T0399292

The Origins and Early History of Conjugated Organic Polymers

The Origins and Early History of Conjugated Organic Polymers

Organic Semiconductors, Synthetic Metals, and the Prehistory of Organic Electronics

SETH C. RASMUSSEN

Department of Chemistry and Biochemistry
North Dakota State University

OXFORD
UNIVERSITY PRESS

OXFORD
UNIVERSITY PRESS

Oxford University Press is a department of the University of Oxford.
It furthers the university's objective of excellence in research, scholarship,
and education by publishing worldwide. Oxford is a registered trademark of
Oxford University Press in the United Kingdom and in certain other countries.

Published in the United States of America by Oxford University Press
198 Madison Avenue, New York, NY 10016, United States of America.

Library of Congress Cataloging-in-Publication Data
Names: Rasmussen, Seth C., author.
Title: The origins and early history of conjugated organic polymers :
organic semiconductors, synthetic metals, and the prehistory of organic
electronics / Seth C. Rasmussen.
Description: New York, NY : Oxford University Press, [2025] |
Includes bibliographical references and index.
Identifiers: LCCN 2025002089 (print) | LCCN 2025002090 (ebook) |
ISBN 9780197638163 (hardback) | ISBN 9780197638187 (epub) |
ISBN 9780197638194
Subjects: LCSH: Conjugated polymers—History.
Classification: LCC QD382.C66 R37 2025 (print) | LCC QD382.C66 (ebook)
| DDC 547/.70457—dc23/eng/20250403
LC record available at https://lccn.loc.gov/2025002089
LC ebook record available at https://lccn.loc.gov/2025002090

DOI: 10.1093/9780197638194.001.0001

Printed by Integrated Books International, United States of America

Contents

Preface

My introduction to conjugated polymers began in 1990 when, as a new graduate student at Clemson University, I was tasked with the research goal of generating metal-coordinated polyacetylenes in order to develop more effective photoconductive materials. Although I had a good background in inorganic chemistry and the synthesis of coordination compounds, I knew nothing about conjugated and conducting polymers at the time. Thus, like many graduate students, I dug into the literature and began learning what I could about these materials. In the process, I also became familiar with the accepted historical narrative that these materials began with research into polyacetylenes and the eventual doping of polyacetylene films by iodine or bromine vapor in 1977. By the time I had defended my doctoral dissertation, I had successfully produced some metal-coordinated conjugated polymers but still really approached these materials with a focus on the photophysics of the bound metal complexes. With a desire to learn more polymer chemistry and a better understanding of conjugated polymers, I then accepted a postdoctoral position under James Hutchison at the University of Oregon with a focus on polythiophenes. While Jim was also new to conjugated polymers, together we continued to learn more about the synthesis and properties of these materials, and I began to interact with other researchers more established in the field. Once I began my independent career at North Dakota State University (NDSU) in 1999, I intended to apply my newfound understanding of conjugated polymers to return to the research of metal-coordinated polymers. However, early successes in the generation of low-bandgap polymers led to a shift in my research focus, and I ultimately developed some recognition for my work on the synthesis and study of these materials.

It was also during my time at Clemson that I developed an interest in science history, although at no point did I ever intend to actively engage in history as a scholarly pursuit. Nevertheless, I began reading every book on the history of chemistry available in the Clemson library, as well as starting to build my own personal collection of books on the topic. Once I began teaching some chemistry courses at Oregon, I incorporated some of my growing knowledge of history into my lectures, primarily correcting historical vignettes and stories included in the textbooks assigned for the courses. These historical sidebars became popular with the students, some of whom asked if I would consider teaching a course focused on the history of chemistry. This ultimately led me to the development of a survey course on the history of chemistry that I taught twice at Oregon. When I moved to NDSU, I brought the course with me, where I have continued to teach it every other year throughout my career. In efforts to fill what I felt to be holes in the historical narrative I was covering in the course, I began researching select topics, but without any intent to do more than just improve my history course. Still, these efforts resulted in what I felt to be new insight into the history of silica glass, which ultimately resulted in my first history publication in 2008.[1] This work was well received, and I continued some scholarly work in history, while maintaining my primary focus on the research of conjugated polymers. While the bulk of this early work in history focused on the history of silica glass or the application of history to chemical education, it is not surprising that I ultimately turned my historical efforts to the topic of conjugated and conducting polymers.

As the result of an invitation to participate in a symposium on the history of plastics by Tom Strom in 2009, I then began research into the history of conjugated polymers, focusing on reports of conducting polymers prior to 1977. These efforts were then published as a book chapter on the topic in 2011,[2] which began my ongoing efforts to provide a more accurate history of the development of these materials. As others in the field then became aware of these efforts, some began to share stories or papers that might merit further investigation. Over time, I had accumulated a significant amount of research and a number of publications on various aspects of this overall history. Of course, each new publication would typically reveal new layers to this history, sometimes requiring me to revise or modify various details or even aspects of the overall narrative of this history. In that aspect, historical research can be similar to chemical research, in which new data can require us to change our accepted viewpoints of a particular relationship or theory. This current book now collects all of my previous published work on this history, as well as adding additional detail and unpublished research. In the process, I attempt to provide the most up-to-date and complete history of conjugated and conducting polymers, from their origins in the 1830s up through their rapid expansion in the 1980s and early 1990s. At the same time, however, I must point out that this history should not be considered complete and that additional details and stories are sure to continue to come to light. Still, it is my hope that the history outlined in this book will change how the field views the development of these materials, as well as their placement in the general history of polymers and synthetic dyes, while also providing a framework for others to contribute to our growing knowledge of their history.

References

1. Rasmussen, S. C. Advances in 13th Century Glass Manufacturing and Their Effect on Chemical Progress. *Bull. Hist. Chem.* **2008**, *33*, 28–34.
2. Rasmussen, S. C. Electrically Conducting Plastics: Revising the History of Conjugated Organic Polymers. In *100+ Years of Plastics. Leo Baekeland and Beyond*; Strom, E. T.; Rasmussen, S. C., Eds.; ACS Symposium Series 1080, American Chemical Society: Washington, D.C., 2011, pp. 147–163.

Seth C. Rasmussen
Fargo, North Dakota

Acknowledgments

I would first and foremost like to thank the Department of Chemistry and Biochemistry at North Dakota State University (NDSU) for supporting my efforts in the history of chemistry in addition to my traditional laboratory research on the synthesis and study of conjugated polymers. In addition, I must acknowledge the American Chemical Society's Division of the History of Chemistry (HIST) in providing the environment, encouragement, and mentorship that allowed my growth into an established chemist-historian. Finally, I would like to thank the Interlibrary Loan Department of NDSU, who went out of their way to track down many elusive and somewhat obscure sources, as well as my brother, Kent Rasmussen, for various discussions relating to linguistics and consultation during some of the translations of French documents. My historical work would be considerably more challenging without these various critical resources.

Considering how long I have been working on this project, as well as its overall scope, a large number of friends and colleagues have assisted by sharing sources and stories over the years, as well as providing critical discussions. Thus, in no particular order, I would like to thank Alan Rocke (Case Western Reserve University) for discussions on issues relating to molecular formulas of organic species during the 19th century; Tony Travis (Hebrew University of Jerusalem) for sharing some of his publications on Lightfoot, Caro, and aniline black, as well as for a number of helpful discussions; John Reynolds (Georgia Tech) for initially bringing the work of Smith and Berets to my attention; David Lewis (University of Wisconsin - Eau Claire) for discussions on the relationship between Fritzsche and Zinin; Vera Mainz (University of Illinois Urbana-Champaign) for help in tracking down various sources; Giorgio Nebbia (Università di Bari) for sharing his knowledge of the life of Riccardo Ciusa; Robert Weiss and Brain A. Bolto (Commonwealth Scientific Industrial and Research Organisation) for their help with collecting information on the life and work of Donald Weiss; Yasu Furukawa (Nihon University) for assistance is collecting biographical information on Hatano and Ikeda; Choon Do for his help in locating Hyung Chick Pyun and his assistance in collecting much of his biographical material; Dennis Cooley (NDSU) for help with philosophy sources relating the nature of discovery; and Richard Kaner (UCLA) for providing his personal image of polyacetylene films. If I have forgotten anyone, please forgive me.

In addition to all mentioned, I would like to give an extra special thanks to Art Diaz and Fred Wudl, who both shared with me their personal memories of their life and work. They both provided me with insight that helped me make important historical connections that would not have been possible otherwise. Finally, I also need to thank the following current and former members of my materials research group for reading various drafts of this manuscript and providing critical feedback: Christopher L. Heth, Kristine Konkol, Eric J. Uzelac, Evan Culver, Spencer Gilman, Wyatt Wilcox, Elsa Buyck, and Trent Anderson.

Chronology

Important Events in the History of Conjugated Polymers

1825   Michael Faraday (1791–1867) isolates benzene from an oil that separated out of illuminating gas.

1826   Otto Unverdorben (1806–1873) discovers *Crystallin* (aniline).

1834   F. Ferdinand Runge (1794–1867) discovers *Kyanol* (aniline) and *Pyrrol* (pyrrole).

1834   Runge reports the first oxidative polymerization of aniline using metal salts. Demonstrates the ability to use the product as a cotton dye, making polyaniline the first known aniline dye.

1836   Edmund Davy (1785–1857) reports the preparation of bicarburet of hydrogen (acetylene).

1840   Carl Julius Fritzsche (1808–1871) discovers *Anilin*.

1842   Benzene isolated from coal tar.

1842   Nikolai Nikolaevich Zinin (1812–1880) synthesizes aniline from nitrobenzene, naming it *Benzidam*.

1843   August Wilhelm Hoffmann (1818–1892) shows Crystallin, Kyanol, Anilin, and Benzidam to all be the same compound, becoming known as simply aniline.

1847   Charles Mansfield (1819–1855) improves the process of fractional distillation of benzene from coal tar, which became the primary source of benzene.

1854   Pierre Jacques Antoine Béchamp (1816–1908) improves Zinin's reduction of nitrobenzene to allow the large-scale production of aniline.

1857   Frederick Crace Calvert (1819–1873) and coworkers develop green and blue polyaniline dyes for the coloring of cotton, naming them *emeraldine* and *azurine*.

1860   Marcelin Berthelot (1827–1907) reports the preparation of *acétylène* by passing organic vapors through a red-hot tube.

1860   Crace Calvert and coworkers patent *emeraldine* and *azurine*, technically making this the first patent of a conjugated polymer. The printers Wood and Wright then commercialize these dyes in late 1860.

1860   Hugo Schwanert (1828–1902) reports the first synthesis of pyrrole.

1860   Black aniline dyes from the oxidative polymerization of aniline were independently introduced by Wood and Wright, John Lightfoot (1831–1872), and Heinrich Caro (1834–1910). All of these became known as *aniline black*.

1862   Berthelot reports the preparation of acetylene via electric discharge.

1862   Henry Letheby (1816–1876) reports the first electropolymerization of aniline.

1865   August Kekulé (1829–1896) proposes the structure of benzene.

1866   Berthelot reports the first polymerization of acetylene via thermal methods.

1872	Heinrich Rheineck introduces the term *nigraniline* (Latin *nigrum* "black" + aniline) for aniline black.
1877	Rudolf Nietzki (1847–1917) isolated quinone from the oxidative degradation of aniline black, providing the first hint of its structure.
1882	Victor Meyer (1848–1897) discovers thiophene.
1885	Carl Paal (1860–1935) reports the first effective synthesis of thiophene.
1892	Thomas L. Wilson (1860–1915) developed a process for the large-scale production of calcium carbide and acetylene.
1896	Heinrich Caro isolates *p*-amidodiphenylamine as an intermediate in the formation of aniline black.
1909	Richard Willstätter (1872–1942) proposes that aniline black is comprised of linear structures of a minimum of eight aniline units that have been oxidized to generate quinoid content.
1912	Arthur G. Green (1864–1941) and Arthur E. Woodhead finalize a refinement of Willstätter's structural model to introduce five oxidation states, giving them the names *leucoemeraldine*, *protoemeraldine*, *emeraldine*, *nigraniline*, and *perinigraniline*.
1915	Angelo Angeli (1864–1931) reports the production of *nero di pirrolo* (*pyrrole black*) from the oxidation of pyrrole.
1929	Kathleen Lonsdale (1903–1971) confirms the cyclic, planar structure of benzene.
1946	Herbert E. Rasmussen (b. 1915) and coworkers develop a vapor-phase synthesis of thiophene from butane and sulfur, allowing the availability of thiophene in commercial quantities.
1949	George Goldfinger (1911–1984) reports the first synthesis of polyphenylene by treating *para*-dichlorobenzene with sodium.
1955	Giulio Natta (1903–1979) reports the first preparation of linear polyacetylene via Ziegler–Natta catalysts.
1960	Peter Kovacic (1921–2022) reports the first oxidative polymerization of benzene.
1960	Tod W. Campbell (1919–1968) reports the first preparation of poly(phenylene vinylene) (PPV).
1961	Berlin and coworkers report the first conductivity of aniline black.
1961	Masahiro Hatano (b. 1930) reports that the conductivity of polyacetylene increases with polymer crystallinity.
1963	Mortimer M. Labes reports the treatment of neutral polyphenylene with iodine vapor, causing the conductivity to increase from 10^{-11} S cm^{-1} to as high as 4×10^{-5} S cm^{-1}. This is the first doping of a neutral conjugated polymer.
1963	Donald Weiss (1924–2008) and coworkers report the thermal polymerization of tetraiodopyrrole to give a cross-linked and partially oxidized polypyrrole with conductivities as high as 0.09 S cm^{-1}.
1964	Marcel Jozefowicz (b. 1934) and Rene Buvet (1930–1992) introduce the modern name *polyaniline*.
1967	Hideki Shirakawa (b. 1936) and Hyung Chick Pyun (1926–2018) accidently discover a method for the production of polyacetylene films.

1967   Donald J. Berets (1926–2002) and Dorian S. Smith (1933–2010) report effects on the conductivity of polyacetylene powder by treating it with a number of different vapors. This includes the first report of the oxygen doping of polyacetylene.

1968   Researchers at the University of Parma report the first electropolymerization of pyrrole to give a freestanding, plastic film with a conductivity of 7.54 S cm^{-1}.

1969   Gerhard Kossmehl reports the first preparation of poly(thienylene vinylene).

1969   Jozefowicz and Buvet report conductivities as high as 100 S cm^{-1} for emeraldine sulfate.

1976   Shirakawa and Chwan K. Chiang perform the first doping of polyacetylene with bromine vapor to give a conductivity of 0.5 S cm^{-1}.

1977   Takakazu Yamamoto (1944–2014) introduces the preparation of polyphenylene via Kumada cross-coupling.

1977   Alan Heeger (b. 1936), Alan MacDiarmid (1927–2007), and Shirakawa report the bromine and iodine doping of polyacetylene to give conductivities up to 38 S cm^{-1}. A second report improved the I_2-doped conductivity up to 160 S cm^{-1} and showed conductivities up 220 S cm^{-1} were possible via AsF_5.

1979   Heeger, MacDiarmid, and Shirakawa report conductivities up to 2150 S cm^{-1} for stretch-oriented films of doped polyacetylenes.

1980   Arthur Diaz optimizes the electropolymerization of polypyrrole to give conductivities up to 200 S cm^{-1}.

1980   Yamamoto reports the first preparation of polythiophene, accomplished using Kumada cross-coupling.

1980   MacDiarmid and Heeger develop polyacetylene-based batteries.

1980   V. L. Afanas'ev and coworkers in the Soviet Union report the first electropolymerization of polythiophene. Diaz reports more optimized conditions the following year.

1981   Kossmehl reports the first preparation of polythiophene via chemical oxidation.

1982   B. R. Weinberger and coworkers report the first polyacetylene-based photovoltaic device.

1984   Heeger and Fred Wudl (b. 1941) report polyisothianaphthene (PITN), the first low-bandgap polymer, with a bandgap of 1.0 eV.

1984   Gérard Tourillon and Francis Garnier report the first photovoltaic device from polythiophene.

1985   Ron L. Elsenbaumer reports the first soluble poly(3-alkylthiophene)s.

1987   Herbert Naarmann reports the production of unstretched, I_2-doped polyacetylene films with conductivities of 5000 S cm^{-1}. These could then be increased up to 170,000 S cm^{-1} upon stretch orientation.

1990   Richard H. Friend (b. 1953), Andrew B. Holmes (b. 1943), and coworkers discover electroluminescence in PPV, allowing the development of polymer-based organic light-emitting diodes (polymer OLEDs or PLEDs).

Chapter 1
Introduction

1.1 Polymers and plastics

Modern society can easily be characterized as a plastic-based culture in which organic polymer-derived plastics are more ubiquitous than older common materials such as ceramics, metals, or glass.[1] Because of this, some have suggested that there is sufficient justification to refer to the period beginning with the 20th century as the *Age of Plastics*.[2] Although semisynthetic organic plastics such as Parkesine and celluloid date back to the mid-1800s,[3] it was the introduction of Bakelite and the establishment of the General Bakelite Corp. by Leo Baekeland (1863–1944) in 1910[4] that initiated a new age in fully synthetic plastics, fibers, and elastomers for a wide multitude of uses, permanently changing technology and society in the process. Since that time, a wide variety of polymers and plastics have become mainstay technological materials, a number of which are shown in Figure 1.1.

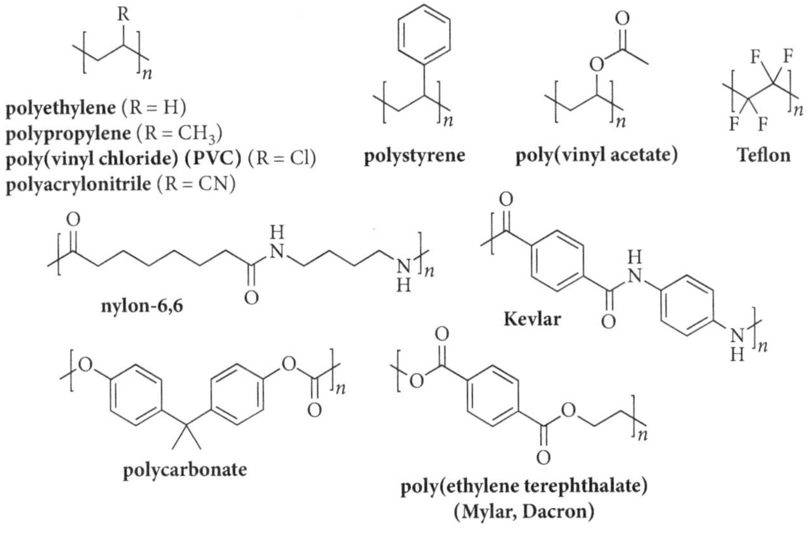

polyethylene (R = H)
polypropylene (R = CH$_3$)
poly(vinyl chloride) (PVC) (R = Cl)
polyacrylonitrile (R = CN)

polystyrene

poly(vinyl acetate)

Teflon

nylon-6,6

Kevlar

polycarbonate

poly(ethylene terephthalate)
(Mylar, Dacron)

Figure 1.1 Representative technological polymers and plastics.

The Origins and Early History of Conjugated Organic Polymers. Seth C. Rasmussen, Oxford University Press.
© Oxford University Press (2025). DOI: 10.1093/9780197638194.003.0001

In a historical account of polymer science up to 1977, Bengt G. Rånby (1920–2000) of the KTH Royal Institute of Technology presented the growth and development of polymers as three generations of traditional technological materials.[5] The first generation of these materials comprises those introduced before 1950. These include simple polymers and plastics such as polystyrene, poly(vinyl chloride) (commonly known as PVC), and low-density polyethylene, as well as synthetic fibers such as nylon. The second generation is characterized by those materials with higher mechanical strength that were introduced during 1950–1965. This includes such materials as high-density polyethylene, isotactic polypropylene, polycarbonates, polyurethanes, and epoxy resins. The final generation of traditional polymeric materials includes specialty polymers developed since 1965. These materials are characterized by more complex chemical structures, high mechanical strength, high softening temperatures, and high chemical resistance. Included in this generation of polymers are materials such as Kevlar and Teflon.

Most organic polymers or plastics are of interest as structural materials and find little application in terms of electronics. This is primarily due to the fact that such common plastics are insulating materials and thus have little utility for most electronic applications. In fact, it is very common to use plastics such as PVC to encapsulate electrical cords in order to isolate the metallic conducting wires from other conducting materials. This all changed, however, with the discovery that the electrical conductivity of conjugated organic polymers (Figure 1.2) can be controlled through oxidation or reduction processes, resulting in what has been referred to as the *fourth generation of polymeric materials.*[5–7]

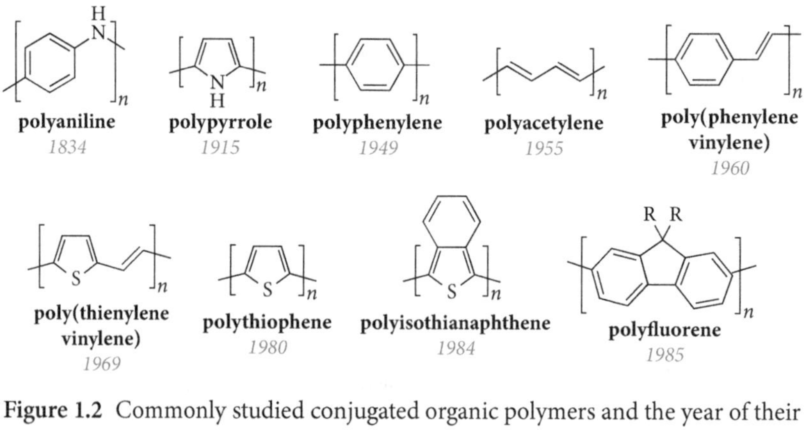

| polyaniline | polypyrrole | polyphenylene | polyacetylene | poly(phenylene vinylene) |
| 1834 | 1915 | 1949 | 1955 | 1960 |

| poly(thienylene vinylene) | polythiophene | polyisothianaphthene | polyfluorene |
| 1969 | 1980 | 1984 | 1985 |

Figure 1.2 Commonly studied conjugated organic polymers and the year of their first report.

Unlike typical saturated organic polymers, conjugated polymers consist of chains of neighboring multiple bonds such that the backbone is comprised of overlapping p orbitals (Figure 1.3) that allow electrons in these orbitals to be delocalized along the polymer chain. As a consequence, conjugated polymers are a class of organic semiconducting materials that exhibit enhanced electronic conductivity (quasi-metallic in some cases) in their oxidized or reduced states.[8,9] As such, these materials combine the conductivity of classical inorganic systems with many of the desirable properties of organic plastics, including mechanical flexibility and low production costs. This combination of properties has led to considerable fundamental and technological interest over the last several decades, resulting in the current field of organic electronics and the development of a variety of modern technological applications, including sensors, electrochromic devices, organic photovoltaics (OPVs), organic light-emitting diodes (OLEDs), and field-effect transistors (FETs).[8-13] In addition, the flexible, plastic nature of the organic materials used as the active layers in such electronic devices has led to the realistic promise of flexible electronics in the near future.[11-13]

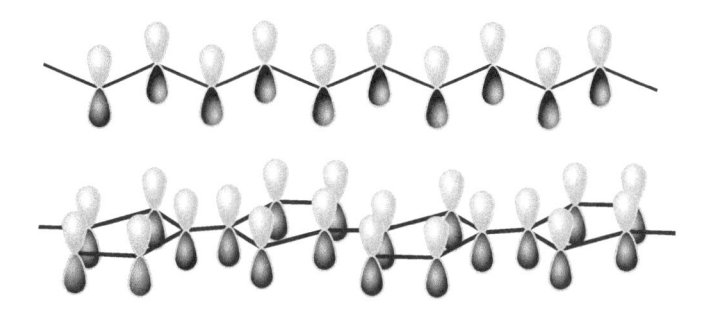

Figure 1.3 Conjugated backbones consisting of overlapping p orbitals.

It must be pointed out, however, that while Rånby's historical account provides a neat, linear picture of the development of polymeric materials, it is not consistent with the known historical record. Not only did initial reports of conducting polymers predate the introduction of Rånby's third generation of polymers, but one of the common early examples, polyaniline, has now been shown to be the earliest known example of a fully synthetic polymer and actually predates polystyrene, the earliest member of the Rånby's first generation of polymers. Of course, while the early forms of polyaniline are essentially identical with the modern conducting forms, its electronic properties were not recognized until the mid-1960s. It is this disconnect between the known historical record and the many commonly presented histories of conjugated and conducting polymers that will be addressed throughout the current volume, with the goal of correcting many misconceptions in the current primary literature.

1.2 Doped conjugated polymers and the birth of synthetic metals

As introduced earlier, conjugated polymers exhibit enhanced electronic conductivity upon oxidation or reduction. As a consequence, the terms *conjugated polymers* and *conducting polymers* are often used interchangeably, although this is not really technically correct. Neutral conjugated polymers are generally semiconducting in nature, with some examples really better described as insulators. In order for these materials to exhibit significant conductivity, the neutral parent polymers require additional oxidative or reductive processes, and thus conjugated polymers only truly become conducting polymers in their nonneutral redox states, typically referred to as *doped* conjugated polymers.

The most common examples of conducting polymers involve the oxidation of these materials. The oxidation of π-conjugated backbones generates positive charge carriers (i.e., holes, Figure 1.4) and an increase of p-type character.[14–17] Such oxidized polymers are thus referred to as p-doped in analogy to p-doped inorganic semiconductors (e.g., gallium-doped silicon). Unfortunately, this terminology can lead to common misconceptions, as unlike classical doping, which is a substitutional procedure (i.e., some silicon atoms are replaced with gallium), the oxidation of conjugated materials removes electrons but does not change the elemental composition of the polymer itself. Still, as the p-doped product is now a polycationic material, the incorporation of anions into the polymer film is required to maintain charge neutrality. In order to differentiate this process from classical doping, it is sometimes referred to as *oxidative doping*.[14]

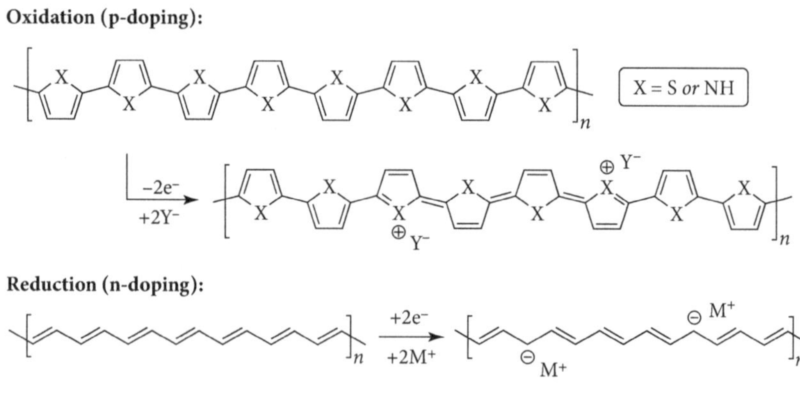

Oxidation (p-doping):

Reduction (n-doping):

Figure 1.4 Redox doping of conjugated polymers.

In a similar manner, the oxidation process itself is referred to as p-doping, which can be accomplished either by treating the polymer with an oxidizing agent or via electrochemical oxidation. When an oxidizing agent (i.e., I_2)

is used to accomplish p-doping, the oxidizer is reduced, which results in the formation of the anions (i.e., I^- or I_3^-) that then become incorporated into the film as counterions. In cases using electrochemical oxidation, the polymer film instead incorporates counterions from the supporting electrolyte used in the electrochemical cell.[15-17]

Although not nearly as common, some examples of conjugated polymers are capable of undergoing reduction, or n-doping. In these cases, reduction results in negative charge carriers (i.e., electrons, Figure 1.4) and an increase of n-type character.[14-17] As with oxidative doping, reduction of conjugated materials can be achieved either electrochemically or via the use of a reducing agent (i.e., Na), both of which require the incorporation of cationic species into the film for charge neutrality. This form of doping is sometimes referred to as *reductive doping*,[14] while both oxidative and reductive doping can also be collectively referred to as *redox doping*.

Finally, it should be noted that the counterions incorporated into the materials during either p-doping or n-doping are often referred to as *dopants*. Again, this is an unfortunate term in that it often results in confusion in the same way as the application of the general term doping to conjugated polymers. Thus, unlike dopants in traditional doped semiconductors, the counterion itself does not cause the enhanced conductivity, although they are necessary to allow the oxidized or reduced forms of the conjugated materials that do provide the resulting conductive nature.[15-17]

The conjugated polymer polyaniline introduces some additional complications, as it is the only member of the common conjugated polymers to exhibit both protonated and free-base forms, with the protonated forms exhibiting significantly higher conductivity than the corresponding neutral bases. This effect of protonation has been referred to as *protonic acid doping*,[18-20] which is sometimes incorrectly described to be a form of doping in which no redox processes are required to make the material conducting.[18] While acid doping does result in significant increases in conductivity of polyaniline samples, it first requires that the sample treated in this way be partially oxidized (typically to what is known as its emeraldine form). As such, polyaniline requires both oxidation and protonation to achieve its maximum conductivity, not just protonation alone.

As many doped conjugated polymers can exhibit electrical conductivities in the metallic range ($>10^2$ S cm^{-1}, Figure 1.5),[21,22] these doped materials have also been referred to as *synthetic metals*.[19,20,23-25] The term *synthetic metals* was first coined by Herbert N. McCoy (1870–1945) at the University of Chicago in 1911.[24,25] As reported in *Science*,[26] McCoy described attempts to produce a "synthetic metal" from the electrolysis of ethanol solutions of tetramethylammonium ($(CH_3)_4N^+$) salts. These efforts were initially reported by McCoy and William C. Moore earlier that same year, in which they utilized an electrolytic chamber comprised of silver-plated platinum gauze for the anode and a mercury electrode as the cathode. Electrolysis of the salt resulted in the production of a crystalline solid of metallic luster, closely resembling sodium amalgam, which was believed to be a mercury amalgam of ammonium radicals with the general formula $HgN(CH_3)_4$.[27] Although this organic amalgam was not very stable, it exhibited an electrical conductivity consistent with a metal (ca. $7–9 \times 10^3$ S cm^{-1}).[26,27] Ultimately, McCoy conlcuded:[26]

> The facts just reviewed, though few in number, seem to me to lend support to this hypothesis, and to lead to the conclusion that it is possible to prepare composite metallic substances, which may be termed synthetic metals, from constituent elements, some of which at least are nonmetallic.

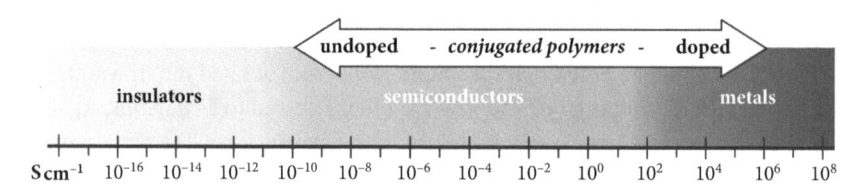

Figure 1.5 Relative conductivity of conjugated polymers.

The term was not used again until 1969, when Alfred René Ubbelohde (1907–1988) of Imperial College used it to describe a new class of materials based on intercalated graphite.[28-32] These carbon-based species exhibited conductivities up to 2.5×10^5 S cm^{-1} and thus provided the first practical and stable example of the type of material described by McCoy. A variety of additional new materials were then discovered through the early 1970s that also exhibited metallic conductivity, including organic charge-transfer salts, metal chain compounds, and the inorganic polymer poly(sulfur nitride), $(SN)_x$.[25] The materials described as synthetic metals were then further expanded to include doped polyacetylenes,[33] following the reported high conductivity of these polymers by Alan J. Heeger (b. 1936), Alan G. MacDiarmid (1927–2007), and Hideki Shirakawa (b. 1936) in 1977.[34-37] Although the initial polyacetylene papers of 1977–1978 had never specifically used the term, MacDiarmid published a review of synthetic metals in 1979.[38] The introduction of this review began with the statement:

> This report is directed toward the very new area of materials science which is concerned with the preparation and characterization of synthetic metals, many of which contain no atoms of any metallic element in their chemical constitution. The three main presently known classes and their potential technological significance will be described.

He then went on to state that these three classes consisted of metallic compounds derived from poly(sulfur nitride), polyacetylene, and graphite. MacDiarmid had also previously used the term in a 1977 radio address to describe poly(sulfur nitride).[39]

By October 1979, Elsevier launched a new journal, aptly titled *Synthetic Metals,* that was dedicated to this class of materials (Figure 1.6).[40] In the introduction of the first issue,[41] Editor F. Lincoln Vogel (b. 1922) described this publication as

> ... a new international journal for the publication of research and engineering papers on graphite intercalation compounds, transition metal compounds, and quasi one-dimensional conducting polymers.

To date, this is still the only journal dedicated to organic conducting materials. By 1980, the term *synthetic metals* was starting to be used more and more often to describe doped polyacetylene, and as the field of conducting polymers continued to grow, the term was further expanded to include other doped conjugated polymers. By 1991, MacDiarmid and Arthur Epstein (1945–2019) included polyparaphenylene, poly(phenylene vinylene), polypyrrole, polythiophene, and

Figure 1.6 The cover of the first issue of *Synthetic Metals,* published October 1979.

Reproduced with permission from *Synthetic Metals,* Volume 1, Issue 1, © Elsevier (1979).

polyaniline in a review of conducting polymers as synthetic metals.[42] Ultimately, the term has come to be more associated with conjugated polymers than any other type of organic conductors.

1.3 2000 Nobel Prize in Chemistry and the history of conductive polymers

On October 10, 2000, The Royal Swedish Academy of Sciences announced that the 2000 Nobel Prize in Chemistry was being awarded to Professors Alan J. Heeger (University of California at Santa Barbara), Alan G. MacDiarmid (University of Pennsylvania), and Hideki Shirakawa (University of Tsukuba) "for the discovery and development of electrically conductive polymers."[43,44] Following the initial announcement, Professor Bengt Nordén (b. 1945), of Chalmers University of Technology and Chairman of the Nobel Committee for Chemistry, presented the significance of the discovery of conductive polymers. The fact that conductive polymers changed the way that one viewed plastics was outlined in the initial press release:

> We have been taught that plastics, unlike metals, do not conduct electricity. In fact plastic is used as insulation round the copper wires in ordinary electric cables. Yet this year's Nobel Laureates in Chemistry are being rewarded for their revolutionary discovery that plastic can, after certain modifications, be made electrically conductive.

The scientific contributions recognized by this award highlighted their collaborative work on highly conducting polyacetylene via p- and n-doping.[34–37] This collaborative work began in the mid-1970s, and as illustrated by the quote, most discussions of the history of this field begin with these efforts, although investigations of electrically conductive conjugated polymers actually date back to the early 1960s, nearly 15 years before the polyacetylene work recognized by the award. While it is sometimes acknowledged that there were other reports prior to the polyacetylene work of Heeger, MacDiarmid, and Shirakawa, this earlier work is usually dismissed with statements such as that given in the advanced information released for the 2000 Nobel Prize:[44] "However, polyacetylene was the conductive polymer that actually launched this new field of research." Similar statements that doped polyacetylene was the first conducting polymer or was at least the material that launched the field of conducting polymers can be found in various discussions of the history of the field. This includes the historical accounts referenced in the Nobel materials, such as a review of conductive polymers by Mercourl G. Kanatzidis (1957–) of Michigan State University in 1990:[14] "However, the conductive polymer that actually launched this new field of research, was polyacetylene $[(CH)_n]$." Similar statements can also be found in the previously discussed historical account of polymer science by Rånby in 1993:[5] "As introduction and background to this symposium my lecture is a short account of important discoveries and inventions made until 1977 when electrically conducting polymers were first prepared by doping of polyacetylene by A.G. MacDiarmid, A.J. Heeger, and H. Shirakawa." More recent publications echo these statements, as can be seen by the historical retrospective published by Chemical Communications in 2003:[45] "A quarter of a century ago, Chemical Communications published a seminal paper called 'Synthesis of Electrically Conducting Organic Polymers: Halogen Derivatives of Polyacetylene, $(CH)_x$.' A new field of chemistry was born."

If one looks to statements by the Nobel Award winners themselves, they too claim the discovery of conducting polymers. As stated by Heeger in the opening of his Noble Lecture,[6,7] "In 1976, Alan MacDiarmid, Hideki Shirakawa, and I, together with a talented group of graduate students and postdoctoral researchers discovered conducting polymers and the ability to dope these polymers over the full range from insulator to metal." While MacDiarmid does not openly claim discovery, he does reinforce the commonly held view that polyacetylene was the first conducting polymer and that additional conducting polymers commonly studied today followed afterwards:[19,20]

> Since the initial discovery in 1977, that polyacetylene $(CH)_x$, now commonly known as the prototype conducting polymer, could be p- or n-doped either chemically or electrochemically to the metallic state, the development of the field of conducting polymers has continued to accelerate at an unexpectedly rapid rate and a variety of other conducting polymers and their derivatives have been discovered.

Shirakawa also points to polyacetylene as the first conducting polymer and the doping experiments of 1976 marking the start of the era of conducting polymers:[46,47]

> At that time, to my regret, I did not recognize that this carbocation could be a charge carrier and thus polyacetylene could be the first conducting polymer. To open an era of conducting polymers, we had to wait until we intentionally carried out the doping experiment with bromine at the University of Pennsylvania, on Tuesday 23rd, of November 1976, and then later also successively with iodine.

All of these views, however, are somewhat at odds with the fact that several reviews of conductive polymers were published prior to the polyacetylene work of 1977, with the first such review published in 1965.[48-53] To add to these works, three books entitled *Organic Semiconductors* were published during the 1960s, all of which included conductive polymers.[54-56] This included an edited volume by James J. Brophy (1926–1991) and John W. Buttrey in 1962,[54] followed by books from Yoshiyuki Okamoto and Walter Brenner (1923–2017) in 1964,[55] and Felix Gutmann (b. 1908) and Lawrence E. Lyons in 1967.[56] Finally, a research report was also published in 1962 by the U.S. Department of Commerce entitled *Organic Semiconductors—Their Technological Promise*.[57] While this report and the books mentoned include significant discussion on semiconducting molecular crystals, the topic of semiconducting organic polymers is also found throughout all four publications, and it is clear that the conduction involved in these materials is electronic, rather than ionic.

Surveying the reviews published on conductive polymers in the years following the reports by Heeger, MacDiarmid, and Shirakawa, it is quite clear that doped polyacetylene is now included and it is acknowledged as a current topic of focus, but these reviews do not portray the doped polymer as unique and is, rather, given as just one of many examples of conductive materials.[58,59] As such, the polyacetylene results are not presented as particularly revolutionary or the clear start of a new field. In fact, a 1982 review by Karlheinz Seeger (1927–2008) of the University of Vienna states:[59]

> All polymers to be discussed below are in fact well known for many years or even decades. Also the technique of increasing the conductivity by doping with halogens or sulfur is not new. It was not known, however, that a metallic conductivity would be obtained at high doping levels.

In terms of the historical record documented by the primary scientific literature, Seeger's quote is really a fairly accurate description of the place of the early contributions of Heeger, MacDiarmid, and Shirakawa with doped polyacetylene. Their studies produced quite exciting results, generating the highest reported conductivity to date for a one-dimensional organic polymer (including the first such example fully in the metallic range) and thus had far-reaching impact on the growth of the fields of conjugated polymers and organic electronics. However, their publications of 1977–1978 detailing the conductivity of doped polyacetylene do not represent the first example of conductive organic polymers. Even in terms of the more specific subclass of conjugated organic polymers, reports of the conductive nature of oxidized conjugated polymers date back to the early 1960s.[15-17] Notable examples include the early 1963 work of both Mortimer M. Labes on polyphenylene[60] and Donald Weiss (1924–2008) on polypyrrole in Australia,[61-63] as well as the 1968 work on electropolymerized polypyrrole at the University of Parma,[64] and the reports of René Buvet (1930–1992) and Marcel Jozefowicz (b. 1934) on polyaniline beginning in 1965.[65-72] Of course, it could be argued that the conductivities reported by Labes and Wiess were still quite low (ca. 4×10^{-4} and 0.1 S cm^{-1}, respectively)[60,63] and Wiess himself states that he did not fully understand the role of doping on the polymer conductivity at the time.[73,74] However, the same could not be said of the work of Buvet and Jozefowicz, who had achieved conductivities as high as 100 S cm^{-1} by 1969 with polyaniline.[72]

The viewpoint expressed by Seeger, however, seemed to change in a relatively short period of time. Thus, by 1985, we find the following statement in a review by W. James Feast (1938–) of Durham University, which downplays the previous efforts in this area:[75]

> By contrast, intrinsically conducting polymers were for many years the subject of intermittent theoretical speculation and a relatively low level of research activity, as judged by the amount of published experimental data. This situation changed dramatically with the publication of a procedure for preparing self-supporting films of polyacetylene, and the past ten years have shown an almost frenetic worldwide programme of research and publication concerned with theoretical and experimental aspects of conducting polymers.

Two years later, Pamela S. Zurer of *Chemical & Engineering News* (American Chemical Society) makes no mention whatsoever of previous work and gives full credit to MacDiarmid and Heeger:[76] "MacDiarmid and Alan J. Heeger, now at the University of California, Santa Barbara, were first to prepare polymers that conduct electricity like metals." Of course, the quote by Zurer could be interpreted a number of different ways. If this is meant to refer to the preparation of organic polymers with conductivities in the metallic regime, this would be technically correct. No matter how one interprets this statement, however, it is unclear why Shirakawa is not credited here.

That the previous studies of Weiss, Buvet, and Jozefowicz, as well as a wealth of other early reports, are overlooked in most discussions of the history of conjugated polymers is unfortunate. In fact, the work of Weiss and coworkers was not referenced in the literature until 1997, nearly 35 years after their initial publication. Perhaps it is no surprise that the first ones to recognize the work of Weiss in the literature were researchers and publications in his home country of Australia.[77,78] Since that time, the initial papers of Weiss and coworkers have been referenced by several dozen authors, but this is still an extremely small fraction of the conjugated polymer literature. In order to make sense of this early work in terms of the commonly accepted history that this field essentially started in the 1970s, some authors have credited the initial discovery of conducting polymers to Weiss in 1963, followed by credit of their rediscovery to the Nobel awardees in the 1970s.[79] However, this still is an oversimplification of the complex history of these materials and implies that these discoveries were isolated events rather than part of the normal progression of research and discovery that forms an unbroken path from the initial synthesis of conjugated polymers in the early 19th century up through their application to organic electronics today.

In 1968, the American sociologist Robert K. Merton (1910–2003) introduced what he referred to as the *Matthew Effect*,[80] in which eminent scientists get disproportionately more credit for their contributions to science, while relatively unknown scientists tend to get disproportionately little credit for comparable contributions. Furthermore, he points out that the history of science has many examples of foundational works published by comparably unknown scientists, only to be neglected for years. This is certainly true in the field of conjugated and conducting polymers, with most discussions of their history typically presenting their development as a somewhat recent advancement, generally beginning with the collaborative work of Heeger, MacDiarmid, and Shirakawa.[14,45,81–86] As a consequence, most researchers in this ever-growing field are unaware of the previous contributions to the field and know little about the historical progression of the research detailing this class of materials.

In attempts to rectify this Matthew Effect in the field of conjugated materials, I have worked to present various aspects of their history in a series of publications over the last decade, with particular emphasis on giving due credit to many early researchers largely unknown to many in the field.[15–17,24,25,73,74,87–92] In addition to my own contributions, a few other authors have also tried to shed light on some of the previous contributions in the field.[92,93] In an effort to continue these collective efforts, the current volume aims to update and expand my previous reports in order to present the first detailed and complete history of the origin and early research of conjugated organic polymers from the earliest reports in the 19th century up through the rapid growth and expansion of the field throughout the 1980s. In the process, it is hoped to highlight the extent of work carried out over this time period and, in the process, properly give credit to all of the various figures, both the well-known and the obscure, that contributed to the origin and development of this important field of electronic materials.

References and Notes

1. Seth C. Rasmussen, Ed. *Chemical Technology in Antiquity*. ACS Symposium Series 1211. American Chemical Society: Washington, DC, 2015.
2. Strom, E. T.; Rasmussen, S. C.; Eds. *100+ Years of Plastics. Leo Baekeland and Beyond*. ACS Symposium Series 1080. American Chemical Society: Washington, DC, 2011, pp. 1–3.

3. Rasmussen, S. C. From Parkesine to celluloid: The birth of organic plastics. *Angew. Chem. Int. Ed.* **2021**, *60*, 8012–8016.

4. Morawetz, H. *Polymers. The Origins and Growth of a Science.* John Wiley & Sons: New York, 1985, pp. 58–59.

5. Rånby, B. Background—polymer science before 1977. In *Conjugated Polymers and Related Materials: The Interconnection of Chemical and Electronic Structure.* Salaneck, W. R.; Lundström, I.; Rånby, B.; Eds.; Oxford University Press: New York, 1993, pp. 15–26.

6. Heeger, A. J. Semiconducting and metallic polymers: The fourth generation of polymeric materials (Nobel Lecture). *Angew. Chem. Int. Ed.* **2001**, *40*, 2591–2611.

7. Heeger, A. J. Semiconducting and metallic polymers: The fourth generation of polymeric materials. *Synth. Met.* **2002**, *125*, 23–42.

8. Reynolds, J. R.; Skotheim, T. A.; Thompson, B., Eds. *Handbook of Conducting Polymers*, 4th ed., 2 volume set. CRC Press: Boca Raton, FL, 2019.

9. Perepichka, I. F.; Perepichka, D. F., Eds. *Handbook of Thiophene-based Materials.* John Wiley & Sons: Hoboken, NJ, 2009.

10. Rasmussen, S. C.; Ogawa, K.; Rothstein, S. D. Synthetic approaches to band gap control in conjugated polymeric materials. In *Handbook of Organic Electronics and Photonics.* Nalwa, H. S., Ed.; American Scientific Publishers: Stevenson Ranch, CA, 2008, Chapter 1.

11. Roth, B.; Sondergaard, R. R.; Krebs, F. C. Roll-to-roll printing and coating techniques for manufacturing large-area flexible organic electronics. In *Handbook of Flexible Organic Electronics.* Logothetidis, S., Ed.; Woodhead Publishing Series in Electronic and Optical Materials, Number 68, Woodhead Publishing: Amsterdam, 2015, pp. 172–198.

12. Sekitani, T.; Someya, T. Stretchable, Large-area organic electronics. *Adv. Mater.* **2010**, *22*, 2228–2246.

13. Logothetidisa, S.; Laskarakis, A. Towards the optimization of materials and processes for flexible organic electronics devices. *Eur. Phys. J. Appl. Phys.* **2009**, *46*, 12502.

14. Kanatzidis, M. G. Conductive polymers. *Chem. Eng. News* **1990**, *68*(49), 36–54.

15. Rasmussen, S. C. Early history of conductive organic polymers. In *Conductive Polymers: Electrical Interactions in Cell Biology and Medicine.* Zhang, Z.; Rouabhia, M.; Moulton, S., Eds.; CRC Press: Boca Raton, FL, 2017, pp. 1–21.

16. Rasmussen, S. C. Early history of conjugated polymers: From their origins to the handbook of conducting polymers. In *Handbook of Conducting Polymers*, 4th ed. Reynolds, J. R.; Skotheim, T. A.; Thompson, B., Eds.; CRC Press: Boca Raton, FL, 2019, pp. 1–35.

17. Rasmussen, S. C. Conjugated and conducting organic polymers: The first 150 years. *ChemPlusChem* **2020**, *85*, 1412–1429.

18. MacDiarmid, A. G.; Epstein, A. J. Polyanilines: A novel class of conducting polymers. *Faraday Discuss.* **1989**, *88*, 317–332.

19. MacDiarmid, A. G. "Synthetic metals": A novel role for organic polymers (Nobel Lecture). *Angew. Chem. Int. Ed.* **2001**, *40*, 2581–2590.

20. MacDiarmid, A. G. Synthetic metals: A novel role for organic polymers. *Synth. Met.* **2002**, *125*, 11–22.

21. Gutmann, F.; Lyons, L. E. *Organic Semiconductors.* John Wiley & Sons, Inc.: New York, 1967, p. 449.

22. Conductivity (σ) is the reciprocal of resistivity, $\sigma = \rho^{-1}$. As the unit for resistivity is the ohm (Ω), the unit for conductivity is Ω^{-1}. In the modern literature, this is represented by the unit Siemens ($S = \Omega^{-1}$). However, an older unit sometimes also seen is the mho (i.e., ohm spelled backwards).

23. MacDiarmid A. G.; Epstein, A. J. Conducting polymers: Past, present, and future... *Mat. Res. Soc. Symp. Proc.* **1994**, *328*, 133–144.

24. Rasmussen, S. C. On the origin of "synthetic metals." *Mater. Today* **2016**, *19*, 244–245.

25. Rasmussen, S. C. On the origin of "synthetic metals": Herbert McCoy, Alfred Ubbelohde, and the development of metals from nonmetallic elements. *Bull. Hist. Chem.* **2016**, *41*, 64–73.

26. McCoy, H. N. Synthetic metals from non-metallic elements. *Science* **1911**, *34*, 138–142.

27. McCoy, H. N.; Moore, W. C. Organic amalgams: Substances with metallic properties composed in part of non-metallic elements. *J. Am. Chem. Soc.* **1911**, *33*, 273–292.

28. Ubbelohde, A. R. Charge transfer effects in acid salts of graphite. *Proc. Roy. Soc. A* **1969**, *309*, 297–311.

29. Murray, J. J.; Ubbelohde, A. R. Electronic properties of some synthetic metals derived from graphite. *Proc. Roy. Soc. A* **1969**, *312*, 371–380.

30. Ubbelohde, A. R. Electronic anomalies in dilute synthetic metals. *Proc. Roy. Soc. A* **1971**, *321*, 445–460.

31. Bach B.; Ubbelohde, A. R. Synthetic metals based on graphite/aluminium halides. *Proc. Roy. Soc. A* **1971**, 325, 437–445.

32. Bach B.; Ubbelohde, A. R. Chemical and electrical behaviour of graphite-metal halide compounds. *J. Chem. Soc. A* 1971, 3669–3674.

33. Pietronero, L. Charge transfer and maximum conductivity of carbon based synthetic metals. *Phys. Scr.* **1982**, *T1*, 108–109.

34. Shirakawa, H.; Louis, E. J.; MacDiarmid, A. G.; Chiang, C. K.; Heeger, A. J. Synthesis of electrically conducting organic polymers: Halogen derivatives of polyacetylene, $(CH)_x$. *Chem. Soc., Chem. Commun.* **1977**, 578–580.

35. Chiang, C. K.; Fincher Jr., C. R.; Park, Y. W.; Heeger, A. J.; Shirakawa, H.; Louis, B. J.; Gau, S. C.; MacDiarmid, A. G. Electrical conductivity in doped polyacetylene. *Phys. Rev. Lett.* **1977**, *39*, 1098–1101.

36. Chiang, C. K.; Druy, M. A.; Gau, S. C.; Heeger, A. J.; Louis, E. J.; MacDiarmid, A. G.; Park, Y. W.; Shirakawa, H. Synthesis of highly conducting films of derivatives of polyacetylene, $(CH)_x$. *J. Am. Chem. Soc.* **1978**, *100*, 1013–1015.

37. Chiang, C. K.; Park, Y. W.; Heeger, A. J.; Shirakawa, H.; Louis, E. J.; MacDiarmid, A. G. Conducting polymers: Halogen doped polyacetylene. *J. Chem. Phys.* **1978**, *69*, 5098.

38. MacDiarmid, A. G. Metallic and semiconducting materials derived from non-metallic elements. *Microstruct. Sci., Eng., Technol.* **1979**, 13.1–13.8.

39. MacDiarmid, A. G. *Synthetic metals. Men and molecules*, no. 212. [Audio cassette] American Chemical Society: Washington DC, 1977.

40. Synthetic metals. http://www.journals.elsevier.com/synthetic-metals/ (accessed February 15, 2021).

41. Vogel, F. L. Editorial introduction. *Synth. Met.* **1979**, *1*, 1.

42. MacDiarmid, A. G.; Epstein, A. J. "Synthetic metals": A novel role for organic polymers. *Makromol. Chem., Macromol. Symp.* **1991**, *51*, 11–28.

43. Nobelprize.org. Nobel Media AB (2014) Press Release: The 2000 Nobel Prize in Chemistry. http://www.nobelprize.org/nobel_prizes/chemistry/laureates/2000/press.html (accessed February 15, 2021).

44. Nobelprize.org. Nobel Media AB (2014) The Nobel Prize in Chemistry 2000—Advanced Information. http://www.nobelprize.org/nobel_prizes/chemistry/laureates/2000/advanced.html (accessed February 15, 2021).

45. Hall, N. Twenty-five years of conducting polymers. *Chem. Commun.* **2003**, 1–4.

46. Shirakawa, H. The discovery of polyacetylene film: The dawning of an era of conducting polymers (Nobel Lecture). *Angew. Chem. Int. Ed.* **2001**, *40*, 2574–2580.

47. Shirakawa, H. The discovery of polyacetylene film. The dawning of an era of conducting polymers. *Synth. Met.* **2002**, *125*, 3–10.

48. Weiss, D. E.; Bolto, B. A. Organic polymers that conduct electricity. In *Physics and Chemistry of the Organic Solid State*. Fox, D.; Labes, M. M.; Weissberger, A., Eds.; Interscience Publishers: New York, 1965, Vol. II, Chapter 2.

49. Labes, M. M. Conductivity in polymeric solids. *Pure. Appl. Chem.* **1966**, *21*, 275–285.

50. Trivedi, P. D. Electrically conductive polymers. *Pop. Plast.* **1968**, *13*(9), 25–29; **1968**, *13*(10), 30–35.

51. Rembaum, A. Conductive polymers. *Encycl. Polym. Sci. Technol.* **1969**, *11*, 318–337.

52. Lupinski, J. H. Conductive polymers. *Ann. N. Y. Acad. Sci.* **1969**, *155*(2), 561–565.

53. Goodings, E. P. Conductive polymers. *Rep. Prog. Appl. Chem.* **1970**, *55*, 53–65.

54. Brophy, J. J.; Buttrey, J. W., Eds. *Organic Semiconductors. Proceedings of an Interindustry Conference.* The Macmillan Company: New York, 1962.

55. Okamoto, Y.; Brenner, W. *Organic Semiconductors.* Rheinhold: New York, 1964

56. Gutmann, F.; Lyons, L. E. *Organic Semiconductors.* John Wiley and Sons Inc.: New York, 1967.

57. Office of Technical Services, U.S. Department of Commerce. *Organic Semiconductors - Their Technological Promise.* U.S. Government Research Report, PB 181037.

58. Mort, J. Conductive polymers. *Science* **1980**, *208*, 819–825.

59. Seeger, K. The morphology and structure of highly conducting polymers. *Angew. Makromol. Chem.* **1982**, *109/110*, 227–251.

60. Originally published as Mainthia, S. B.; Kronick, P. L.; Ur, Hana; Chapman, E. F.; Labes, M. M. Electronic conductivity of complexes of poly-p-phenylene. *Polymer Preprints* **1963**, *4*(1), 208–212. Data was later republished in Labes, M. M. Conductivity in polymeric solids. *Pure Appl. Chem.* **1966**, *12*, 275–285.

61. McNeill, R.; Siudak, R.; Wardlaw, J. H.; Weiss, D. E. Electronic conduction in polymers. *Aust. J. Chem.* **1963**, *16*, 1056–1075.

62. Bolto B. A.; Weiss, D. E. Electronic conduction in polymers. II. The electrochemical reduction of polypyrrole at controlled potential. *Aust. J. Chem.* **1963**, *16*, 1076–1089.

63. Bolto, B. A.; McNeill, R.; Weiss, D. E. Electronic conduction in polymers. III. Electronic properties of polypyrrole. *Aust. J. Chem.* **1963**, *16*, 1090–1103.

64. Dall'Olio, A.; Dascola, G.; Varacca, V.; Bocchi, V. Résonance paramagnétique électronique et conductivité d'un noir d'oxypyrrol électrolytique. *C. R. Seances Acad. Sci. Ser. C* **1968**, *267*, 433–435.

65. Jozefowicz, M.; Belorgey, G.; Yu, L.-T.; Buvet, R. Oxydation et réduction de polyanilines oligomères. *C. R. Seances Acad. Sci.* **1965**, *260*, 6367–6370.

66. Combarel, M. F.; Belorgey, G.; Jozefowicz, M.; Yu, L.-T.; Buvet, R. Conductivité en courant continu des polyanilines oligomères: Influence de l'état acide-base sur la conductivité électronlque. Note. *R. Seances Acad. Sci., Ser. C* **1966**, *262*, 459–462.

67. Jozefowicz, M.; Yu. L. T. Relations entre propriétés chimiques et électrochimiques de semi-conducteurs macromoléculaires. *Rev. Gen. Electr.* **1966**, *75*, 1008–1013.

68. Yu, L. T.; Jozefowicz, M. Conductivité et constitution chimique de semi-conducteurs macromoléculaires. *Rev. Gen. Electr.* **1966**, *75*, 1014–1018.

69. De Surville, R.; Jozefowicz, M.; Yu, L. T.; Perichon, J.; Buvet, R. Electrochemical chains using protolytic organic semiconductors. *Electrochim. Acta* **1968**, *13*, 1451–1458.

70. Cristofini, F.; De Surville, R.; Jozefowicz, M.; Yu, L.-T.; Buvet, R. Propriétés élecrochemiques des sulfates de polyaniline. *R. Seances Acad. Sci., Ser. C* **1969**, *268*, 1346–1349.

71. Labarre, D.; Jozefowicz, M. Polymères conducteurs organiques fimogènes à base de polyanilines. *C. R. Seances Acad. Sci., Ser. C* **1969**, *269*, 964–966.

72. Jozefowicz, M.; Yu, L. T.;Perichon, J.;Buvet, R. Proprietes nouvelles des polymeres semiconducteurs. *J. Polym. Sci. Part C: Polym. Symp.* **1969**, *22*, 1187–1195.

73. Rasmussen, S. C. Electrically conducting plastics: Revising the history of conjugated organic polymers. In *100+ Years of Plastics. Leo Baekeland and Beyond*. Strom, E. T.;Rasmussen, S. C.; Eds.; ACS Symposium Series 1080, American Chemical Society: Washington, DC, 2011, pp. 147–163.

74. Rasmussen, S. C. Early history of polypyrrole: The first conducting organic polymer. *Bull. Hist. Chem.* **2015**, *40*, 45–55.

75. Feast, W. J. Synthesis and properties of some conjugated, potentially conductive, polymers. *Chem. Ind.* **1985**, 263–268.

76. Zurer, P. Organic polymer exhibiting copperlike conductivity developed. *Chem. Eng. News* **1987**, June 22, 20–21.

77. Allen, N. S.; Murray, K. S.; Fleming, R. J.; Saunders. B. R. Physical properties of polypyrrole films containing trisoxalatometallate anions and prepared from aqueous solution. *Synth. Met.* **1997**, *87*, 237–247.

78. Truong, V.-T. Conducting polymers: A novel alternative material. *Chem. Aust.* **1997**, *64*, 20–22.

79. Tran, H. D.; Li, D.; Kaner, R. B. One-dimensional conducting polymer nanostructures: Bulk synthesis and applications. *Adv. Mater.* **2009**, *21*, 1487–1499.

80. Merton, R. K. The Matthew effect in science. *Science* **1968**, *159*, 56–63.

81. Street, G. B.; Clarke, T. C. Conducting polymers: A review of recent work. *IBM J. Res. Dev.* **1981**, *25*, 51–57.

82. Bott, D. Electrically conducting polymers. *Phys. Technol.* **1985**, *16*, 121–126.

83. Kaner, R. B.; MacDiarmid, A. G. Plastics that conduct electricity. *Sci. Am.* **1988**, *258*(2), 106–111.

84. Scott, C. History of conductive polymers. In *Nanostructured Conductive Polymers*. Eftekhari, A., Ed.; John Wiley & Sons: Chichester, UK, 2010, pp. 3–17.

85. Hargittai, I. Risking reputation. Conducting polymers. In *Drive and Curiosity: What Fuels the Passion for Science*. Prometheus Books: Amherst, NY, 2011, pp. 173–190.

86. Li, Y. Conducting polymers. In *Organic Optoelectronic Materials*. Lecture Notes in Chemistry 91. Y. Li, Ed.; Springer: Heidelberg, 2015, pp. 23–50.

87. Rasmussen, S. C. The path to conductive polyacetylene. *Bull. Hist. Chem.* **2014**, *39*, 64–72.

88. Rasmussen, S. C. The early history of polyaniline: Discovery and origins. *Substantia* **2017**, *1*(2), 99–109.

89. Rasmussen, S. C. *Acetylene and Its Polymers. 150+ Years of History*. Springer Briefs in Molecular Science: History of Chemistry. Springer: Heidelberg, 2018.

90. Rasmussen, S. C. The early history of polyaniline - revisited: Russian contributions of Fritzsche and Zinin. *Bull. Hist. Chem.* **2019**, *44*, 123–133.

91. Rasmussen, S. C. New insight into the "fortuitous error" that led to the 2000 Nobel Prize in Chemistry. *Substantia* **2021**, *5*(1), 91–97.

92. Rasmussen, S. C. The early history of polyaniline II: Elucidation of structure and redox states. *Substantia* **2022**, *6*(1), 107–119.

93. Inzelt, G. *Conducting Polymers. A New Era in Electrochemistry*. Monographs in Electrochemistry. F. Scholz, Ed.; Springer-Verlag: Berlin, 2008, pp. 265–269.

94. Elschner, A.; Kirchmeyer, S.; Lovenich, W.; Merker, U.; Reuter, K. *PEDOT. Principles and Applications of an Intrinsically Conductive Polymer*. CRC Press: Boca Raton, FL, 2011, pp. 1–20.

Chapter 2
Polyaniline

2.1 Introduction

Our discussion of the history of conjugated polymers begins with the oldest known example of these materials, the polymeric species now known as polyaniline.[1-4] This material has a very long history[5-11] and dates back to 1834 with a report by the German chemist F. Ferdinand Runge (1794–1867), in which he treated protonated aniline salts with various oxidants to generate green and black materials.[12] Furthermore, Runge's paper predates the 1839 report of Eduard Simon (1789–1856) that describes the generation of the material now identified as polystyrene.[13] As such, this gives polyaniline the distinction of being the oldest known fully synthetic organic polymer.[5,14] Of course, the modern name *polyaniline* was not introduced until the 1960s,[15-17] prior to which these materials were referred to by various color-based names, the most common of which was *aniline black*. This emphasis on color was due to the fact that the primary application of early polyaniline materials was as green, blue, and black dyes for cotton, with the commercialization of these dyes dating back to 1860.[5-7,11] The fact that these materials were known by other names for the first circa 125 years of their history has led many modern scientists to completely underestimate the amount of early research carried out on these materials. Other significant milestones in the history of polyaniline include its electropolymerization in 1862,[5,18] thus making it the oldest known conjugated polymer produced electrochemically, as well as it being the first conducting polymer to exhibit conductivities as high as 100 S cm^{-1}.[5] While the early report of electropolymerization is generally highlighted in the modern literature, the later milestone in conductivity is generally ignored, with most sources emphasizing later, and less substantial, conductivity studies in the 1980s.[1,2] Of course, by placing reports of polyaniline's conductivity in the 1980s, this further reinforces the general view that conducting polymers began with polyacetylene and that other examples followed afterward.

Polyaniline is synthesized almost exclusively by oxidative polymerization,[1-3,19-23] which is a form of step-growth polymerization[14] that can be accomplished either through the use of chemical oxidizing agents or electrochemically via the application of anodic potentials. Regardless of how oxidation

The Origins and Early History of Conjugated Organic Polymers. Seth C. Rasmussen, Oxford University Press.
© Oxford University Press (2025). DOI: 10.1093/9780197638194.003.0002

is induced, the general mechanism is the same and is initiated by the removal of an electron from the π-system of aniline to form the corresponding radical cation, which can exist in multiple resonance forms (Figure 2.1).

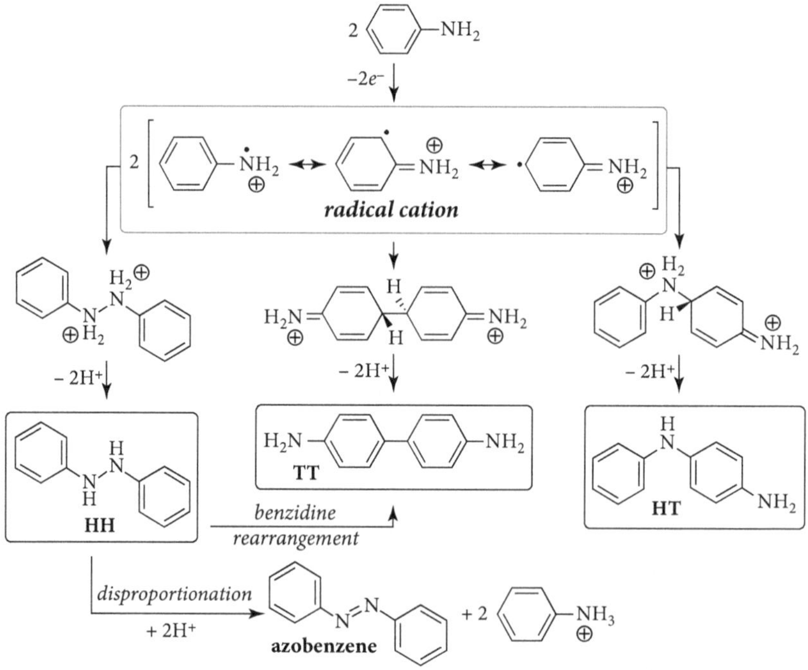

Figure 2.1 Initial dimerization of aniline.

Modern spin density studies predict that the unpaired electron is localized on either the aniline nitrogen or the *para*-carbon of the benzene ring, with near equal distribution between the two resonance forms.[23] As such, this can result in three possible couplings between radical cations: nitrogen–nitrogen (head-to-head, HH); nitrogen–arene (head-to-tail, HT); and arene–arene (tail-to-tail, TT).[19-23] This initial coupling is then followed by deprotonation to generate the neutral dimer. Of the three possible products, diphenylhydrazines formed by HH coupling are not stable, particularly under acidic conditions where the HH dimer is converted to the TT dimer via the benzidine rearrangement.[19,24-26] It is also possible for two equivalents of the HH dimer to disproportionate in the presence of acid to give azobenzene and two equivalents of aniline (Figure 2.1).[24,25] Because of this additional reactivity, HH units do not contribute to polyaniline production,[20] and the initial coupling gives either TT or HT dimers. Of these two possibilities, TT coupling is favored over HT coupling at the high radical cation concentrations typical of most polymerization conditions (i.e., large excess of oxidant and low pH),[21,22] although HT coupling is also possible depending on the exact conditions applied.

Further polymerization then continues through oxidation of the neutral dimers to form new radical cations (Figure 2.2). These oligomeric radical cations again undergo coupling, either with simple monomeric radical cations or radical cations of other oligoanilines, thus generating still larger oligoaniline species after deprotonation, which in turn can be oxidized yet again. The overall step-growth process thus propagates via sequential oxidation, coupling, and deprotonation steps to ultimately give polymeric products.[19,24–26]

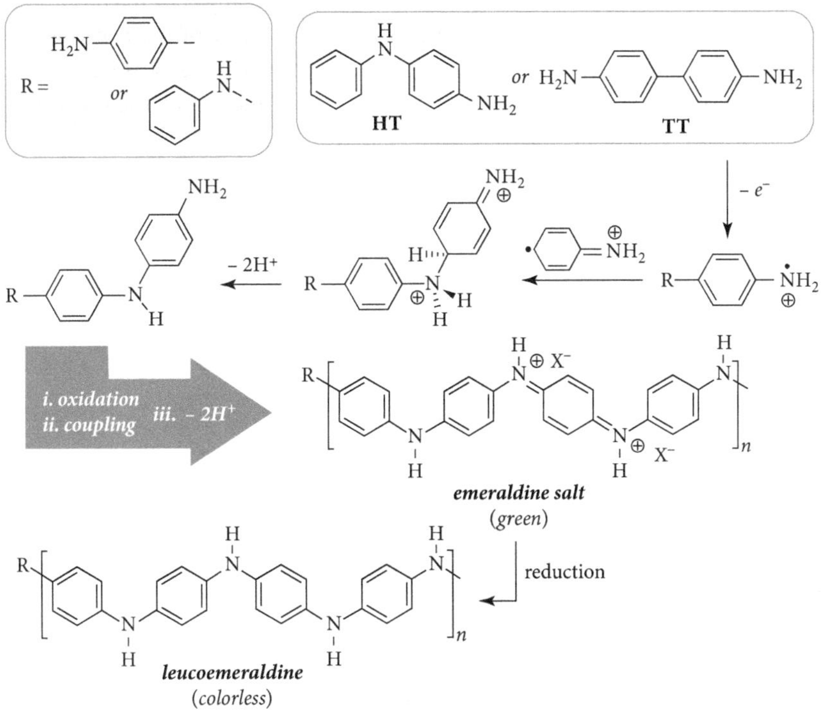

Figure 2.2 Polymerization mechanism from the initially produced dimer intermediates.

As the polymeric materials produced undergo oxidation at lower potentials than either aniline or shorter oligoanilines, the products resulting from oxidative polymerization are initially formed in their oxidized state. Thus, in order to isolate the neutral polymer, a final reduction step is required. In the case of polyaniline, its most common oxidized form is the half-oxidized *emeraldine*, which can exist as either the violet-blue base or the protonated green salt (Figure 2.2). However, as aniline is most commonly polymerized in strongly acidic media, the typical product initially generated is the emeraldine salt.

2.2 A brief history of aniline

The history of aniline is an interesting case, as it was independently "discovered" by multiple researchers,[10,27,28] beginning with the 1826 report by the German chemist Otto Unverdorben (1806–1873, Figure 2.3) of an oil isolated via the dry distillation of indigo.[29] Indigo also has a long and colorful history, with documented reports as far back as 27 BCE.[30] Utilized as a dye and pigment throughout antiquity, the color of the indigo dye originates from an organic species also commonly known as indigo (Figure 3.3). In addition to being the primary coloring agent of the dye isolated from the indigo plant *Indigofera tinctorial*, the compound indigo is also largely responsible for the color of the dye isolated from the plant woad, *Isatis tinctorial*.[31,32] The structure of indigo was not known in Unverdorben's time, however, and was not determined until 1883 by Adolf Baeyer (1835–1917).[33]

Figure 2.3 Otto Unverdorben (1806–1873).

By subjecting indigo to dry distillation, Unverdorben first distilled off water, followed by an oil, after which distilled a resin-oil mixture.[29] The oil was a colorless, alkaline liquid that was determined to be heavier than water. When combined with various acids, the liquid produced crystallizable salts, which led to Unverdorben giving it the name *Crystallin*.[34] Of such crystalline salts, he described the preparation and properties of the sulfuric acid and phosphoric acid derivatives. Finally, he noted that Crystallin oxidized in air to produce first a yellow and then a red species, the latter of which gave a yellow solution in water.

Eight years later, a volatile oil was then isolated from coal tar distillates by another German chemist, F. Ferdinand Runge (1794–1867, Figure 2.4). Runge began the investigation of distillates of coal tar in 1834, which he separated into discrete acidic and basic substances.[35] Runge initially treated coal oil by shaking it with lime and water, from which an aqueous lime-oil solution was isolated. This yellow-brown solution was then distilled to give a thick oil consisting of a mixture of species. Two organic bases could then be isolated from this initial oil by sequential distillations. Thus, distillation was first performed from an excess of HCl, the product of which was then distilled from an excess of NaOH. The isolated product was then distilled again from an excess of acetic acid, which resulted in the isolation of a mixture of two acetic acid salts. These salts were then converted to oxalic acid salts, which allowed separation via crystallization. Neutralization of the first of these salts gave a volatile oil characterized by a scarcely noticeable, but peculiar, odor. As illustrated by the separation methods applied

Figure 2.4 Friedlieb Ferdinand Runge (1794–1867).

Reproduced courtesy of the Edgar Fahs Smith Memorial Collection, University of Pennsylvania (CC0 1.0).

by Runge, this oil was basic in nature and formed colorless salts when treated with various acids. Furthermore, when treated with chlorine of lime (calcium hypochlorite, $Ca(OCl)_2$), both the oil and its salts became aquamarine. Because of this color response, he gave this oil the name *Kyanol*, from a combination of the Greek *kuanós* ("blue") and the Latin *oleum* ("oil").[35]

A third study was then reported in 1840 from Russia by the German chemist Carl Julius Fritzsche (1808–1871, Figure 2.5).[36-38] By treating indigo with a hot, strongly basic solution, Fritzsche isolated a salt mass of a reddish-brown color. If this salt was then heated in a retort, it was converted to an oily material during which aqueous ammonia was simultaneously distilled. Further distillation of the brown, oily material resulted in the isolation of a colorless oil, leaving a brown, resinous body remaining in the retort. Fritzsche gave this product the name *Anilin* after *anil*, which was an older name for the indigo plant introduced by the Portuguese, but which can ultimately be traced back to Sanskrit origins.[10]

Figure 2.5 Carl Julius Fritzsche (1808–1871).

Anilin was characterized as an oxygen-free base that formed light, highly crystalline salts when treated with acids, reporting both the HCl and oxalic acid salts as examples. In its purest state, the base strongly refracts light and exhibits a strongly aromatic, but unpleasant, odor. The specific gravity of Anilin was determined to be 1.028, its boiling point to be 228°C,[39] and combustion analysis led to the formula $C_{12}H_{14}N_2$ (a doubling of the modern formula C_6H_7N).[40]

Although no one had seemed to initially make the connection between Crystallin and Kyanol, the same could not be said for Crystallin and Anilin, perhaps due to the fact that they were both isolated from indigo. In the initial German publication of Fritzsche's report,[37] the editor of the *Journal für praktische Chemie*, Otto Erdmann (1804–1869), added a postscript that gave a side-by-side comparison of the known properties of Anilin and Crystallin, as well as giving a list of unknowns not addressed by Fritzsche's publication:[41]

> These and other questions that have to be imposed remain undecided. However, the last of these can almost certainly be answered. Anilin is most probably no other body than Krystallin described by Unverdorben already 14 years ago. . . . Unverdorben's description of Krystallin is not complete. However, the agreement between the properties of Krystallin and of Anilin given by him is so great that Herr Fritsche, if there is a difference, had the obligation to prove it by specific experiments.

When Fritzsche's paper was then republished in *Annalen der Chemie und Pharmacie*,[38] Justus Liebig (1803–1873) added his own postscript, which repeated Erdmann's paraphrased comments along with his full agreement.[42] Liebig, however, went even further to state,[42]

> Herr Erdmann . . . must not be surprised at the methods of Herr Fritzsche. Herr Fritzsche is one of those who mines by robbery; when he learns that some chemist is engaged in an investigation promising him valuable results, he undertakes not to help or render him any services, or to help carry the burden, but, like the corsairs, tries to unburden him in a quite particular way.

It is interesting to note here that neither Erdmann nor Liebig mentions Runge's Kyanol, and confirmation that Crystallin, Kyanol, and Anilin were all the same species would not come until later. In the meantime, however, Fritzsche's Russian colleague Nikolai Nikolaevich Zinin (1812–1880, Figure 2.6) reported the reduction of nitrobenene with hydrogen sulfide (H_2S) in 1842.[43-45] By adding H_2S to nitrobenzene in ammonia-saturated ethanol, Zinin obtained a mixture of elemental sulfur and yellow needles. The initial mixture was then boiled and the solution decanted from any solid sulfur, after which the isolated liquid was distilled to give a yellowish oil that was heavier than water. The oil was characterized as an oxygen-free base that was insoluble in water, but miscible in alcohol or ether, with a boiling point of circa 200°C, and combustion analysis led to a formula of $C_{12}H_{14}N_2$ (again, a doubling of the modern C_6H_7N).[40] The oil readily formed salts, several of which he characterized and reported (sulphate, HCl, and mercuric chloride). Based on its determined composition, Zinin named the oil *Benzidam*.[43-45]

Figure 2.6 Nikolai Nikolaevich Zinin (1812–1880).

After the initial report, Fritzsche was quick to point out the similarities of Benzidam and Anilin, publishing a short note following the publication of Zinin's work in the *Journal für praktische Chemie*:[46]

> To the most interesting treatise of Mr. Zinin, I must add the remark that the base designated as new under the name of Benzidam is nothing but Anilin. In its properties, as well as in its composition and the composition of the salts, Benzidam agrees so perfectly with Aniline that there can be no doubt about its identity.

One can certainly imagine that Fritzsche's own rebukes from Erdmann and Liebig were most likely on his mind as he wrote this. Still, it is interesting to note that Fritzsche only refers to his own Anilin here and includes no mention of either Crystallin or Kyanol. Nevertheless, all of these debates were finally put to rest in 1843, when August Wilhelm Hoffmann (1818–1892) published a comparative study on the analysis and reactivity of all of these species, as well as their salts and derivatives. In the process, he concluded that Crystallin, Kyanol, Anilin, and Benzidam were indeed all the same compound.[47] Interestingly, of the various names given for the base, Hoffmann felt that only the original name Crystallin might be retained, although he favored instead the name *Phenamid*.[47] Nevertheless, Zinin was also using the name Anilin by 1845,[48] and it ultimately was the name that endured as the modern aniline, the preferred International Union of Pure and Applied Chemistry (IUPAC) name for this aromatic amine.

A little over a decade later, the French scientist Pierre Jacques Antoine Béchamp (1816–1908, Figure 2.7) then improved on Zinin's reduction of nitrobenzene in 1854.[49] While investigating the reduction of various nitro species with iron reagents, Béchamp found that ferrous acetate proved to be a very simple and effective reducing agent for the conversion of nitrobenzene to aniline. This could be accomplished either by the use of ferrous acetate solutions or the combination of iron filings in concentrated acetic acid, after which the aniline acetate salt can be isolated from the reaction by distillation and then converted to the free base via treatment with KOH. The efficiency and cost effectiveness of this method was highlighted by Béchamp as follows:

> From the point of view of the applications of which aniline might become the object, it is perhaps not unhelpful to point out how inexpensive the process is. In several experiments I obtained an amount of aniline, which is almost ¾ of the nitrobenzene used. One kilogram of nitrobenzene can produce 750 grams of aniline. Consequently, with commercial benzene, aniline could be obtained at a price of about 20 francs per kilogram.

Figure 2.7 Pierre Jacques Antoine Béchamp (1816–1908).

Not surprisingly, Béchamp's method then became the primary route to the large-scale preparation of aniline, providing access for its utilization in various chemistries and products, including the polymeric materials of discussion here.[27] However, the initial study of aniline oxidations predated the ready access of the aniline of the 1860s and dates back nearly to its early discovery. While Unverdorben reported insoluble black byproducts during his isolation of Crystallin in 1826, he did not study or characterize these materials, and thus we can only speculate if these may have contained polyanilines.[29] As such, the first to purposely oxidize aniline was Runge during his study of Kyanol,[12] which allows us to date the beginnings of polyaniline to 1834.[5-7,11]

2.3 Runge and the oxidation of aniline

Friedlieb Ferdinand Runge was born on February 8, 1794, in the village of Billwärder, now an area of Hamburg, Germany.[11,50-53] Known by his middle name,[51] Ferdinand was the son of Pastor Johann Gerhard Runge, and the third of seven children.[11,50-52] His mother died in 1806, after which his father remarried, with an additional son from the second marriage.[51] His family was quite poor and could not financially support the children's education beyond the local primary school.[11,51,52] As a consequence, Runge was apprenticed to his uncle at age 15, studying pharmacy from 1810 to 1816 at the Ratsapotheke in Lübeck, on the Baltic coast.[11,51-53] He then entered the University of Berlin in October 1816 in order to study medicine.[11,51] Less than two years later, in the spring of 1818, he moved to the University of Göttingen, where he attended lectures by Friedrich Stromeyer (1776-1835).[11,51,52] Runge left Göttingen after only one semester, however, this time transferring to the University of Jena, where he studied analytical chemistry under Johann Wolfgang Döbereiner (1780-1849).[11,51-53] There, he was finally awarded a Dr. med. (Doctor of Medicine) degree on May 21, 1819, with a dissertation on belladonna.[11,50-52] While still in Jena, Runge wrote a book in which he summarized his results on the biologically active ingredients of plants. This was published as two volumes in 1820/1821 and included investigation of the extracts of coffee beans (caffeine), cinchona barks (quinine), safflower, and aloe. As a result, Runge is generally credited as the first to discover caffeine.[50-52]

Runge returned to Berlin at the end of 1819 with the desire to become *Privatdozent* in chemistry.[11,50-52] First, however, he needed to acquire a Dr. phil. degree, which he completed in in the summer of 1822 with a dissertation on indigo.[11,50-52] Runge then lectured at Berlin until October of 1823, after which he spent a couple years traveling Europe.[52] This began with a visit of Paris where he worked in the laboratory of J. B. Quesneville, who had a factory for the manufacture of chemical products.[51,52] Runge then continued his travels through a number of countries as the companion of Carl Milde, the son of a manufacturer in Breslau (now Wroclaw, Poland), in order to visit various industrial plants.[51] Upon their return in 1826, Carl invited Runge to move to Breslau, where Runge became associated with his father's factory and became familiar with the process of textile dyeing.[51] That same year he also joined the University of Breslau as *Privatdozent*, while additionally continuing his laboratory studies.[11,52] Runge was then promoted to extraordinary professor of technical chemistry in November of 1828.[11,50–53] However, without the security of a full professorship, his salary was somewhat limited, and he felt the conditions unfavorable for development of his practical ideas. As a result, he decided to end his academic career at Breslau in 1831 and moved back to Berlin.[11,51,52]

While in Berlin, Runge was offered a position in the chemical works recently acquired by the Königliche Seehandlungs Societät (Royal Maritime Trade Society), which was located in Oranienburg (a small town 35 km north of Berlin).[11,50–53] He accepted this position in either 1832[51,52] or 1833,[53] and it was here that he began perhaps the most fruitful period of his career, ultimately becoming technical director of the chemical plant in 1840.[50] Still, he had problems with the plant's business manager, Ernst Eduard Cochius, who ultimately purchased the factory from the Prussian state on September 4, 1850.[51,52] After this purchase, Cochius dismissed Runge in 1851,[51,53] with his last day of employment as December 31, 1852.[51,52] Runge was granted a small pension, however, providing that he remained in the local area where he could be consulted if needed.[11,51–53] This was not the end of his troubles, however. The business was not succeeding, and Cochius committed suicide in 1855, after which his widow ordered Runge out of the house provided as part of his pension, along with stopping further payments.[51,52] Luckily, he was able to move into a neighboring house belonging to a friend, and the King of Prussia, Friedrich Wilhelm IV, granted him a small royal pension.[51,52] Ultimately, at the age of 74, Runge died after a short illness on March 25, 1867.[11,50–53]

It was in 1833, shortly after the start of his new position at Oranienburg, that Runge began the investigation of distillates of coal tar, including the previously discussed isolation of his Kyanol (modern aniline)[51,52] that he published in 1834.[35] In a continuation of his aniline research published later that same year,[12] Runge found that the treatment of aniline with various metal salts resulted in the formation of several colored species. The effectiveness of these reactions was dependent on the metal salt applied, with silver nitrate exhibiting the least reactivity. At room temperature this initially seemed to cause little change, although a black-brown precipitate formed after an extended period of time, which could be accelerated with heating. These efforts were continued with the application of a HCl solution of gold oxide to a porcelain plate, followed by a drop of aqueous aniline. Heating this combination to 100°C then resulted in a purple-colored spot, which rapidly acquired blue margins to become blue-gray upon drying.[12] If the HCl solution of the gold oxide was heated with a sufficient excess of aqueous aniline, a purple liquid formed, rather than just a spot, which was not blued by the action of alkalis.

Some of the most striking results, however, were achieved through the action of copper. Much like the previous reactions with gold, an HCl solution of copper oxide was applied to a hot porcelain plate and allowed to dry. The addition of a drop of aniline nitrate solution heated to 100°C then resulted in the formation of a dark green-black spot.[12] It was also found that the same result could be obtained by using copper oxide in nitric acid and aniline hydrochloride. In order to confirm that it was the metal species that was causing the color change, Runge confirmed that heating either aniline nitrate or aniline hydrochloride alone gave no decomposition or reaction. The addition of any copper salt, however, then caused the generation of the black material.[12] In the same way, heating aniline hydrochloride in the presence of nitric acid alone had no effect, but the addition of a copper salt again formed a black spot.

In the process of these studies, Runge noted that if sufficient amounts of the aniline salts could be produced, the colored products resulting from their reaction with these metal species could provide practical applications.[12] To illustrate this, he first showed that the addition of aniline hydrochloride to a hot porcelain plate coated with potassium dichromate ($K_2Cr_2O_7$) produced a dark black spot similar to the previously described reactions with metal salts. Then he treated cotton with the analogous lead chromate ($PbCrO_4$), which caused the cotton to take on a yellow hue. Aniline hydrochloride was then printed onto the fabric, which gave green patterns over a span of 12 hours. The use of more concentrated aniline hydrochloride solutions resulted in the development of black patterns rather than green, and both the green and black patterns remained unchanged when rinsed in water.[12,54] It is important to note that these green and black dyes represent the very first synthetic aniline dyes, predating other aniline-based dyes such as mauve (aniline purple) and magenta (fuchsine) by more than 20 years.[14,55-58] Thus, while mauve is often stated to be the "first aniline dye"[55,58] or the "first synthetic dye,"[57,58] it is more correctly the first commercial example of such dyes.[14]

Runge had a strong interest in dyeing and the coloring of fabrics, dating back to his time working at the Breslau factory of Carl Milde's father in 1826.[51] This interest is further demonstrated by his publication of a three-volume series entitled *Farbenchemie* (Color Chemistry).[59-61] Published over the time span of 1834–1850, these volumes summarized the state and technical methods of the field of dyeing at the time.[50] Due to both this interest and the various colored species that he had successfully produced from coal tar species such as aniline, Runge had written to his superiors in 1836 to propose that the chemical works at Oranienburg could produce such synthetic dyes from coal tar.[50,52,53] Much to his supreme disappointment, however, such efforts were never approved. Unknown to Runge, all of his letters to the heads of the Königliche Seehandlungs Societät eventually came to the desk of Cochius for review and recommendation, as they considered him the person ultimately in charge at Oranienburg.[52] Unfortunately for Runge, Cochius did not recommend his proposals as worth pursuing, and thus a significant opportunity was missed, as evidenced by the highly successful coal tar dye industry that developed in the 1860s. Nevertheless, Runge moved onto other studies, and it was only much later that the merits of his discoveries with synthetic dyes were finally recognized. Following Runge's initial efforts, further studies on the oxidation of aniline did not appear until 1840 with the work of yet another "discoverer" of aniline, Julius Fritzsche.[36-38]

2.4 Fritzsche and continued studies of oxidation products

Carl Julius Fritzsche was born in Neustadt, Saxony (now part of Germany), on October 29, 1808.[10,11,62-66] Neustadt was located near the town of Stolpen, where his father, a physician, served as the district medical officer.[10,63] His mother came from the prominent Struve family. Although his given name was Carl, Fritzsche appeared to go by Julius, his middle name. Among his many publications, neither Carl nor the initial C is given, and the majority of these he authored as simply J. Fritzsche (although some publications give the alternate spelling of Fritsche).[10]

As there was no local gymnasium, Fritzsche was educated through private lessons until the age of 14, after which he moved to nearby Dresden in order to spend five years as an apprentice in pharmacy at the Salomons-Apotheke under his uncle Friedrich Adolph August Struve (1781–1840).[10,63,64] Completing his apprenticeship, he then moved to Berlin to manage the pharmacy laboratory of Johann Gottfried August Helming (1770–1830).[63,64] Although this position was not strictly scientific, it did enable him to acquire a position at the University of Berlin as assistant to Eilhard Mitscherlich (1794–1863) in 1830.[10,11,62-66]

It was at Berlin that Fritzsche is said to have developed his passion for science over the next two and a half years, largely due to his close relationship with Mitscherlich. It is also believed that it was Mitscherlich who most likely persuaded Fritzsche to enroll at the university in 1831,[10,63,64] where he had already been attending lectures the previous year. Fritzsche then acquired the Dr. Phil.

degree in 1833, with a dissertation on plant pollen.[10,11,63-66] As the subject of his doctorate was botany, not chemistry, Mitscherlich is generally credited with all of Fritzsche's chemical training.[10,11,65] Fritzsche clearly expressed his deep appreciation of Mitscherlich in his dissertation, stating,[63] "In these times, I express the greatest affection for Mitscherlich. With the deepest gratitude I will remember him to the grave. With paternal precaution, he led my occupations and gave me the opportunity to complete my knowledge." In 1834, Fritzsche then emigrated to Russia,[10,11,62-66] moving to St. Petersburg in order to become the head of Struve's Institute of Artificial Mineral Waters.[10,11,62] This was just one of a number of such institutes established by his uncle Friedrich, who was a pioneer in the commercialization of artificial mineral water.[10] In St. Petersburg, he also dedicated himself to scientific pursuits, with his name appearing in the *Mèmoires des savants étrangers* of the St. Petersburg Academy of Sciences for the first time in 1836.[10,63,64] From that point on, all of his papers first appeared in the publications of the Academy of Sciences, of which he became an adjunct member on August 24, 1838.[10,11,62-64,66] Fritzsche was then granted status as an extraordinary member of the Academy in 1844 and appointed an academician (full member) in either 1852[11,63,64,66] or 1853.[62] Besides his scientific activity in the Academy, he also contributed time to the Russian government as a member of the Imperial Commission for the Research and Utilization of the Caucasus Mineral Waters, as a chemist to the Medical Department, and as a consulting member of the Medical Council of the Minister of the Interior.[10,63,64] He also held occasional administrative positions within the Academy itself and spent three years as a member of its Administrative Committee.[63]

During his nearly 33 years of scientific activities within the Academy, he authored more than 60 papers, most covering various topics within organic chemistry.[63,64] The bulk of this research, including the aniline work of focus here, was all carried out in a small, modest laboratory next to his residence.[10,11,63] This was largely due to the fact that the laboratory of the St. Petersburg Academy was very primitive, with very little funds for its support.[67] The Academy eventually constructed a new and spacious chemical laboratory either in 1866[11,63,66] or 1867,[64,67] where Fritzsche then occupied shared facilities with Nikolai Zinin.[10,63-66]

Throughout his life, Fritzsche had experienced excellent health until 1869, when he had a stroke.[10,11,63-66] Although he did recover to some degree, he continued to suffer from paralysis on one side, and his speech and memory suffered.[10,11,63,64,66] His demeanor seemed suddenly changed, with his condition leading him to suggest that he preferred death to such a life.[10,63,64] Still, he did continue some scientific work,[63-65] even if he rarely attended academy sessions after that point.[63] Seeking physical and spiritual support, he eventually returned home to Germany in 1870, where he had a circle of family and friends.[10,63,64,66] Nevertheless, his health continued to decline, and he finally died in Dresden on June 20, 1871.[10,11,63-65]

It was early in his time at St. Petersburg that Fritzsche reported the isolation of his Anilin from indigo,[36-38] only two years after being granted status as an adjunct member of the Academy. In addition to its isolation and basic characterization, Fritzsche also went on to study the oxidation products of aniline, something previously done only by Runge before him. As previously noted by Unverdorben in 1826,[29] Fritzsche observed the air oxidation of aniline to give a yellow species. Fritzsche, however, found that this yellow product was just an initial intermediate, with longer exposure times leading to further products. While Unverdorben had reported a red species with longer exposure (although still yellow in solution),[29] Fritzsche observed the further transition to a brown species, ultimately leading to the production of a dark resinous mass.[36-38]

Fritzsche then extended the study to the intentional addition of oxidizing agents, beginning with the addition of nitric acid to generate blue or green materials.[36-38] The material produced was dependent on the specific reaction conditions, but Fritzsche concluded that it did not appear to be indigo. However, its further study was limited by the small amounts of material formed, and the fact that it continued to react with nitric acid, ultimately resulting in decomposition.

Fritzsche then continued his studies by dissolving aniline salts in chromic acid (H_2CrO_4, usually as a H_2SO_4 solution), which resulted in the formation of a dark green precipitate that ultimately became a dark blue-black.[36-38] Unlike the previous application of nitric acid, the use of chromic acid reproducibly produced these colored solids under a variety of conditions, even in fairly dilute solutions. Combustion analysis revealed that significant amounts of chromium were retained in these products, even for samples obtained from acid solutions.

Finally, he also investigated the application of potassium permanganate ($KMnO_4$).[36-38] Here, treatment of aniline salts with $KMnO_4$ caused the production of a brown precipitate containing manganese oxide. Fritzsche concluded with an acknowledgment that he had not yet been able to investigate these various reactions of aniline and its salts in much detail, and thus he planned to return to their further study in later publications.

He then reported another paper on aniline oxidation in 1843.[68,69] While this publication did not include further discussion of most of the previously reported reactions, it did return to the treatment of aniline with chromic acid. Here, he reported that he was able to reproducibly obtain the previously reported green product via the treatment of aniline with chromic acid, but he was unable to obtain products of a consistent composition, with analysis revealing that both the amount of carbon and chromium varied significantly from sample to sample. He did seem to recognize, however, that the product composition was affected by both the amount of chromic acid used and the amount of any other acids involved, even if he didn't understand how these variables caused the determined differences in the products generated. As Fritzsche put it,[69]

> Apart from the fact that the products are different in appearance, as one uses more or less chromic acid, or applies a greater or less excess of another acid, even apparently similar products give very different results in analysis, to which I still am missing the key.

In the same paper, Fritzsche also reported the treatment of aniline salts with potassium chlorate ($KClO_3$).[68,69] By slowly adding an HCl solution of $KClO_3$ to the aniline salt in a water-alcohol mixture, a fine blue precipitate formed to give the combined mixture a quite viscous nature. If the blue precipitate was then isolated by filtration and washed with alcohol, its color turned green, with the filtrate brown-red in color. After drying, the isolated solid became dark green, and analysis of its composition led to an empirical formula $C_{24}H_{20}N_4Cl_2O$. Based on Fritzsche's discussion of this formula, however, this appears to be a doubled formula in the same way as his previously reported formula of Anilin.[36-38] Still, this is in near perfect agreement with the structure of the emeraldine salt given in Figure 2.2 ($X = Cl^-$), which would give an empirical formula of $C_{12}H_{10}N_2Cl$.

Lastly, Fritzsche also stated that he had produced analogous products via the successful application of H_2SO_4 solutions of either potassium bromate ($KBrO_3$) or potassium iodate (KIO_3). Unfortunately, though, he did not provide any details on these other materials. Nevertheless, Fritzsche seemed to have limited his study of the oxidation products of aniline to just these two papers, and he left unanswered the various remaining questions he had raised.

2.5 The birth of commercial aniline dyes

As introduced in Section 2.3, Runge demonstrated the first aniline dyes for the coloring of cotton via the oxidation of aniline with lead chromate ($PbCrO_4$) in 1834.[12] While he had proposed to his superiors to mass-produce and commercialize such dyes, that unfortunately did not come to pass. Of course, a significant limitation to such a commercial enterprise in the 1830s was the limited availability of aniline at the time, which still had to be isolated from natural sources. Of course, this changed with the synthetic production of aniline via the reduction of nitrobenzene, first by Zinin in 1842[43-45] and then by the more efficient methods of Béchamp in 1854.[49] As such, the stage was now set for the commercial production of aniline dyes, beginning with the production of the purple dye mauve (aniline purple) by British chemist William Perkin (1838-1907) in 1856.[14,55,70]

Generated as a minor byproduct during the oxidation of aniline sulfate with potassium dichromate, the purple material was isolated by washing the resulting black precipitate with methanol. This was then followed up with the production of magenta (fuchsine) by the French chemist François-Emmanuel Verguin (1814–1864) in 1858.[14,71] Based on the initial commercial success of both mauve and magenta, it was thus only a matter of time before someone revisited the colored materials of Runge and Fritzsche for commercial applications. The earliest of these efforts was led by the British chemist Frederick Crace Calvert (1819–1873) during the final years of the 1850s.[72]

Frederick Crace Calvert (sometimes given as Crace-Calvert, Figure 2.8) was born on November 14, 1819, near London.[11,73–75] By the age of 16, he left England and moved to France.[76,77,79] There, he went on to study chemistry under Auguste Gerardin at the University of Rouen. After two years, he then moved to Paris, where he continued studies in the natural sciences at the Jardin des Plantes, École de Medecine, the Sorbonne, and the College de France.[11,73,74] At the age of 21, he then became the manager of the chemical works of Robiquet and Pelletici, but he soon left to become assistant to Michel Eugène Chevreul (1786–1889) at the National Museum of Natural History in 1841.[11,73–75]

Figure 2.8 Frederick Crace Calvert (1819–1873).

Reproduced courtesy of Wellcome Collection. Attribution 4.0 International (CC BY 4.0).

At the age of 28, he then returned to the country of his birth in late 1846, although his years in France left him with a French accent.[76] Upon his return, he was appointed to the chair of the honorary professorship of chemistry at the Royal Institution, and later as a lecturer on chemistry at the Manchester Royal School of Medicine on Pine Street.[73,74] By 1847, he was publishing papers on the bleaching of fabrics and went on to work on a wide range of topics, although most with some form of application to manufacture. Due to his various contributions, he became a Fellow of both the Royal Society and the Chemical Society, as well as a member of other societies both at home and abroad. His successful career was then cut short when he was struck with a fatal illness in the summer of 1873,[73,74,76] resulting in his death in Manchester on October 24, 1873.[11,73-75]

Crace Calvert first isolated a satisfactory amount of phenol from coal tar in 1847, using it to develop picric acid as a yellow dye in 1849,[75] and then set up the partnership of F. C. Calvert and Co. in Manchester specifically for its manufacture in 1857.[70] In the same year, Crace Calvert claimed that he and his co-workers, Samuel Clift and Charles Lowe, had obtained purple and red dyes by oxidizing aniline with manganese dioxide, potassium dichromate, or nitric acid, which could then be precipitated with tannin. Acetic acid solutions of the resulting tannates could then be used to dye cotton, silk, and wool, samples of which he had exhibited early in 1858.[72] However, these were viewed to be too expensive to be competitive.[70]

This seemed to have changed by 1860, by which time Crace-Calvert and his coworkers had obtained the blue-green color of Fritzsche by reproducing his reaction of aniline hydrochloride with potassium chlorate ($KClO_3$), with the ultimate goal of producing the same color on cloth fibers.[54] Sometime prior to the summer of 1860, they had successfully developed green and blue aniline dyes for the coloring of cotton, for which they filed a joint patent on June 11, 1860.[77-79] It should be highlighted that this patent is technically the first patenting of a conjugated polymer.

The final optimized methods were quite similar to those initially reported by Runge in 1834 and involved the printing of an aniline salt (either the hydrochloride or tartrate) onto cotton previously treated with $KClO_3$, which resulted in a green color after 12 hours. This green color was given the name *emeraldine*,[54,77-81] which later became the name used to refer to the most common form of polyaniline (Figure 2.2). If the initially produced green-dyed fabric was further treated with potassium dichromate ($K_2Cr_2O_7$), the green color could be converted to an indigo blue color, which was given the name *azurine*.[77-81] Although such dichromate reagents had been previously shown to effectively oxidize aniline, the conversion of emeraldine to azurine was also shown to occur with the treatment of various alkaline species or when boiling with sufficiently basic soap solutions,[54,78,79] indicating that the primary action of the dichromate here was as a base rather than a further oxidant. Crace Calvert presented these results as part of an address before the Society of Arts on February 5, 1862, which was then published in the *Journal of the Society of Arts*.[77] Later that same year, samples printed by these methods were exhibited as part of the chemical section of the London International Exhibition of 1862.[81]

Crace Calvert and his coworkers then contracted the printers Wood and Wright to produce their green and blue colors late in 1860.[80,81] Wood and Wright considered the colors good enough for commercialization and further optimized the methods to result in a dark shade that could be considered black (Figure 2.9).[80,81] The improvements introduced were achieved by either adding iron salts or other oxidizing agents to the $KClO_3$, after which the fabric was printed with aniline hydrochloride and then treated with either a weak solution of potassium dichromate or bleaching powder (calcium hypochlorite or $Ca(OCl)_2$) to give a black color.[54,80,81] Alternately, a mixture of copper nitrate $(Cu(NO_3)_2)$ and aniline hydrochlorate, without the addition of $KClO_3$, could be printed directly onto the fabric to gradually give a dark green or black. The blacks in these cases were really either a green or dark blue, but ones that were so dark that they could be viewed as black.[80,81] About the same time, aniline-based black dyes were independently introduced by both John Lightfoot (1831–1872) and Heinrich Caro (1834–1910).

(a)   (b)

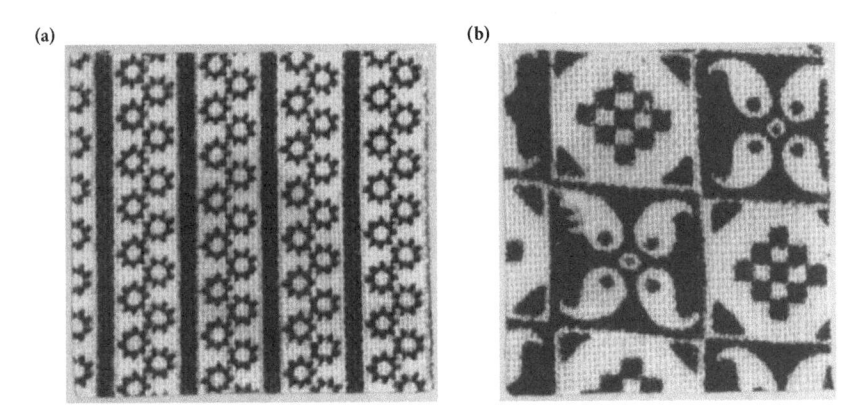

Figure 2.9 Printed emeraldine and aniline black on cotton.

Reproduced from Crace Calvert, F. *Lectures on Coal-Tar Colours, and on Recent Improvements and Progress in Dyeing and Calico Printing.* Henry Carey Baird: Philadelphia, 1863.

The son of Thomas Lightfoot (1811–1866), John Lightfoot Jr. was born in 1831 and named after his paternal grandfather, John Lightfoot Sr. (1774–1820).[75,82] Both his father and grandfather, as well as his uncle John Emanuel Lightfoot (1802–1893), were all colorists and connected with the Broad Oak Printworks of Hargreaves, Dugdale & Co. in Accrington, a town in Lancashire, about 20 miles north of Manchester.[75,79,82] During the 1800s, Accrington was a center of the cotton and textile machinery industries, with cotton mills and dye works providing work for the town's inhabitants. Here, Broad Oak had grown to become the largest printworks in the northwest of England by the mid-19th century.[75]

Although little is known about Lightfoot's upbringing and education, it is known that his family had excellent connections in continental Europe, with Lightfoot traveling to France and Alsace as part of his practical education in 1854.[75] He then followed the family tradition and began working at Broad Oak in circa 1855.[82] Working at Broad Oak until about 1868, Lightfoot played a leading role there in the use of madder, archil (or orchil), and indigo colors, while also beginning to experiment with the semisynthetic colorant murexide, the behavior of aniline colors, methods of fixing new dyes to wool and cotton, and imitations of madder and aniline colors. However, it was his development of a highly successful aniline black while at Broad Oak that is of focus to the current discussion. Lightfoot then moved to a new position in either late 1867 or early 1868, to become general manager at the Lower House print works in Burnley. At Lower House, Lightfoot continued with experiments on aniline black, notably the efficacy of different metals.[82]

According to Lightfoot, it was in November of 1859 when he discovered a fast and brilliant black for the printing of cotton through the combination of aniline hydrochloride and $KClO_3$.[78,82] His initial efforts applied aniline hydrochloride and $KClO_3$, which was combined in a starch paste and printed onto cotton with a wood block. Unfortunately, this generated little to no color, even after 24 hours. However, when the same formulation was printed using a machine with copper rollers, a green color was produced within 12 hours.[78,79] Believing that copper was a critical factor, he then added nitromuriate of copper (a mixture of $CuCl_2$ and $Cu(NO_3)_2$) to the formulation, which ultimately gives an intense black. Lightfoot noted that the intensity of the black was dependent on both the amount of copper salt added, as well as the length of aging.[78] The initially printed color was a pale green, which gradually developed into an intense green in about 24 hours. When washed in water without either alkali or soap, this color became an intense black.[78]

The conclusion that the copper was a necessary factor is not consistent with either the previous studies of Fritzsche[68,69] or the production of emeraldine by Crace Calvert as discussed earlier,[77-81] both of which produced dark green materials by treating aniline hydrochloride with $KClO_3$ alone. However, in reviewing the formulation given by Lightfoot,[78] it is clear that the ratio of $KClO_3$ to aniline was insufficient to generate the polymeric species given in Figure 2.2. As such, an additional oxidant such as a copper(II) salt was required for him to generate the color in sufficient intensity, particularly if black was the goal.

Of particular interest is that this is the same basic approach utilized by the printers Wood and Wright as discussed earlier,[54,80,81] although they primarily used iron salts as the additional oxidant. Nevertheless, the addition of copper salts to the formulation of aniline black became a standard for the next several decades.

Although Lightfoot recognized that the price of aniline in 1859 was still high enough to be a disadvantage, he continued with experiments in July 1860, and, by 1861, he had produced a few samples.[78,79] He reported his discovery the following year as a letter to the editor in the December 6 issue of *Chemical News and Journal of Industrial Science*, along with some sample swatches.[83] At the end of Lightfoot's letter, the editor added a note that the color was not a pure black, and did not look that good in the solid form, although it showed extremely well in the printed samples. As with the previous cases of both Runge[35] and Wood and Wright,[54,80,81] the true color of the dye was most likely a very dark green or blue, which thus appeared black in sufficient concentration. Still, the black introduced by Lightfoot was generally considered superior to that of Wood and Wright.[82] British and French patents for this process were filed in early 1863,[78,79,82] with the U.S. patent granted in May of the same year.[79,84] In these patents, Lightfoot describes the dyeing of fabric with mixtures of aniline hydrochloride and $KClO_3$, followed by the addition of copper chloride.[78;81,84] The patents point out, however, that the copper chloride can be replaced with other copper salts or even salts of other metals, with iron, nickel, manganese, chromium, bismuth, and antimony given as specific examples.[78,84] Lightfoot then sold the patent rights to Jakob J. Müller-Pack of Basle.[79,82]

A considerable limitation of Lightfoot's printing process was its corrosive nature, leading to severe corrosion of the steel rollers of the printing machines, as well as streaking during the printing.[79,81,82] This problem was solved by Charles Lauth, who registered a British patent in 1964 that specified the use of insoluble copper sulfide (CuS). Due to its insoluble nature, the copper salt was inactive during the printing process but was later converted to soluble copper sulfate ($CuSO_4$) by action of the chlorate salt during development of the aniline black. As such, this dramatically reduced damage to the rollers of the printing machines.[79,81] Müller-Pack then purchased the patent rights to Lauth's process, which was combined with Lightfoot's methods in an effort to dominate the European and U.S. markets.[79]

In addition to the dyes of Crace Calvert and Lightfoot, yet another black aniline dye was produced in 1860 by Heinrich Caro (Figure 2.10). Caro was born February 13, 1834, in Posen, Prussia (now Poznań, Poland).[9,11,85,86] In 1842, the family moved to Berlin, where Caro was educated at the Köllnische Realgymnasium until 1852, after which he attended the Königliches Gewerbeinstitut (Royal Technical Institute) until 1855, which trained students for industry. At the same time, he also attended lectures at Berlin's Friedrich-Wilhelms-Universität (now the Humboldt University of Berlin).[9,11,85–88] It was near the end of his studies that Caro was encouraged to pursue subjects related to printing and dyeing, as discussion had begun about setting up a state school for the training of technicians in this field, which would require teachers.[9,85] Thus, in April of 1855, Caro took an apprenticeship with a calico printing company in Mülheim an der Ruhr, C. & F. Troost, where he primarily carried out analytical work.[9,11,85,86,88] Caro was then sent on a study trip to England in March of 1857, where he was instructed to visit various printing and dyeing factories. One of the factories he visited was Roberts, Dale & Co, having previously met the owner John Dale during a trip to Germany in 1854.[85,87] After Caro's return to Germany, C. & F. Troost built a larger factory and made the decision to transform itself into a corporation, becoming Luisenthaler Aktiengesellschaft of Mülheim.[85] Within this same period, Caro carried out his military service in 1857–1858.[9,11,85]

Figure 2.10 Heinrich Caro (1834–1910).

Caro then decided to try his luck in England, moving there in November of 1859. His initial efforts in England unfortunately failed, but he finally obtained a position with John Dale at the Cornbrook Chemical Works of Roberts, Dale & Co. in Manchester.[85-89] He remained there until October of 1866, after which he gave up his position and returned to Germany.[87,88] It is believed that this was at least partially due to his declining health. Once back in Germany, he took a post at the University of Heidelberg in the laboratory of Robert Bunsen (1811–1899), and began private consultancy work, primarily for the newly formed Badische Anilin- & Soda-Fabrik (BASF).[87] By the end of 1868, his work with BASF had developed to the point that he was hired as a technical director.[85-89] At BASF, he oversaw the development of several new dyes, including artificial alizarin, eosin, methylene blue, and azo dyes, as well as the initial stages of the indigo synthesis.[86-88] In 1877, Caro was awarded an honorary doctorate by the University of Munich in recognition of his various contributions to the dye industry.[88] It was also during this time period that German patent law was being introduced, with Caro involved in its development.[88,89] In 1884, Caro was appointed to the BASF board of directors,[89] but by the end of 1889, the strain of his activities in both the fields of chemistry and patent law, as well as his duties as director, led to an end of his direct involvement in the work at BASF.[88] Afterward, however, he did remain active on the company's supervisory board.[88,89] On September 11, 1910, Caro died in Dresden following a short illness.[11,85,86,89]

It was while working for Roberts, Dale & Co. that Caro developed a process for making aniline purple (mauve) by the oxidation of aniline salts with various salts of copper in 1960. Extraction of the desired purple dye by alcohol then left a dark, oily residue, which provided an excellent fast black dye for printing on cotton.[54,79,87] The residue was then commercialized for sale to printers by Roberts, Dale & Co. in 1862.[54,79,90] Within that same year, Caro's black was printed at the nearby works of Edmund Potter & Co. of Dinting Vale.[90] As with the previously discussed formulation of Lightfoot, this residue caused problems with the printing machinery, but it could be successfully printed with wooden hand blocks.[78] The black was particularly popular on mainland Europe, where it was marketed by Müller-Pack, which had close ties with Roberts, Dale & Co. in 1863.[87,90] These early efforts of Roberts, Dale & Co. helped to establish the advantages of aniline black over other black colorants of the time. Furthermore, it has been proposed that the aniline black work of Roberts, Dale & Co. may have been the motivating factor for Lightfoot to patent his process in 1863.[78] As discussed earlier, the aniline black found more widespread application in England following the introduction of insoluble copper sulfide by Charles Lauth in 1864.[87,90]

By 1871, all of these black dyes from aniline collectively became known as *aniline black*,[78] which later became the first real general term for the modern polyaniline. Furthermore, both of the dye names aniline black and emeraldine have been retained to some degree in the modern nomenclature of polyaniline materials. Following the introduction of these various aniline dyes, the next major innovation came from the physician and chemist Henry Letheby (1816–1876).[18]

2.6 Electrolysis of aniline: Letheby, Coquillion, and Goppelsroeder

Henry Letheby was born in 1816[91,92] or 1817,[93] in the port city of Plymouth, England. He received his early education in Plymouth,[93] before he commenced chemical studies in the laboratory of the Royal Cornwall Polytechnic Society in Falmouth, on the south coast of Cornwall, England.[91,93] There, he eventually became assistant, while also doing some lecturing.[93] In 1837, he entered the Aldersgate Medical School in East London, where he joined the chemistry class of Jonathan Pereira (1804–1853). That same year he also became a licentiate of the Society of Apothecaries.[91] He later became assistant to Pereira, who was professor of chemistry at the London Hospital.[91,93]

Letheby received his MB (Bachelor of Medicine) from London University in 1842[91,93] or 1843,[94] after which he succeeded Pereira as the chair of chemistry and toxicology at the London Hospital in 1846.[91,93] Already serving as the chief gas examiner of London,[92,93,95] he was elected as the city's medical officer of health in 1855,[91–93,95] while also serving as the city's public analyst.[91,92,94,95] In addition, he also served as a consulting chemist to the Great Central Gas Company.[92,95] He then went on to receive an MA and Ph.D. from an unknown German university in 1858.[91,94] Letheby was a fellow of both the Linnean Society of London and the Chemical Society of London[91,92] and was a founding member of the Society of Metropolitan Medical Officers, serving as its president in 1874.[91] Due to failing health, he resigned his city posts in February of 1874.[91] For some weeks in 1876, he had been unwell with what was believed to be inflammation of the lungs.[94] Although it was expected that he would recover, he died quite suddenly on March 28, 1876, at the age of 60.[91–95] He was then buried in Highgate Cemetery on March 30.[91]

After he was required to investigate two cases of fatal poisonings by nitrobenzene, Letheby began investigations into developing chemical tests for the detection of aniline, as he had found that nitrobenzene was reduced to aniline after ingestion. This thus led to a reported 1862 study that detailed the generation of blue or purple colors from the treatment of acidic solutions of aniline with various oxidizing agents.[18] These efforts began with describing the effects of oxygen on aniline salts, before showing that similar results could be made by treating dilute solutions of aniline in sulfuric acid with manganese dioxide (MnO_2), potassium dichromate ($K_2Cr_2O_7$), hydrogen peroxide, barium peroxide (BaO_2), chlorine, or chloride of lime ($Ca(OCl)_2$).

Letheby then continued his study by showing that similar deep blue to bluish-green materials could be generated via the electrochemical oxidization of sulfuric acid solutions of aniline.[18] This was accomplished using a strip of platinum connected to the positive pole of a small Grove cell (3 × 2 inches). The Grove cell (Figure 2.11) was an early high-current battery introduced by William Robert Grove (1811–1896) in 1839[96-98] as an improvement of the previous Daniell cell.[99] The basic Grove cell utilized two half-cells separated by a diaphragm, with one chamber containing an inert platinum electrode in HNO_3 and a second chamber containing a zinc anode in HCl or H_2SO_4.[97-99] Such a configuration would thus generate a cell potential of circa 1.7 V, which Letheby was able to use to generate a deep blue to bluish green pigment that adhered to the platinum strip as a fine powder.[18] As with the other previous reports of chemical oxidation, the exact color observed depended on the concentration of the aniline solution used.

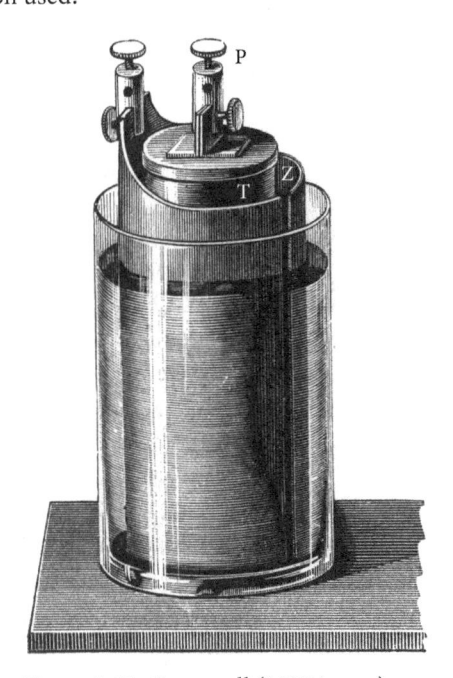

Figure 2.11 Grove cell (1897 image).

Letheby then prepared the material on a larger scale, using greater quantities of aniline and two larger Grove cells (6 × 4 inches) that, according to Letheby, were "arranged for intensity."[18] As he refers to intensity rather than capacity, it is assumed that the batteries were connected in series, which would double the cell potential. Aniline in H_2SO_4 was placed in a beaker into which a porous cell containing just dilute H_2SO_4 was also placed. A large sheet of platinum (4×6 in) was placed in the aniline solution and brought into contact with the positive pole of the battery, while a second platinum sheet was placed within the porous cell without aniline and attached to the negative pole. Using this setup, a thick layer of dirty bluish green pigment quickly covered the platinum sheet connected to the positive pole, which was then removed, washed with water, and dried to give a bluish black powder. This product was insoluble in water, alcohol, ether, or ammonia, and was only soluble in H_2SO_4. Immersion in ammonia did not dissolve the material, but this did cause it to acquire a brilliant blue color.

When dissolved in concentrated H_2SO_4, the solution formed was either blue, green, or violet depending on the concentration. Diluting these various acid solutions with water resulted in the precipitation of a dirty emerald green powder, which could be made blue with treatment of concentrated ammonia. Treatment of the green powder with concentrated H_2SO_4 again generated blue to purple solutions. The blue pigment generated via treatment with ammonia could also be partly decolorized by various reducing agents, which also increased its solubility.[18]

It should be highlighted that Letheby is often credited in the literature with the earliest production of polyaniline. Although this is certainly not correct, his report is still particularly noteworthy, as it does represent the earliest known report of the electropolymerization of a conjugated polymer.[5–7,11] Following Letheby's innovation, the electrochemical oxidation of aniline was not investigated again until further independent reports by Jacques Coquillion (b. 1836) and Friedrich Goppelsroeder (1837–1919) in 1875.

Of the French scientist Jacques J. Coquillion, very little is known. He was born in 1836 in Bellenot, Cote-d'Or,[100] and worked in the Laboratoire de Chimie à la Faculté de médecine in the École de Médecine of Paris during the 1870s.[101] Charles Wurtz (1817–1884) was the director of the Laboratoire de Chimie, but Coquillion does not seem to have been a Wurtz student. As of 1881, he was a professor at the Lyceum in Charleville.[100] Interestingly, he is listed as a civil engineer,[100] although all of his published work is chemistry.

Nearly all of his published work seems to be from his time at the École de Médecine, including reports on the production of aniline black from the electrolysis of aniline salts beginning in 1875.[102] His intention was to show that aniline black could be produced without the use of any metal species, as typically used for the oxidation of aniline salts to aniline black. To accomplish this, Coquillion used two Bunsen cells (Figure 2.12). Such Bunsen cells were essentially Grove cells in which the platinum electrode in the HNO_3 half-cell was replaced with a carbon rod, an innovation introduced by Robert Bunsen (1811–1899) in 1841[103] in an effort to reduce the cost of the cell.[99] As neither the previous platinum electrode nor the current carbon rod is part of the HNO_3 half-cell reaction, this did not have any real effect on the cell potential of Grove versus Bunsen cell. Although Coquillion does not directly specify that the two cells are coupled, it is assumed that these were used in the same way as previously discussed for Letheby. This is further supported by the fact that while he does not cite Letheby directly, he does mention him in passing (although misspelled as Litheby).[102] Also like Letheby before him, Coquillion attached platinum electrodes to both poles of the cell, which are then used to interact with the electrolysis solution.

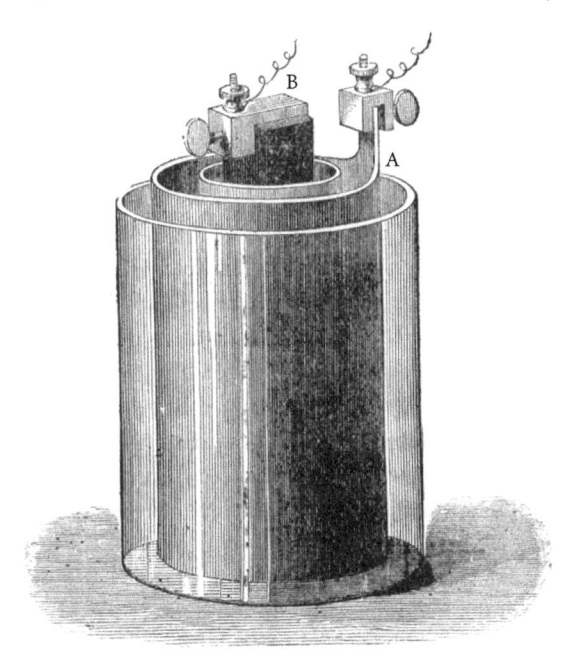

Figure 2.12 Bunsen cell.

Using this setup to electrolyze a concentrated solution of aniline sulfate, Coquillion noted that the bluish green film of Letheby is not seen but, rather, that a black mass was formed on the platinum electrode that attached to the positive pole after 12–24 hours.[102] He noted that this mass was sufficiently adhered to the electrode, yet it was easy to detach. After washing with ether and alcohol, the mass was dried in an oven to yield an amorphous substance that was black with some green reflections. The product was insoluble in most solvents, but upon treatment with sulfuric acid, it would take on a greenish color. This reaction was found to be reversible, and treatment with alkalis would again give a velvety black.

To verify that the production of the black mass was due to the electrolysis and not the result of the platinum serving as the electrode, Coquillion then used carbon electrodes.[102] These carbon electrodes were connected to the Bunsen cells via platinum wires, but without allowing the platinum wires any contact with the solution to be electrolyzed. This modified assembly resulted in identical results to those previously obtained via the use of platinum wires placed directly into the aniline sulfate solution.

Coquillion then repeated the electrolysis with other various aniline salts, including aniline nitrate, aniline hydrochloride, aniline acetate, and aniline tartrate.[102] Of these, the nitrate gave very similar results to the sulfate, with the exception that treatment with sulfuric acid resulted in a brown material that was ascribed to decomposition. The hydrochloride gave what was described as "a black, lumpy product," with the poorer quality believed to be due to competing oxidation of chloride anions. Of the two organic salts, the acetate gave a sticky black substance, while the tartrate gave no product. Although he did not attempt to determine the composition of any of these products, he concluded that he should be able to do so via further analysis.

He then followed this with an additional report in 1876, in which he began by describing the extensive purification of the carbon rods used as his primary electrodes.[104] In order to verify that these rods were free of metals or other contaminates, Coquillion first treated the carbon for three hours under a flow of chlorine in a strongly heated porcelain tube. The rods were then boiled in nitric acid, treated again with chlorine, and washed with distilled water. The rods treated in this way were 10 centimeters long, in which the upper parts were wound with platinum wires in order to connect them to the two Bunsen cells, leaving the lower ends of the rods to be immersed in the aniline salt solutions.

Of the various aniline salts that he had electrolyzed in this way, Coquillion concluded that the hydrochloride and sulphate were the only ones that he viewed suitable to give aniline black industrially. In both of these cases, he was able to obtain a suitable pasty mass at the positive pole after 24 hours of electrolysis, which when washed and dried gave a purplish black material. This material was soluble in concentrated sulfuric acid, which when diluted with water caused the precipitation of a greenish mass.

He then went on to discuss two additional aniline salts not previously reported, aniline arsenate and aniline phosphate, the second of which he acknowledged was really a mixture of phosphates.[104] In both cases, the generation of product was slow and difficult with these salts, with only small quantities of aniline black after 12 hours. The materials generated, however, were similar to those obtained from the hydrochloride and sulphate, although he noted that they did not appear to be identical. Still, he did not view these salts as suitable for the generation of aniline black industrially. As with his original report, Coquillion did not report any data concerning the constitution of these materials, but he did state that he planned to do so soon. Unfortunately, he did not seem to ever report any further studies as promised. However, a significant number of additional reports on the electrochemical oxidation of aniline were then reported by Friedrich Goppelsroeder, with the first of these published between the two reports by Coquillion.[105]

Christoph Friedrich Goppelsroeder (Figure 2.13) was born born April 1, 1837, in Basel, Switzerland.[106–110] His parents were Georg Friedrich Goppelsroeder (1801–1837), a banker, and Emma nee von Speyr (1812–1891), the daughter of a banker.[106,107] Going by his middle name, Friedrich attended the gymnasium in Basel before attending the higher "Les Auditoires" school of the Académie de Neuchâtel.[108,110] It was at the latter that he enjoyed the chemistry and physics lessons from Charles Kopp.[108,110] This experience influenced him to enter the University of Basel in the winter of 1855,[106–108,110] where he studied chemistry under Christian Friedrich Schönbein (1799–1868), a friend of the family.[106,107,109,110] He then moved to the University of Berlin in October of 1856,[106–108,110] where he received education in analytical chemistry through internships with Franz Leopold Sonnenschein (1817–1879) and Heinrich Rose (1795–1864).[106,108,110] He then moved to the University of Heidelberg in the winter of 1857, where received his doctorate (Dr. phil.) in chemistry, physics, and mineralogy under Robert Bunsen (1811–1899) in 1858.[106–110]

Figure 2.13 Christoph Friedrich Goppelsroeder (1837–1919).

Reproduced from W. Ostwald, Zu Ehren Friedrich Goppelsroeder's. *Kolloid-Zeitschrift* **1912**, *10*, 1.

Goppelsroeder then found employment with the Indienne fabric manufacturer Köchlin, Baumgartner & Cie. in Lörrach, working at their Wiesenthal bleaching, dyeing, and printing works. He remained there until 1860, when he became first the deputy of the public chemist in Basel, and then his successor in 1861.[106-110] That same year, he also became a private lecturer in chemistry at the University of Basel.[106,108-110] Following Schönbein's death in 1868, he was appointed professor extraordinarius of chemistry at Basel in 1869.[106-110] From 1872 to 1880, he served as director and professor of chemistry of the Mulhouse standing high school for chemistry.[106-110] However, his prosperity enabled him to resign from teaching in the spring of 1880 in order to work as a private scholar. Initially, he remained in Mulhouse but returned to Basel in 1898,[106-110] where he set up a private laboratory next to his house in order to focus on his research.[109] Throughout his career, he made important contributions to analytical chemistry, electrochemistry, the manufacture of dyes, and the study of capillary analysis, a precursor of paper chromatography.[107-109] After a brief illness, he ultimately died in Basel on October 14, 1919.[106-110]

Goppelsroeder's initial report on the electrolysis of aniline salts came in 1875, following the first paper of Coquillion.[105] Geoppelsroeder claimed that he had been investigating the electrolysis of aromatic species since late 1874 and was motivated by Coquillion's related work on aniline black to give an initial communication of his own efforts on aniline. Goppelsroeder provided no real experimental conditions, however, and just briefly discussed results, stating that he had observed the formation of aniline black via the direct oxidation of aniline and that "it is endowed with a metallic luster, like the colors of aniline in general, and gives, on paper, a complete black coloration."[105]

He then went on to point out that the electrolysis of organic bodies comes with difficulties and thus precautions need to be taken. In addition, he highlighted that temperature, concentration, and the pressure under which the electrolysis is performed all play a role, which led him to insist on the importance of making in-depth tests. Although he spends much of the communication discussing the difficulties of the technique, he also finds future potential, stating "I am convinced that one day we will succeed in taking advantage of electrolysis for dyeing and printing."[105]

Goppelsroeder then followed this up with a more extensive report,[111] which was published shortly after Coquillion's second paper. While this paper contains significantly more experimental detail, it does not describe the type of electrochemical cell used in the way that Letheby or Coquillion did, only stating that he used platinum electrodes with both weak and strong "galvanic current." In addition, he seems to have separate solution vials for the positive and negative poles, as he discusses the fact that the material used to conduct current from one vial to the other (cotton, filter paper, wool or silk) all become colored in the same way as the positive pole. In terms of the aniline salts studied, he found that aniline hydrochloride, sulfate, and nitrate all gave a green deposit at the positive pole, which passes through violet, violet blue, to dark indigo blue colors. In contrast, the tartrate, oxalate, and acetate salts only gave a brown deposit, with little green content.

While Goppelsroeder refers to effects of current strength, temperature, and concentration, this largely seems to only effect the rate of the product formation.[111] However, he only gives details for effects of concentration, stating that one milligram of the hydrochloride salt in 60 mL of water gave green material after a few hours. Increasing the concentration to one milligram in 30 mL of water gave not only green, but also blue and purple, products. Further increasing the concentration to 2.5 milligrams in 30 mL gave a blue-violet color, with some green after only two hours, with a very noticeable green material with additional time.

In terms of characterization of the products, he found the initial green deposit to be unaltered by treatment with ozone, but it becomes blue-green and then blue when treated with ammonia gas.[111] This second reaction is reversible, however, with the material becoming green again after removal of the ammonia. Furthermore, the green product becomes dark purplish blue on heating with a solution of potassium dichromate, which again becomes green with the addition of strong acid. The dark indigo blue deposit formed after longer periods was concluded to be a mixture of different dyes, primarily consisting of aniline black. It was found that the other dyes of the mixture could be separated through treatment with various solvents, including water, alcohol, ether, benzene, acids, and alkalis.

After its purification with various solvents, the resulting black material is described as a beautiful crystalline black with a metallic luster, which he generally referred to as *electrolytic black* to differentiate it from normal aniline black. Goppelsroeder went on to state, "Optical examination of electrolytic black showed it to be blacker than other aniline blacks that I have compared it to."[111] The material could not be sublimed and was found to be insoluble in water, alcohols, or benzene. Weak acids had no effect, even when boiling, but boiling with concentrated acetic acid did cause some greening. In a similar manner, the black material was resistant to the action of a variety of reducing and oxidizing agents. Lastly, he found that heating the black material to red in a combustion tube with a mixture of lime and soda gave off white vapors, which have the odor of aniline and turned brown. Ammonia was detected with even stronger heating.

Goppelsroeder concluded that he would be reporting the elementary analysis results of the purified bodies soon but then planned to leave the further study of electrolytic black to Coquillion. This was due to the fact that he felt Coquillion had priority on the topic, as he had been the first to report on it to the Academy.

Of course, it is again interesting that no reference to Letheby is ever made, which Coquillion at least mentioned in passing.

Goppelsroeder then turned to the electrolysis of various aniline derivatives,[112] including toluidine (methylaniline), pseudotoluidine, diphenylamine, and naphthylamine, before ultimately returning to the elementary analysis of electrolytic black later in 1876.[113] The material analyzed was generated electrolytically as previously described and then purified by means of successive treatments with water, alcohol, ether, and benzene, after which it was again washed with alcohol. This resulted in a velvety black powder, which was then dried at 110°C. Elemental analysis then led to the empirical formula $C_{24}H_{21}N_4Cl$. Boiling the purified product with a solution of caustic potash (KOH) removed all chlorine content to give a black material with a crystalline appearance and a metallic sheen. Analysis of this secondary product gave an empirical formula of $C_{24}H_{20}N_4$. The combination of these results then led to the conclusion that the original electrolytic product was an HCl salt of the formula $C_{24}H_{20}N_4 \cdot$ HCl. It should be noted that this formula is very similar to that previously reported by Fritzsche in 1843 for the chemical oxidation of aniline salts (see Section 2.4),[11,68,69] although with slightly higher H content and lower Cl content.

Following his report of the elemental analysis of his electrolytic black, Goppelsroeder stayed somewhat true to his word that he would leave the further study of electrolytic black to Coquillion. As such, he reported no further papers on the production or characterization of aniline black via electrolysis. He did, however, publish additional papers on the products of further reactions of the electrolytic product the following year, focusing largely on the various color changes involved with the treatment of his electrolytic black with a large number of different chemical reagents.[114,115] In addition, he reported the results of the effectiveness of various vats of such solutions for the dyeing of fibers.

2.7 Other early efforts to determine the composition and structure of aniline black

While Goppelsroeder contributed to the analysis of the composition of aniline oxidation products, he was not really the first,[9] with a particularly noteworthy report by Heinrich Rheineck dating back to 1872.[116] Although Rheineck's work is an important factor in the overall history of polyaniline, very little is actually known about him. Originally from Neckarsulm, Germany (near Stuttgart), Rheineck enrolled in the University of Tübingen at in the fall of 1860 in order to study pharmacy.[117] During his time at Tübingen, he carried out research on the action of sodium on allantoin under Adolph Strecker (1822–1871), resulting in a paper published in 1865.[118] Afterward, he then worked in Hohenheim for several years as an assistant chemist in agricultural chemistry.[119] By 1871, however, he had moved to Hagen, Germany, in North Rhine-Westphalia (near Dusseldorf),[119–121] where he studied aspects of inorganic dyes such as Prussian blue.[121] By the following year, he had already moved to nearby Elberfeld (ca. 30 km to the southwest),[116,122] likely employed by one of the smaller aniline dye works located there. It was here that he then published his seminal work on aniline black,[116] which attracted significant interest and was republished several times.[122–124]

Rheineck began his study with a discussion that explained that aniline black, as with other aniline dyes, is produced by the oxidation of aniline, which results in the molecular condensation of repeating aniline units. In addition, he stated that these oxidation products retain the basic nature of aniline, with aniline black representing a specific base. He then proposed that this base should be given the name *nigraniline* (Latin *nigrum* "black" + aniline) in analogy to the name *rosaniline* already used for aniline red.[116]

In order to illustrate the basis for his conclusions, he produced aniline black following commonly applied conditions for its production on fabrics (aniline hydrochloride, potassium chlorate, and copper chloride). He then allowed this aqueous mixture to evaporate in a porcelain dish and repeatedly moistened it until a dry, velvety black powder was finally produced, after which the powder was washed with hot water. As it was already well known that this material appeared dark green on fabric and turned dark blue-violet after treatment with alkalis, Rheineck proposed that the blue-violet species was a free base, with the initial green material its corresponding hydrochloride salt. To support this conclusion, he used either soda or ammonia to remove hydrochloric acid from the initial aniline black in order to generate what he viewed to be the free base.[116] He then showed that this base could be used to successfully remove acid from aniline salts. This was accomplished by first producing the isolated free base on a piece of cotton to give a blue-violet color. If the dyed cotton was then treated with aniline salts in the absence of oxidant, the fabric immediately turned green, consistent with the protonation of the fabric dye. Treating the fabric again with alkali then returned the fabric to its original blue-violet color. He then went on to show that the green hydrochloride salt could be treated with sulfuric acid to release HCl fumes and produce a violet solution. Upon dilution, this solution again gave a black-green precipitate, which he was confident corresponded to the sulfuric acid salt.[116]

Rheineck believed that these combined results confirmed his view that the blue-violet material was a free base (nigraniline), which gave green salts when treated with acid. Still, he decided not to follow through with further studies, stating,[116] "For lack of opportunity and facilities, I have to refrain from further elaboration on this interesting subject in scientific terms, and make other chemists aware of it, to which a well-established laboratory is available." As such, this remained his only publication on aniline black, and it was left to others to expand upon his initial efforts. One such example of such further efforts was the work of the German chemist Rudolf Nietzki (1847–1917), who reported various studies on aniline black starting in 1876.[125,126]

Rudolf Hugo Nietzki (Figure 2.14) was born in Heilsberg, Prussia (now Lidzbark Warmiński, Poland), on March 9, 1847.[127–130] His father was Karl Johann Emil Nietzki, a Protestant pastor, rector, and writer.[127,128] The family later moved to the Prussian town of Zinten (now Kornevo, Russia) in 1854, where his father served as the town's pastor.[129] Nietzki was initially educated

Figure 2.14 Rudolf Hugo Nietzki (1847–1917).

Reproduced courtesy of the University of Basel, Attribution-ShareAlike 4.0 International (CC BY-SA 4.0).

by his father, after which he was sent to the gymnasium in Königsberg (now Kaliningrad, Russia), as Kinten had no schools for higher education.[127-129] Nietzki was unhappy with his education at the gymnasium, however, and he left before completion to begin the study of pharmacy,[127,129] first as an apprentice in Zinten, followed by an internship in Kreuzburg, Prussia (now Kluczbork, Poland).[128,129]

Passing the assistant examination in 1865,[127,128] Nietzki took a position at a pharmacy in nearby Hirschberg (now Jelenia Góra, Poland),[128,129] before pursuing additional studies at the University of Berlin in 1867.[127-129] He passed the state pharmacy examination in 1870,[129] after which he was called to serve as a military pharmacist in the Franco-Prussian War.[127-129] After the end of the war in 1871, he became private assistant to August Wilhelm Hofmann (1818–1892) at the University of Berlin.[127-129] During his spare time, he also managed to complete some research in plant chemistry,[129] such that he obtained his Dr. phil. at the University of Göttingen in 1874.[127-130] As he had left his gymnasium education before completion, he did not have a certificate of maturity and thus could not take the degree in Berlin.[129]

After obtaining his doctorate, he then worked as a chemist for Matthes & Weber in Duisberg.[127,129] He soon moved to the University of Leiden, however, to become first assistant of the laboratory there in 1876,[129] working under Antoine Paul Nicolas Franchimont (1844–1919)[127-129] and Jacob Maarten van Bemmelen (1830–1911).[129] He moved again in 1879 to work as a research chemist in the laboratory of Kalle & Co. in Biebrich, Rhineland-Palatinate[127-129] but left in 1883 with the desire to carry out independent industrial and scientific studies, renting space in the laboratory of Dr. Schmidt, in Wiesbaden.[129] He moved yet again the following year, this time to Basel, Switzerland, in order to introduce processes he had developed to the dyestuffs company J. R. Geigy.[128,129]

While in Basel, he also obtained space in the laboratory at the University of Basel, where Jules Piccard (1840–1933) quickly recognized the value of having a specialist in dye chemistry and thus offered to establish Nietzki as a *Privatdozent*.[129] Completing his habilitation at the University of Basel on June 30, 1884, Nietzki was then appointed professor extraordinarius in 1887.[127-129] In 1895, he was made professor of chemistry,[127-129] but as the university laboratory did not have enough spaces to meet student demands, he established an organic chemistry laboratory in a private house located in the Rue du Rhin. As the lab was established at his own expense, it initially accommodated only 20 students but was later expanded to allow 36 students. During his third decade at Basel, rapidly developing arteriosclerosis limited his ability to work and made laboratory manipulations difficult.[129] He thus resigned as professor in March of 1911,[127-129] retiring the following year to Freiburg im Breisgau, in Southern

Germany.[129] Following an extended illness, he died on September 28, 1917, at the Neckargemund Sanatorium.[127–130]

It was while at Leiden, in 1876, that Nietzki began reports of his investigations into the nature of aniline black,[125,126] with his first paper published shortly before Goppelsroeder's report of the elementary analysis of electrolytic black.[113] Beginning with efforts to produce high-purity samples for study,[125] Nietzki then moved on to the analysis of its chemical composition, leading to the empirical formula $C_{18}H_{15}N_3 \cdot HCl$. Again, this formula is very similar to those previously reported by both Fritzsche[68,69] and Goppelsroeder,[125] although with slightly higher H content than the two previous reports, and a Cl content that falls in between the two. As both Nietzki and Goppelsroeder had extensively purified their aniline black samples, it is quite possible that they had been partially reduced in the process, thus inadvertently reducing the cationic content and thus the amount of Cl^- counterions. Two additional papers on the analysis of aniline blacks then followed in 1876[126] and 1878,[131] but these reports focused primarily on the blue material generated by boiling aniline black in excess aniline, which Nietzki viewed to be a type of phenylated material. In the 1878 paper, however, he does conclude that the amount of Cl found during analysis is dependent on both the purification and drying of the sample.[131]

It was during the period between the second and third of these analysis papers, however, that Nietzki reported a discovery that would have far more impact on the eventual elucidation of aniline black's structure. Beginning in 1877,[132] he found that heating aniline black with potassium dichromate ($K_2Cr_2O_7$) in sulfuric acid produced copious amounts of quinone (1,4-benzoquinone). To investigate this unexpected reaction, he suspended aniline black in dilute sulfuric acid and gradually added the dichromate salt. This resulted in the consumption of the black material, while also giving rise to a strong odor of quinone. He then separated the brown, liquid mixture by steam distillation to give a considerable amount of quinone, although not enough to be consistent with the quantity of aniline black used. The remaining content of the still was then evaporated to give a colorless crystalline material that was identified as hydroquinone. As a result, Nietzki concluded that aniline black was initially decomposed to generate hydroquinone, which was then oxidized to quinone (Figure 2.15). As aniline black is generated through the oxidation of aniline, Nietzki proposed that it should be able to produce hydroquinone or quinone directly from aniline, in which aniline black was thus just an intermediate oxidation product. The oxidation of aniline with dichromate in dilute sulfuric acid confirmed this view to be correct, with the ratio of hydroquinone to quinone dictated by the amount of oxidant applied.

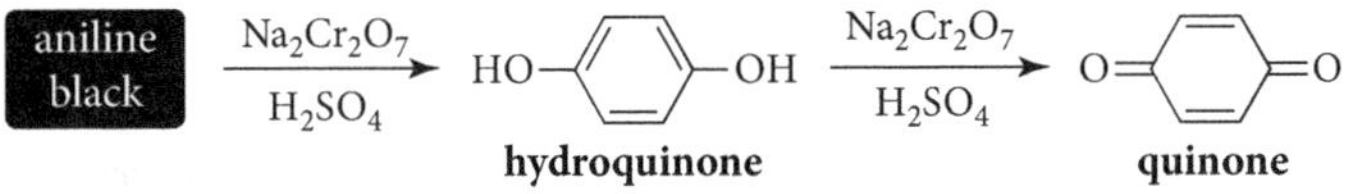

Figure 2.15 Nietzki's proposed oxidative degradation of aniline black.

This initial report was then followed with a second paper in 1878, which focused on optimizing the isolated yield of quinone.[133] While hydroquinone could be produced in good yield, efforts to generate the final quinone always occurred in significantly lower yields. The only exception to this was when attempts were carried out at very small scales. This led him to suspect that the quinone was possibly undergoing further reactivity under the combination of both heat and oxidation. To test this, he steam-distilled samples of pure quinone in the presence of oxidant, which resulted in the production of hydroquinone along with a resinous product. Repeating the process without the oxidant, however, gave him similar results, which led to the conclusion that quinone was undergoing condensation to give the resinous product under heat. Ultimately, he found that near quantitative amounts could be generated by the slow addition of potassium dichromate to cooled sulfuric acid solutions of aniline, after which the quinone was isolated by ether extraction.

Of course, the ability to produce quinone and hydroquinone via the oxidative degradation of aniline black provided some clues as to the structural nature of the initial material. This, however, did not seem to be a substantial focus for Nietzki, who seemed much more interested in this route as an easy and effective method for the generation of quinone. The next major step toward revealing the structures involved then came from one of the original developers of commercial black aniline dyes, Heinrich Caro.

In 1896, toward the end of his career, Caro published studies that followed up on a previous statement by Nietzki describing the generation of a yellow species via the oxidation of aniline in cold, aqueous alkaline solutions.[134,135] Caro went on to show that the addition of a permanganate solution to an alkaline, aqueous solution of aniline gave a green solution coupled with precipitation of a solid.[134] Filtration of the solid, which was determined to be mixture of MnO_2 and azobenzene, left a yellow solution (Figure 2.16). Caro's final optimized methods utilized a 2% solution of potassium permanganate to an NaOH solution of aniline with vigorous stirring.[135]

The yellow filtrate produced in this way was then treated with iron sulfite to give a colorless solution, which was then filtered, and evaporated to remove unreacted aniline.[134,135] Upon cooling of the remaining solution, a black tar separated, which was then boiled with dilute H_2SO_4 to result in the crystallization of a colorless sulfate salt that was found to be insoluble in cold

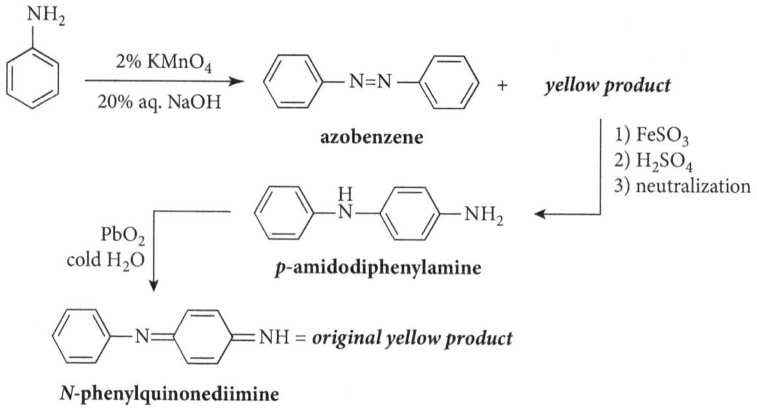

Figure 2.16 Caro's synthesis and identification of N-phenylquinonediimine as an intermediate oxidation product.

water. This salt was neutralized to give the corresponding base, which was then crystallized from ligroin to give colorless, flat needles that were found to be the known species p-amidodiphenylamine (Figure 2.16).[134,135] Oxidation of this base with PbO_2 in cold water led to the isolation of yellow crystals. These crystals were ultimately determined to be N-phenylquinonediimine, and Caro believed this product to be the identity of the original yellow species.[135] Caro recognized that N-phenylquinonediimine and azobenzene were isomeric species, both of which were formed during the alkaline oxidation of aniline. The exact mechanism involved, however, still remained unknown.[135] Further efforts were then continued in 1906 by the German chemist Richard Willstätter (1872–1942).[136]

2.8 Willstätter, Green, and the elucidation of structures

Richard Willstätter (Figure 2.17) was born in Karlsruhe, Germany, to Jewish parents on August 13, 1872.[137–139] After two years of education at the gymnasium at Karlsruhe, the family moved to Nürnberg in 1883, where he entered the Realgymnasium with aspirations of a commercial career.[138] In October 1890, he began studies at the University of Munich, while also attending lectures at the Technische Hochschule.[138,139] Even though the chair at Munich was Adolf Baeyer (1835–1917), the bulk of Willstatter's studies were really under Eduard Buchner (1860–1917), Johan Rupe (1866–1951), and Eugen Bamberger (1857–1932). In 1893, he then took the predoctorate examination and was assigned to Alfred Einhorn (1856–1917) as a research student.[138,139]

Figure 2.17 Richard Willstätter (1872–1942).

Reproduced courtesy of the Edgar Fahs Smith Memorial Collection, University of Pennsylvania (CC0 1.0).

Willstätter began his independent career in 1894, after which he became *Privatdozent* in 1896.[138–140] In 1902, he was then made professor extraordinarius and head of the Organic Section.[138,140] In 1905, he left Munich to accept the chair of chemistry at the Eidgenössische Polytechnische Schule in Zurich (now ETH Zurich).[137–141] He moved again in 1912, this time to Berlin, where he was appointed director of the new Kaiser-Wilhelm-Institut for Chemistry in Dahlem (now the Max Planck Institute for Chemistry).[137–141] Succeeding Baeyer in 1915, he then returned to Munich to become professor and director of the State Chemical Laboratory. That same year, he was also awarded the Nobel Prize for Chemistry for his work on chlorophyll and plant pigments.[138–141] Regrettably, escalating antisemitic views in Germany finally led to his resignation on June 24, 1924,[138,139] although he did retain a room to continue some limited research.[138,139] A handful of attractive offers from other institutions followed, but all of these were declined.[139,141] He also continued to receive various recognitions for his scientific contributions, including the Davy Medal of the Royal Society of London in 1932 and the Willard Gibbs Medal of the Chicago Section of the American Chemical Society in 1933.[138–140] Finally, the Gestapo searched his house in late 1938, and he was later ordered to leave the country.[138–141] As a result, he moved to Muralto-Locarno in southern Switzerland in 1939, where he later died of a heart attack on August 3, 1942.[139–141]

Willstätter's investigation of aniline black and its production began by revisiting Caro's oxidation of p-amidodiphenylamine with PbO_2 as discussed earlier (Figure 2.16). Analyzing the isolated oxidation product, Willstätter found that the nitrogen content was lower than expected, leading to the conclusion that performing the reaction in water resulted in hydrolysis of some of the oxidation product to the corresponding N-phenylquinone monoimine (Figure 2.18), which cocrystallizes with the diimide.[136] However, he found that pure N-phenylquinonediimine could be obtained if the oxidation was carried out in ether solution using dry silver oxide.

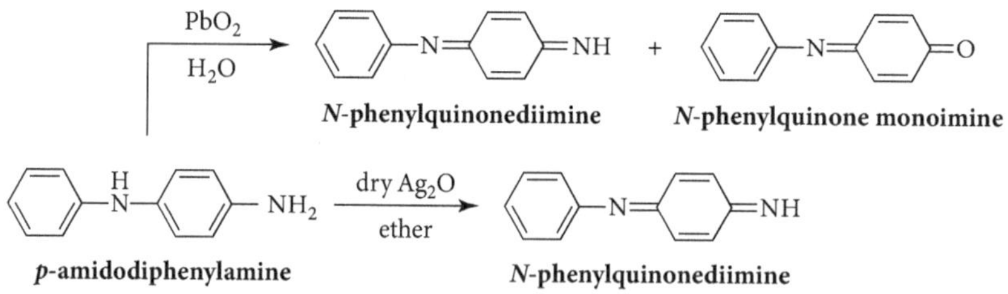

Figure 2.18 Oxidation products of p-amidodiphenylamine.

Now that he was satisfied that he was generating pure N-phenylquinonediimine, Willstätter began efforts to study its polymerization via treatment with acid.[136] Upon protonation, the yellow species first turned red-brown, after which it eventually gave a dark green product. This green material was referred to as *emeraldine* in reference to the commercial name of the green aniline-based dye developed by Frederick Crace-Calvert and coworkers in 1860 (see Section 2.5). Based on a suggestion from Nietzki, Willstätter found that the same material could be produced more easily by oxidizing p-amidodiphenylamine with iron chloride in acidic media. Similar results were also found using hydrogen peroxide with a catalytic amount of $FeSO_4$.[136]

From these green materials, Willstätter isolated a blue species that was viewed to be the emeraldine base, also known as the blue aniline-based dye azurine. This species was crystallized and analyzed to give a formula of $C_{24}H_{20}N_4$. From this formula, it was viewed that the blue compound had been formed from the coupling of two molecules of p-phenylquinonediimine.[136,142] The blue base was then dissolved in benzene and oxidized with PbO_2 to give a red species with two fewer hydrogens ($C_{24}H_{18}N_4$). Both the blue and red materials could be reduced to the same colorless leuco base ($C_{24}H_{22}N_4$), which could not be reduced any further.[136] As with the previous the oxidation of *p*-amidodiphenylamine in water, oxygen-terminated byproducts were also found in the three of the isolated species here.

In an attempt to determine the various structures of the blue, red, and leuco base species, Willstätter began by considering what type of bond might link the two phenylquinonediimine units. Of the possible linkages considered, the only one viewed to be consistent with the observed results was the formation of a new C–N bond between a terminal nitrogen of one unit and a phenyl unit of the other. Still, even by limiting the linkage to a simple C–N bond, there still remained various possible structural motifs as illustrated in Figure 2.19.[9,136] Of the three possible structures, it was thought that the two branched structures (II and III) should be easily converted into azine species (IIb and IIIb). As such, efforts to find evidence of any azine content in the blue species were pursued. However, as no evidence of azine content could be found, it was determined that the most reasonable structure for the blue species was the linear structure (I),[136,142] which was also consistent with the initial formation of *p*-amidodiphenylamine from aniline. Based upon this conclusion, the structures of the three species and their transformations via oxidation or reduction could be summarized as shown in Figure 2.20.[9,136]

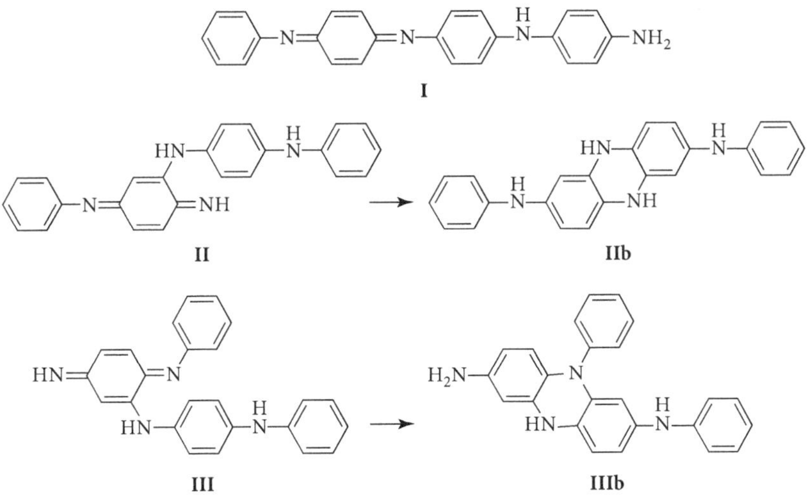

Figure 2.19 Possible structural motifs for the isolated blue product.

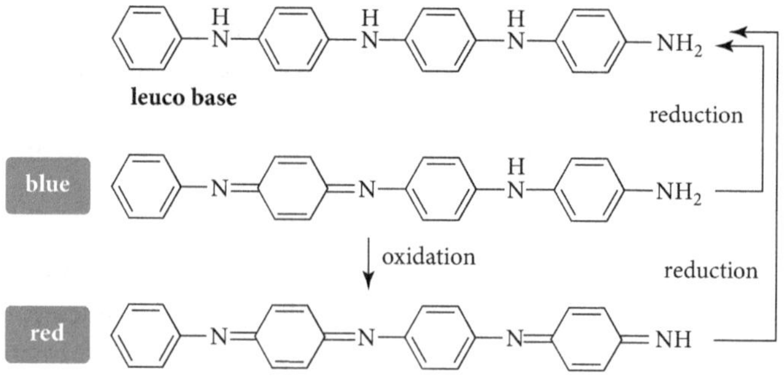

Figure 2.20 Structures and interconversions of Willstätter's isolated species.

Although Willstätter viewed the red and blue species as intermediates in the formation of aniline black, it was found that of the three species given in Figure 2.20, only the red species could be transformed directly to an insoluble black material.[136] Thus, upon heating or treatment with dilute acids, the isolated red product was found to undergo further polymerization to give a black product, which Willstätter referred to as *polymerization black*. In his reports, Willstätter acknowledged that the properties of the polymerization black prepared in this way did depend on the exact conditions applied, but the analysis of samples that produced a black material without byproducts gave a composition of $(C_6H_{4.5}N)_x$. Therefore, he felt that this formula best represented the composition of aniline black. Furthermore, as his aniline black was produced from the red intermediate consisting of four units (i.e., $x = 4$), he concluded that the smallest possible structure for aniline black was where x equaled a minimum of 8 (i.e., the coupling of two units of the red species), with the simplest possible formula thus being $C_{48}H_{36}N_8$.[136]

Still, Willstätter recognized that this was not a definitive determination of the structure of aniline black, acknowledging in a subsequent 1909 paper,[142] "The way in which the aniline residues are linked in [aniline] black has not yet been determined." This uncertainty was further illustrated by the dissenting views of the German chemist Hans Theodor Bucherer (1869–1949), who had published a 1907 paper that refuted Willstätter's proposed structures.[143] In Bucherer's view, the significant stability of the aniline black could only be the result of azine-type structures. To provide further support for his proposed linear structures (Figure 2.20), Willstätter subjected samples of aniline black to strong oxidation, which Nietzki had previously illustrated as a route to decompose the black material to quinone.[132,133] In the process, Willstätter further optimized Nietzki's methods by utilizing a combination of bichromate and PbO_2, thus further maximizing the resulting yield of quinone from aniline black.[142] The experimental yields of quinone thus obtained were then compared with the theoretical yields expected from the various proposed structures for aniline black at the time. Such comparisons showed that the yields obtained were much higher than that expected for either branched or azine-based structures, but they were within 95% of that expected for Willstätter's linear structure. At the same time, a more detailed elemental analysis was also performed to give a slightly more precise average composition of $C_{5.97}H_{4.55}N$, although still very close to that previously reported by Willstätter. Based on this composition, as well as the newly demonstrated support for his linear structures, Willstätter proposed that aniline black had a composition consisting of units of the threefold quinoid derivative given in Figure 2.21.[9,142] It should be noted that Willstätter was careful to highlight that the exact position of the two central quinoid units was unknown and their placement was arbitrary.[142]

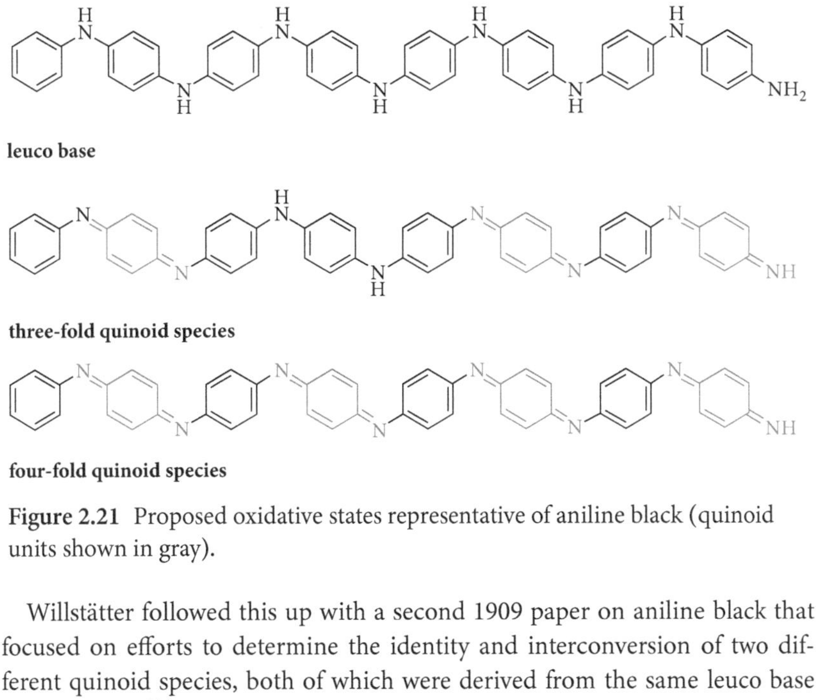

leuco base

three-fold quinoid species

four-fold quinoid species

Figure 2.21 Proposed oxidative states representative of aniline black (quinoid units shown in gray).

Willstätter followed this up with a second 1909 paper on aniline black that focused on efforts to determine the identity and interconversion of two different quinoid species, both of which were derived from the same leuco base (Figure 2.21).[144] In addition to the previously proposed threefold quinoid derivative, he now added the fully oxidized fourfold quinoid analog, which could be produced via further oxidation of the threefold species with hydrogen peroxide. This new species was said to be blue-black, with its salts exhibiting a dark green color. The relative content of these two oxidized species in aniline black was believed to be dependent on the specific oxidation conditions applied. Of the two, Willstätter viewed the threefold quinoid species, which gave a blue-colored base and green salts, to represent the traditional material emeraldine.

In contrast, he viewed the polymerization black that he had previous generated from the red intermediate (Figure 2.20) to be closer to the fourfold quinoid species.

This was then followed up with a 1910 paper in which phenylhydrazine carbamate was used to study the gradual reduction of aniline black.[145] Thus, aniline black, which was viewed to be primarily the threefold quinoid species, was found to progress through three transitions. The initial aniline black was first observed to undergo reduction at 30°C–45°C, resulting in a color change from dark blue to a light blue. Further heating then produced a second transition at circa 80°C to give a gray material, with a final colorless product obtained above 120°C. In the presence of a small amount of ferrous salt, the leuco base produced in this manner converted back to the black material under an oxygen atmosphere.

About the same time as this final paper, the English industrial chemists Arthur G. Green (1864–1941) and Arthur E. Woodhead began reporting competing studies. While the first of the Green and Woodhead papers was reported in 1910,[146] two additional solo works by Green had been reported the previous year.[147,148] As illustrated by these works, Green initially had very different ideas concerning the structure of aniline blacks.

Although little is known about Woodhead, Arthur George Green (Figure 2.22) was born in the town of Ealing, located in West London, in 1864.[149,150] He was the eldest son of William John Green, an architect, and educated at Lancing College, Sussex. In 1880, he then entered University College, London, where in his first year he won the gold medal in the junior practical chemistry class and the Clothworkers' Exhibition in chemistry. The following year he won the gold medal for chemical analysis, and in 1883 the Tuffnel Scholarship.[149,151] At the College, Green carried out research under lecturers Henry Forster Morley (1855–1943) and Richard John Friswell (1849–1908),[150,151] as well as working as a volunteer during college vacations in the laboratory of Messrs. Williams Bros., aniline dye manufacturers of Hounslow.[150]

Green commenced his industrial career in June 1885 with Messrs. Brooke, Simpson & Spiller, Ltd.,[149–151] where he worked as a research chemist at the Atlas Aniline Dye Works in the East London neighborhood of Hackney Wick. It was in these same facilities that he had previously carried out his research with Friswell.[152] He quickly gained success, being made a Fellow of the Chemical Society in 1885 and awarded the silver medal of the Royal Society of Arts in 1891.[149] In 1894, he then moved to become manager at the Clayton Aniline Company, in Manchester.[149,150] By 1901, however, he decided to become independent and thus set himself up as a consultant in London. He then accepted an invitation to the Chair of Chemistry and Dyeing at the University of Leeds in 1903, a position made vacant by the death of John James Hummel (1850–1902).[149] In 1915, Green was elected to the Royal Society.

Figure 2.22 Arthur George Green (1864–1941).

Reproduced from J. Baddiley. Arthur George Green 1864–1941. *Obit. Notices Fellows R. Soc.* **1943**, 4, 251–270.

In March 1916, he decided to resign the chair to became director of research at Levinstein Ltd. in Manchester,[149] while also spending a portion of his time during 1916–1918 at the College of Technology, Manchester. At the college, he worked to establish the Dyestuffs Research Laboratory with the assistance of Frederick Maurice Rowe (1891–1947), one of his past students.[149,150,153] Throughout this period, Green continued to receive recognition for his work. In 1917, he was awarded the Perkin Medal of the Society of Dyers and Colourists and received an honorary MSc from the University of Leeds. He was elected to the Livery of the Worshipful Company of Dyers in 1918, and he received the Dyers' Company gold medal for the third time in 1923, previously receiving it in 1909 and 1914 (with W. Johnson).[149]

In 1919, Levinstein, Ltd., merged with British Dyes to become the British Dyestuffs Corporation, Ltd, with Green remaining director of research. He resigned his position in 1923, however, and returned to his private practice, which was considerable both in Europe and America.[149] Still, he eventually returned in 1936 to become consultant to the Dyestuffs Group of Imperial

Chemical Industries Ltd., which was established in December 1926 from the merger of British Dyestuffs Corporation with three other British companies: Brunner Mond, Nobel Explosives, and the United Alkali Company. Five years later, at the age of 78, he died peacefully in his sleep at his home at Walton-on-Thames on September 12, 1941.[149,150]

Green's earliest publication on the structure of aniline black consisted of a short 1909 paper that presented proposed structures for both the emeraldine and nigraniline materials,[147] along with brief comments on the initial 1907 paper by Willstätter and Moore.[136] A second paper that same year covered much of the same material, although in more detail.[148] Green's proposed structures for emeraldine and nigraniline are shown in Figure 2.23, both given as cyclic species consisting of three aniline repeat units. It is important to note that at no point does he say anything about the nature of the substitution geometry on the various bridging phenyl rings (i.e., *ortho, meta,* or *para*). Furthermore, he states that nigraniline is derived from emeraldine by the removal of hydrogen, and can be regenerated via reduction, but does not address the presence of chloride only in the nigraniline structure. In terms of the previous work of Willstätter and Moore, Green agreed with the empirical formula of $(C_6H_{4.5}N)_x$ but did not accept any of the proposed structures. As stated by Green,[148]

The view expressed by these authors that the product of this so-called "polymerisation" must have a molecule in which x = at least 8, and is to be represented as a complex indamine with a long open chain, appears to me to be very difficult to reconcile with its properties (stability to acids, etc.).

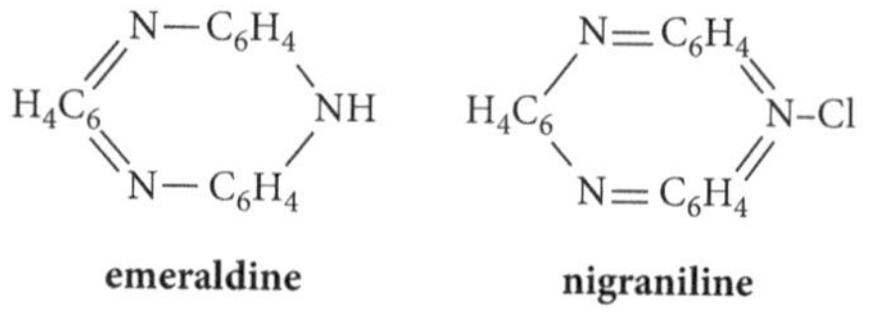

emeraldine   nigraniline

Figure 2.23 Green's initially proposed cyclic structures of emeraldine and nigraniline.

By the publication of his first paper coauthored with Woodhead in 1910,[146] however, this stance seems to have softened to some degree, stating that the work of Willstätter and his coworkers has "added much of value to our knowledge of the complex oxidation products of aniline," but that "the constitution of aniline-black and of its intermediate products still cannot be regarded as completely elucidated." Furthermore, they point out that the results of Willstätter can really only apply to the primary oxidation products of aniline (i.e., emeraldine or nigraniline), while the most stable form of aniline black, known as *ungreenable aniline-black*, must be an azine structure of some type. Based on this view, Green and Woodhead then proposed[146] "that the term "aniline-black" should be restricted to the higher condensation products (ungreenable black), whilst the original names "emeraldine" and "nigraniline" should be retained for the primary oxidation products." Green and Woodhead then went on to reinterpret some of previous results of Willstätter and Moore, as well as providing additional data in order to present a more detailed model of the primary oxidation products of aniline. In providing their revised models, Green and Woodhead retained the general linear octameric structures of Willstätter, even though they felt that the question of linear chains versus ring structures was still undecided. On this topic, they specifically state,[146]

> Assuming the correctness of the eight-nucleal structure for the primary oxidation products, it still remains an undecided question whether the aniline residues are to be regarded as united in an open or in a closed chain, but without attempting to decide this point we shall make use of the open-chain formulae to express provisionally the analytical results.

In the pursuit of these updated models, Green and Woodhead started with the preparation of emeraldine via multiple methods, neutralizing it to the corresponding base, and carefully purifying it to produce an initial material for study. This purified material was then dissolved in either 80% aqueous acetic acid or 60% aqueous formic acid to give yellowish-green solutions. A dilute solution of chromic acid was then added, which resulted in a color change from green to pure blue. The addition of more oxidant then led to a color change to violet, which ultimately gave a violet precipitate.[146] If a weak solution of sodium hydrogen sulfite ($NaHSO_3$) was slowly added to the violet solution, these color changes would occur in the opposite direction, transitioning from violet to blue to green. Stronger reducing agents, such as phenylhydrazine ($PhNHNH_2$), sodium hyposulfite ($Na_2S_2O_3$), or titanium trichloride, would convert the green solution to the colorless leuco base. As the initial green solution corresponded to the emeraldine form, the next sequential colored solution via oxidation was designated nigraniline. The final violet solution was then given the new designation of *pernigraniline* (Latin *per* "through, entirely, utterly" + nigraniline). It should be noted here that the prefix *per-* is commonly used in chemistry to denote the highest known oxidation state of various species (i.e., persulfate or permanganate) and thus is added to nigraniline here to denote the fully oxidized form.[9] At the other end of the series, the parent leuco base was given the name *leucoemeraldine* (i.e., the leuco base of the original emeraldine).

Their analysis then continued with titration of the initial acetic acid solution of emeraldine with titanium trichloride in order to determine the quantity of hydrogen required for the conversion of emeraldine into leucoemeraldine, which led to the conclusion that the emeraldine form corresponded to a diquinoid species.[146] In a similar manner, the acetic acid solution of emeraldine was again titrated, this time with a standard solution of chromic acid in order to determine the quantity of oxygen consumed in the conversion of emeraldine into nigraniline. From such measurements, it was then concluded that the introduction of one additional quinonoid group was required to convert emeraldine to nigraniline. The combination of these various conclusions then led to the series outlined in Figure 2.24.[146] As it had been determined that the difference between emeraldine and leucoemeraldine was two quinoid units, an additional single quinoid species was added to the series, which was given the name *protoemeraldine* (Greek *prōtos* "first" + emeraldine) to denote the first oxidized form. Here it should be noted that for the emeraldine and nigraniline structures, the quinoid units were clustered together at one end of the structure, without any mention if the placement of these units was arbitrary, as previously emphasized by Willstätter.

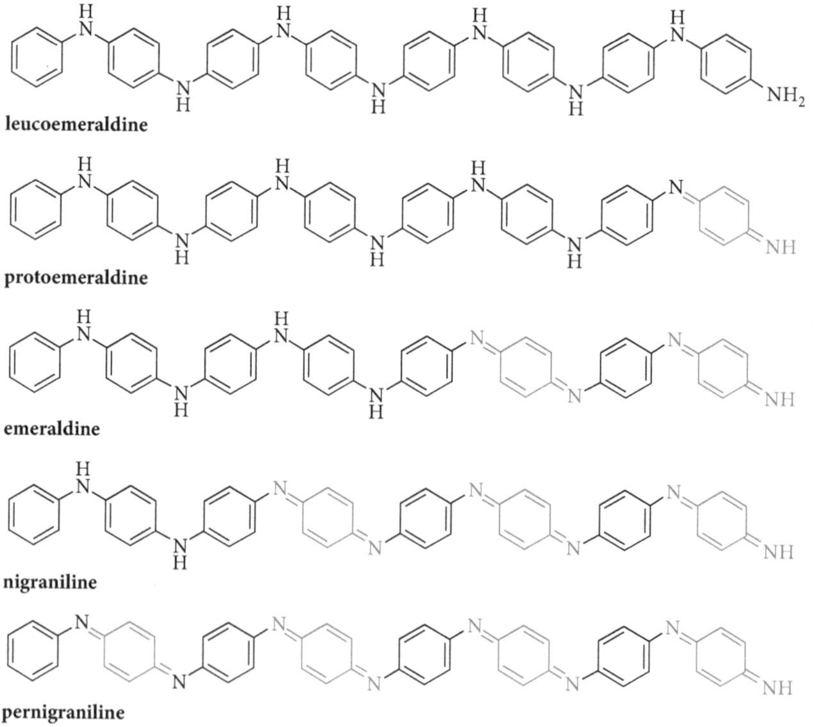

Figure 2.24 The full series of aniline oxidation products proposed by Green and Woodhead in 1910 (quinoid units shown in gray).

Willstätter then responded to the work of Green and Woodhead in the following year with an additional paper.[154] While he pointed out that the proposed renaming of the structures provided no real improvement over his previous descriptive nomenclature, it was the contradicting views on the quinoid content of the emeraldine form that dominated his response. The key issue raised by Willstätter was that from their own previous attempts at reducing materials to the leuco base, the use of titanium trichloride required higher temperatures for full reduction than that reported by Green and Woodhead. As such, he felt that the leucoemeraldine reported by Green and Woodhead was actually the monoquinoid product, not the true leuco base. He felt that this conclusion was also supported by Green and Woodhead describing the leucoemeraldine as a "pale brown amorphous powder," when it should have been completely colorless. Of course, such an assignment would then remove any differences between the studies of the two groups.

Green and Woodhead then provided their own rebuttal to Willstätter in 1912,[155,156] in which they refute Willstätter's claim that full reduction to the leucoemeraldine cannot be accomplished with titanium trichloride at lower temperatures, and highlight various points in Willstätter's work that they viewed as problematic. "In order to place the matter beyond doubt," Green and Woodhead attempted to further reduce their leucoemeraldine samples with titanium trichloride at high temperature as specified by Willstätter. This was accomplished by first preparing the leucoemeraldine from the reduction of emeraldine by phenylhydrazine at a low temperature, after which it was treated with boiling titanium trichloride. As the results were not consistent with the further reduction of a monoquinoid species to the full leuco base, it was concluded that their original conclusions were correct.

After that, it does not seem that Willstätter had anything further to say on the subject, at least not in the chemical literature. As such, the debate was generally considered decided. This is further supported by the fact that no further modifications to these structural models have been reported since, with the final oxidative series as presented by Green and Woodhead still remaining the currently accepted structural forms (other than the specific placement of the quinoid units). Willstätter thus moved on to other research topics, and, although Green continued with a couple more publications on aniline black,[157,158] these focused primarily on various azine-based structures that he felt were representative of ungreenable aniline black.

It should be noted that although the modern literature typically gives Green and Woodhead sole credit for determining the structure and oxidative forms of polyaniline, it is quite clear from the historical record that the primary structure determination was really accomplished by Willstätter.[9] Such a statement is not intended to ignore the very real contributions of Green and Woodhead, who clearly made important corrections and additions to the oxidative series, as well as establishing the traditionally accepted nomenclature, but these advancements were accomplished by refinement of the previous structural models of Willstätter. Throughout Green's work, he was never convinced of the linear nature of these materials and only used the linear structures as a convenient working model. In contrast, Willstätter had carefully eliminated other possible structures through detailed analysis and fully believed in the linear forms of his structures. In addition, although Willstätter largely presented his structures as linear octamers, he made it quite clear that this was only the minimum length necessary to explain the presented results and the actual materials could certainly be larger, thus paving the way for the eventual understanding of these materials as macromolecules.

2.9 Evolution of the conductive nature of aniline products

Although the basic structures of the aniline oxidation products had now been largely worked out, there was still no realization concerning the electronic nature of these species. Exactly what first led researchers to start looking at the electronic properties of aniline black and its various redox states is unclear, but as early as 1961, Berlin and coworkers in Russia had characterized aniline black by electron paramagnetic resonance (EPR) spectroscopy and determined its semiconducting behavior, with measured conductivities of 10^{-13} to 10^{-5} Ω^{-1} cm^{-1} for the base and 10^{-5} to 10^{-2} Ω^{-1} cm^{-1} for the corresponding hydrochloride salts.[159] The following year, Herbert A. Pohl (1916–1986)[160] and E. H. Engelhardt at Princeton University reported similar values of 3×10^{-11} Ω^{-1} cm^{-1} for samples of aniline black.[161] However, it wasn't until later in the mid-1960s that significant advances were made by Marcel Jozefowicz (b. 1934) and others, all working in the laboratory of Rene Buvet (1930–1992) at the École Supérieure de Physique et de Chimie Industrielles de la ville de Paris (ESPCI Paris).[8]

Marcel Jozefowicz was born in Paris on December 14, 1934.[162] His family were Polish Jews, and his father died when he was only seven years old. The bulk of the rest of his family were killed in the gas chambers of Auschwitz during the Second World War, and thus he was recognized a ward of the nation by judgment of the Civil Court of the Seine in 1949.[162] He completed secondary studies at various high schools in Paris before being accepted in 1955 to the l'École nationale supérieure de chimie de Paris (also known as ENSCP or Chimie Paris-Tech), within the University of Paris. He obtained a license in physical sciences in 1958 and then a doctorate in physical sciences under the Faculty of Sciences of the University of Paris in October 1962.[162]

Jozefowicz then became an assistant lecturer at the Faculty of Sciences of Paris until 1966, then a lecturer at the Faculty of Sciences de Nancy-Université until 1969, before being appointed professor of chemistry at the newly created Scientific Center of Saint-Denis-Villetaneuse.[162,163] In 1970, this center was renamed the Scientific and Polytechnic Center (CSP) of the Université Paris XIII (Université Paris Nord), currently known as the Université Sorbonne Paris North as of 2020.[162] During his time at Université Paris Nord, Jozefowicz held a number of administrative positions, including the first head of the chemistry department, the second director of the CSP, member of the university council, and the second president of the university (1973–1978).

Jozefowicz was a strong supporter of the development of relations between the university and industry, particularly in the field of fundamental and applied research. In this capacity, he chaired both the Reflection Group and the Concerted Action Committee on Biomaterials within the framework of the Direction Générale de la Recherche Scientifique et Technique (DGRST). His efforts continued in various successive structures of which he was the founder and director, including the "Hemocompatible polymers" of the National Center for Scientific Research from 1976 to 1984, and the public interest group "Substitute Therapeutics" from 1986 to 2001, the latter of which became the Therapol Limited Company in 2001. Jozefowicz retired in 2002, but he continued to chair and lead Therapol until 2006.[162]

Jozefowicz and Buvet worked together for a period of nearly 10 years, from 1960 until Jozefowicz became professor at Université Paris Nord. After a couple papers on polyparaphenylenes,[164,165] they turned to optimizing methods for the reproducible preparation of emeraldine sulfate samples by using persulfate ($S_2O_8^{2-}$) to oxidize aniline in sulfuric acid solutions.[15] It is important to note that they referred to these products as "polyanilines" and appear to be the first to use the modern name for this polymer in 1964.[15] In addition to emeraldine sulfates, they also prepared the analogous chloride, formate, and acetate salts via the neutralized emeraldine base. In the process, efforts were also made to carefully study the acid-base properties of these materials, including precise control of the level of protonation. An additional 1965 report then reported studies of the redox properties of the various polyaniline materials synthesized in this initial report.[16] In the process, the absorption spectra of emeraldine acetate was measured in acetic acid to give a maximum at approximately 700 nm. Oxidation of the sample resulted in a shift in the absorbance to lower energy.

Between the publication of these first two papers, Jozefowicz and Buvet then reported their first studies of the conductivity of these materials in 1965.[166] The polyaniline samples were compressing into pressed pellets measuring 13 mm in diameter, with a thickness between 0.3 and 2 mm. The conductivity was then determined by injecting a constant current through the pellet and measuring the resulting voltage across the terminals. The resulting conductivities were determined to fall between 10^{-2} and $1 \ \Omega^{-1} \ cm^{-1}$, depending on certain variables in the sample production. These values were considerably higher than previously reported values for aniline blacks,[159-161] and it was found that factors providing the largest effects on the conductivity were generally the water content of the samples and the level of sample protonation. For materials prepared in the same way, values were found to be reproducible within ±5%.

A more detailed study of the conductive nature of polyaniline then followed over the time period of 1966–1971.[17,167-174] In the process, they highlight that revealing a deeper understanding of the relationships between chemical properties and conductivity is complicated by the fact that the structures of these materials were still not completely decided, citing the works of both Willstätter and Green.[167] The conductivities were measured as discussed earlier, with particular focus on the effects of both water content and the pH of the solution used during the preparation of the emeraldine form of the polymer.[17,167,168] While they do point out that the oxidation state of the polymer influences the conductivity, they state that studies on this are still ongoing and limit reported results to only the emeraldine form.[168] The pH dependence was found to provide the most significant effects, with decreasing pH giving a linear increase in conductivity when plotted on a log scale, as shown in Figure 2.25. In this way, it was found that the conductivity of the emeraldine form could be controlled with pH over a range of 10^{-2} to over $1 \ \Omega^{-1} \ cm^{-1}$.[17,167,168] Of significant interest is the fact that this effect is described as *doping*, a term generally attributed to the later work of MacDiarmid, Heeger, and Shirakawa on polyacetylene. Specifically, Jozefowicz states that the "acid absorbed presents doping characteristics for a conductivity that is without possible dispute of electronic origin."[168] The effect of water was more limited, but the conductivity was found to increase with water content. Lastly, they show that the conductivity is temperature dependent and increases with increasing temperature, as typical of semiconducting materials.[168]

By early 1966, they had achieved conductivity values from 10^{-5} to $10 \ \Omega^{-1} \ cm^{-1}$,[171] with Jozefowicz stating,[167] "The conductivity of the polyanilines is very high and classifies these compounds among the best-known organic conductors. This conductivity is, without possible dispute, electronic." This statement is then backed up by direct comparison with the reported conductivities of other organic polymers, with the highest previous value at circa $10^{-3} \ \Omega^{-1} \ cm^{-1}$. Furthermore, he points out that determining what type of polymer structures should result in the most favorable conductivities is severely limited by a lack of sufficient characterization of the chemical and physical properties of the materials being studied. This, he proposes, may be the source of the pessimism about the possibilities of organic semiconductors at the time.[167]

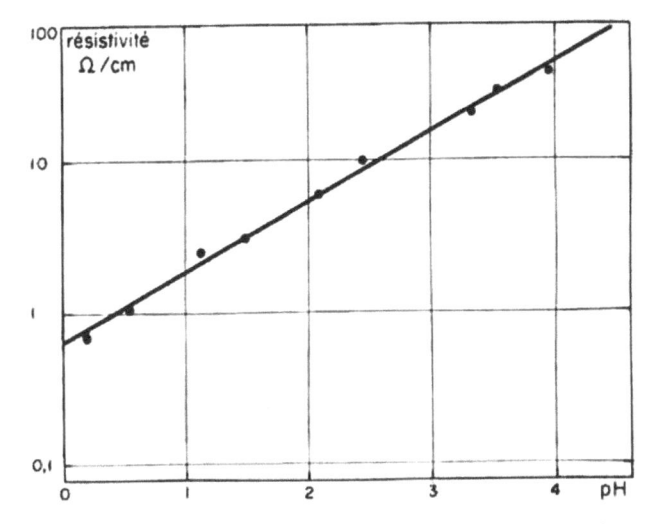

Figure 2.25 The effect of pH on the conductivity of the emeraldine form of polyaniline.

Reproduced from M.-F. Combarel, G. Belorgey, M. Jozefowicz, L.-T. Yu, R. Buvet. Conductivité en courant continu des polyanilines oligomers: Influence de l'état acide-base sur la conductivité électronique. *C. R. Seances Acad. Sci.* **1966**, *262*, 459–462, courtesy of gallica.bnf.fr/Bibliothèque nationale de France.

By 1969, Jozefowicz and Buvet were reporting conductivities as high as $100 \ \Omega^{-1} \ cm^{-1}$ for pressed pellets of protonated polyanilines in their oxidized forms.[173] A couple years earlier, in April of 1967, Buvet presented the electronic characteristics of polyaniline at the 18th meeting of CITCE (Comité International de Thermodynamique et Cinétique Electrochimiques), reporting that its conductivity was electronic in nature and not due to any ionic conduction.[170,175] It has been pointed out by Inzelt, however, that this did not seem to generate much excitement at the time.[175] The reported value of $100 \ \Omega^{-1} \ cm^{-1}$ remained the highest known conductivity for an organic polymer until 1977, when it was finally surpassed by the conductivity of highly doped films of polyacetylene.

2.10 Adams, Diaz, and a return to electrochemical studies

Shortly after the birth of this developing interest in the conductivity of polyaniline, a renewed interest in the preparation of these materials by electrochemical methods also emerged. This began with a 1962 report by Ralph N. Adams (1924–2002) and coworkers at the University of Kansas.[176] Adams was born in Atlantic City, New Jersey, on August 26, 1924.[177] His college education was interrupted by World War II when he was drafted in 1942. Originally drafted into the Chemical Warfare Service, his desire to fly led to him joining the Army Air Corps in 1943, where he earned his lifelong nickname, Buzz. He returned to his education after the end of the war, earning his bachelor's in chemistry from Rutgers University in 1950.[177] He then continued his graduate studies under N. Howell Furman (1892–1965) at Princeton University, receiving his Ph.D. in 1953. He remained at Princeton for two years as an instructor, after which he joined the faculty of the University of Kansas in 1955. He was named a university distinguished professor in 1965. Adams then retired in 1992 and died in 2002.[177]

While Adams cites Letheby's original report, and makes passing mention of Coquillion and Goppelsroeder, he does not seem to be aware of the extent of work on aniline electropolymerization reported in the 1870s. However, his primary interest is in the mechanism of the electropolymerization, which was largely unexplored at this point. For his study, Adams prepared his samples by constant potential electrolysis of a 0.01 M solution of aniline in 1.0 M sulfuric acid using a platinum anode.[176] For comparison, he also prepared both emeraldine and nigraniline samples by chemical oxidation. Comparison of the IR spectra of the electropolymerized product with that of the chemically oxidized species showed that all three samples were structurally similar, with the electropolymerized product slightly more like emeraldine. The differences observed were concluded to be due to the sample being a partial mixture of both emeraldine and nigraniline, although comparison of other properties such as solubility and reactivity led to the conclusion that the electropolymerized material was principally emeraldine.

The rest of the study then focused on the potential mechanism for the electropolymerization. As p-aminodiphenylamine was assumed to be an intermediate in the polymerization, as previously demonstrated by Caro[134,135] (see Section 2.7), Adams compared the electropolymerization of aniline and p-aminodiphenylamine to conclude that, qualitatively, they both gave the same product. In the process, he also found that p-aminodiphenylamine was much more easily oxidized than aniline, which led to the conclusion that all succeeding intermediates should also be much more easily oxidized than aniline. As such, the rate-limiting step of the production of emeraldine from aniline should be the first charge transfer step, that is, the initial oxidation of aniline.

From a kinetics study of the initial oxidation, it was concluded that it involved two electrons and that the reaction was second order with respect to aniline. This led Adams to decide that the initial product of this oxidation was the radical cation of aniline, which was also felt to be the most energetically feasible option. These radical cations would then undergo condensation with the loss of two protons (H^+) to produce p-aminodiphenylamine, consistent with the HT coupling shown in Figure 2.1. In contrast to the modern mechanism, however, Adams felt the sequential oxidations of *p*-aminodiphenylamine transpired via the loss of two electrons and two protons to give quinoid intermediates as shown in Figure 2.26. Interestingly, although each step involves the loss of protons that would produce an acidic environment, Adams presents both his quinoid intermediates and the emeraldine product as the free bases.

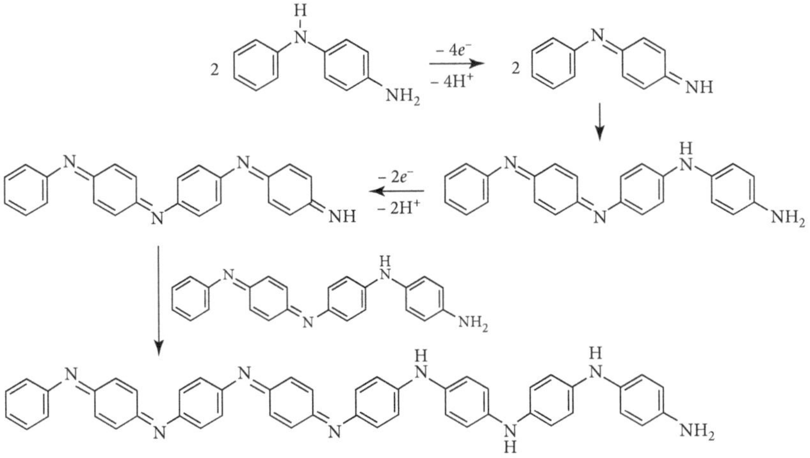

Figure 2.26 Adam's mechanism for the formation of emeraldine starting from *p*-aminodiphenylamine.

A second paper then followed in 1968, which revisited the analysis of the initial oxidation of aniline and the formation of the first coupled product.[22] To further investigate this initial mechanistic step, the oxidation of a number of aniline derivatives was added and compared with the unfunctionalized parent species. As before, these results favored the production of an initial radical cation and coupling to produce the HT product *p*-aminodiphenylamine. However, he does point out that other mechanisms are also possible and that the point remains ambiguous.

There was then a lag in further aniline electropolymerization studies until a 1980 report by Arthur Diaz,[178] over a decade after Adams's last report. Diaz is much better known for his work on the electropolymerization of first pyrrole and then thiophene, but he briefly turned to the electropolymerization of aniline as well. We will return for a more in-depth discussion of his career in Chapter 3 (see Section 3.7).

His first 1980 paper on electropolymerized polyaniline presented optimized methods for the generation of strongly adhered, quality films on the electrode surface.[178] Although previous electropolymerization reports had successfully generated polyaniline, it was found that the products were typically powders that adhered poorly to the working electrode. However, the application of potential cycling between –0.2 and +0.8 V vs. SSCE (saturated sodium calomel electrode), rather than constant potential, was found to generate true films that strongly adhere to the platinum surface. The resulting films were insoluble in all the common solvents, and scanning electron microscopy (SEM) revealed that they were continuous, with a fairly smooth surface.

Cyclic voltammograms of the films showed three clearly defined peaks, with the redox reaction accompanied by a color change of the film.[178] When the electrode potential was positive of +0.2 V, the film exhibited a blue color that changed to a transparent yellow when the potential was negative of 0 V. The observed color changes occurred evenly throughout the film with no evidence of regions exhibiting different reactivity. The absorption spectra of the yellow film held at –0.2 V showed a broad featureless band with a maximum at 420 nm, while the blue film held at +0.6 V exhibited a broad bond with a maximum at 630 nm.

This was then followed with a second paper in 1984 that compared the surface characteristics of polyaniline thin films prepared by plasma and electrochemical polymerizations.[179] The electropolymerized films were prepared as previously described in the 1980 paper,[178] while the plasma-generated films were prepared in a glow discharge chamber using a power of 10–15 W and a frequency of 13.56 MHz and were deposited on glass slides.[179] The films by either method were continuous and adherent to the substrate surface. In general, the films were insoluble in conventional solvents and could not be removed by soaking. However, some of the plasma-polymerized films could be lifted off the substrate with acetone.

Both films were found to have similar surface energies and stoichiometry, with the principal difference being surface roughness.[179] The SEM of the plasma-polymerized films showed that they were very smooth and featureless. In contrast, the electrochemically prepared films, which, while fairly smooth, could exhibit a wide variation in surface topology. In general, the plasma-polymerized films were found to have a greater deficiency of nitrogen content and higher oxygen content than the electropolymerized films. This was thought to probably stem from the fact that the plasma conditions were more severe.

A final paper then appeared in 1988 that investigated the effects of different acid electrolytes on the electrochemical behavior of polyaniline, including H_2SO_4, HCl, HNO_3, perchloric ($HClO_4$), fluorosulfonic (HSO_3F), trifluoromethylsulfonic (HSO_3CF_3), HF, and trifluoroacetic (CF_3COOH) acids.[180] In all cases, the polyaniline films were prepared in an aqueous 2 M acid solution containing 0.1 M aniline, with the voltage applied to the platinum electrode swept between –100 and 900 mV at 50 mV/s. In H_2SO_4, the resulting films exhibited either two (250 and 900 mV) or three (250, 500, and 900 mV) oxidation peaks, depending on control of the anodic voltage limit. By controlling the voltage limit in order to maintain current densities below 0.4 mA/cm^2, films were produced without the central peak at 500 mV. The application of HCl, HNO$_3$, HSO$_3$F, and HSO$_3$CF$_3$ acid electrolytes all resulted in similar results, with exact differences not quantified. Films prepared with HF and CF$_3$COOH were quite different, with polymerization reaction slower and lacking the ability to reduce the anodic voltage limit during preparation. As such, these films always exhibited three peaks at 250, 500, and 800 mV. The presence of the peak at 500 mV was attributed to the presence of *ortho*-coupled defects and/or degradation products. Overall, the choice of electrolyte had little effect on switching rates or film stability, although they did result in a variable degree of fibril formation in structure of the polymer films.

2.11 New applications of polyaniline

It was also in the 1980s that new applications started being proposed for polyaniline, beginning with a 1982 paper on photoelectrode applications from researchers at the Solar Energy Research Institute in Colorado.[181] This study reported the preparation of polyaniline-coated semiconductors, including n- and p-type silicon, as well as p-type gallium phosphide (GaP). The stability of the resulting photoelectrodes was then evaluated to reveal a dramatic enhancement in the photocurrent stability of the n-Si/polyaniline photoanode compared with the original n-Si electrode. Similar stability was observed for the other photoelectrodes, with the conclusion that the polyaniline films enhance stability of semiconductor photoelectrodes by permitting efficient charge exchange with the electrolyte.

Shortly thereafter, Tetsuhiko Kobayashi, Hiroshi Yoneyama, and Hideo Tamura at Osaka University introduced polyaniline-coated electrodes as electrochromic display devices in early 1984.[182] Devices were constructed by depositing a circa 10 nm–thick film of Pt onto a quartz plate via arc plasma to generate an optically transparent electrode. Polyaniline films were then electropolymerized galvanostatically onto the transparent electrode from a 2 M HCl solution containing 1 M aniline. The thickness of the polyaniline film was determined to be circa 100 nm. Depending on the potential applied over the range –0.2–1.0 V versus SCE, it was found that the polyaniline films underwent multiple color changes—from transparent yellow, to green, to dark blue, to black. Unfortunately, the electrochromic behavior of the films was unstable upon multiple cycling of these color changes. However, if the potential region was limited such that only switching between the transparent yellow and green states occurred, very stable electrochromic behavior was exhibited, with repetitions greater than 10^6 times achieved.

Two additional reports were then published as back-to-back papers in the spring of 1984, with the first of these focusing on the electrochemical reactions responsible for the observed color changes.[183] As before, the films exhibited reversible color change, with a transparent yellow color (absorption maximum at 305 nm) at potentials of –0.2–0.0 V. Oxidation via higher potentials of 0.3–0.6 V then caused the color to change to green (absorption maximum at 760 nm). At potentials in the transitional range of 0.0–0.3 V, however, a shoulder was observed at 420 nm, giving rise to an intermediate "yellowish green" state.

As the occurrence of the yellowish green state coincides with anodic and cathodic peaks of the cyclic voltammograms, at 0.18 and 0.08 V, respectively, this potential region was then investigated in more detail. By changing the potential from 0.15 to 0.2 V, a large decrease in the absorption at 305 nm was observed, along with moderate increases at 420 and 740 nm. In the same manner, scanning from potentials of 0.40–0.45 V resulted in further loss at 305 nm, decreased absorption at 420 nm, and significant growth at 740 nm. These collective results then led to the conclusion that the redox currents associated with the anodic peak are largely related with the color change from yellow to the intermediate yellowish green, while the capacitive current observed in the region of 0.3–0.6 V is related to the color change from yellowish green to green. As such, the initial color change was determined to result from the polymer oxidation, with a red shift in absorption resulting from the formation of the quinoid units. In contrast, the second color change was attributed to the insertion of anions into the polymer film, which mainly occurs at potentials of 0.3–0.6 V, and that would be necessary to allow effective protonation of the emeraldine form.

The final paper then focused on study of the oxidative degradation processes of the polyaniline films at higher potentials, thus limiting additional electrochromic switching beyond the green state.[184] While the polyaniline films were found to be fairly stable within the potential ranges of –0.2–0.6 V versus SCE, polarization of the electrodes at potentials greater than 0.7 V caused oxidative degradation of the film properties. Following such oxidative degradation, a pair of new redox waves was found to appear at circa 0.5 V versus SCE, which was attributed to the formation of p-benzoquinone. Such an oxidative degradation would also be consistent with the previous studies of Rudolf Nietzki discussed in Section 2.7.

This foray into electrochromics was then followed by a short 1985 report on the photoresponse of polyaniline films at a liquid junction from Masao Kaneko and Hideki Nakamura at the Institute of Physical and Chemical Research (Riken) in Japan.[185] Here, a polyaniline film was prepared on a 1 cm^2 Pt plate via electropolymerization in an aqueous 0.2 M $LiClO_4$ solution. The resulting film was then dipped into a fresh aqueous solution containing 0.1 M $LiClO_4$ and irradiated with visible light from a 500 W xenon lamp. At applied potentials above 0 V, irradiation induced a low anodic photocurrent of the order of 1 μA cm^{-2}. Decreasing the potential below 0 V resulted in a sharp increase in cathodic photocurrent with decreased voltage, reaching a maximum value of 21 μA cm^{-2} at –0.5 V. The photocurrent was generated reversibly by repeated on–off cycles of the irradiation, and the cycles could be repeated at least 30 times without any significant change of the current response. Increasing the concentration of $LiClO_4$ by an order of magnitude resulted in a circa 30% increase in cathodic photocurrent, while changes in pH caused little change in the photocurrent. At the applied potential of –0.5 V for 5 hours in a 0.5 M $LiClO_4$ solution, a steady photocurrent of 27 μA cm^{-2} could be produced.

Finally, also in 1985, David W. DeBerry of SumX Corporation in Texas published a report on the application of polyaniline coatings for corrosion inhibition of stainless steels.[186] Polyaniline was first electropolymerized onto stainless steels of type 410 (11.6% Cr) and 430 (16.2% Cr), after which samples were removed from the deposition bath in their most oxidized state (dark blue/black in color). Upon placement in the acidic corrosion test solution, the open-circuit voltage of the modified stainless steel decreased fairly rapidly from a maximum of 0.6 V, during which the polyaniline became a dark green. With longer exposure times, a lighter, more metallic green color developed with an open-circuit potential of circa 0.2 V, which was found to be the most stable state of the electrode. While bare type 410 or 430 samples started undergoing corrosion with minutes of being placed in the acidic corrosion test solution at open-circuit potential, polyaniline-coated type 410 or 430 samples were found to remain passive from several hours to as much as 1200 hours after immersion. Overall, it was found that polyaniline-coated type 410 samples showed average equivalent penetration rates of less than 25 $\mu m/yr$ as long as the coating maintained an open-circuit voltage more positive than circa 0.0 V.

2.12 MacDiarmid and polyaniline "rediscovered"

Although Alan MacDiarmid is primarily known for his work on polyacetylene films, he transitioned from polyacetylene to polyaniline in the mid-1980s, which was just a couple years after Alan Heeger left the University of Pennsylvania to move to the University of California, Santa Barbara. While we will return later to MacDiarmid's life and work in Chapter 5 (see Sections 5.9, 5.10, and 5.12), it is worth discussing his early work on polyaniline here. This began with three nearly back-to-back papers published in the journal *Molecular Crystals and Liquid Crystals* in the spring of 1985.[187-189] Oddly, MacDiarmid begins his first paper with the statement that[187] "'Polyaniline' has been described in many papers during the past approximately 100 years in various, usually ill-defined forms such as 'aniline black,' 'emeraldine,' 'nigraniline,' etc., synthesized by the chemical or electrochemical oxidation of aniline, $(C_6H_5)NH_2$." He then supports this statement with references to a small number of polyaniline papers, including one of the earliest papers from Green and Woodhead in 1910.[146] The rest include two papers from Josefowicz and Buvet,[170,171] the earliest paper from Diaz,[178] and the paper from Noufi and Nozik discussed at the start of the previous section.[181] While Green may have been working with ill-defined forms in 1910, he certainly was not by 1913, and all of the other papers cited were clearly working with well-defined materials, all using the modern name polyaniline, not aniline black. As such, he is not only missing about 50 years of polyaniline history, but he seems to be dismissing even the more modern literature of which he has cited an extremely small fraction (4 of the 23 papers since 1964 discussed in the previous sections).

This first paper then goes on to report the constitution of polyaniline, its synthesis by both chemical oxidation and electropolymerization, the effect of protonation on conductivity, and the polymer electrochemistry. Of course, as documented in the previous sections, this was a material that was already well known and had been previously reported by other authors. Yet, it all seems to be presented as if it were a new material, with absolutely no discussion or citation of previous studies. In terms of composition, he proposes that polyaniline can exist in four forms, which using the nomenclature introduced by Green would correspond to the leucoemeraldine (fully reduced) and pernigraniline (fully oxidized) forms as both free bases and their corresponding salts.[187] However, the oxidized forms of the polymer are described as dark blue and dark green, for the base and salt, respectively, colors that are commonly attributed to the emeraldine forms. He then goes on to state that electrochemical studies suggest that it may be possible to further oxidize the samples beyond the oxidized forms presented, although further study is required. This observation would again be consistent with the emeraldine forms of the polymer, not the pernigraniline forms as given. Lastly, it should be highlighted that the "metallic" conductivities claimed in the study here are less than 10 S cm^{-1}, well below the values reported 16 years earlier by Josefowicz and Buvet.[173]

The second paper then discusses the application of polyaniline anodes and cathodes to rechargeable batteries.[188] Here, it is highlighted that Jozefowicz had previously shown related results in H_2SO_4.[170] The bulk of the study focused on cells comprised of polyaniline and Zn, which were found to have an open-circuit potential of 1.4 V and short-circuit currents ranging from circa 150 to 800 mA. Such cells exhibited a discharge voltage of 1.1 V and were predicted to have a capacity of 167 amp hr/kg. A second cell configuration of polyaniline and PbO_2 was also briefly discussed.

The third paper, in collaboration with colleagues from the University of Linköping, then investigates the conductivity of polyaniline in more detail.[189] Pressed pellet conductivity measurements were carried out on both the free base and salt forms of MacDiarmid's oxidized polyaniline, which again are presented structurally as the pernigraniline forms. For the salt forms, conductivities of 0.3 $S cm^{-1}$ were found for the SO_4^{2-} salt, while values greater than 10 S cm^{-1} were determined for the BF_4^- salt. In contrast, the fully deprotonated polymer exhibited a conductivity of less than 10^{-10} S cm^{-1}. In all cases, the polymers exhibited a temperature dependence consistent with a semiconducting material.

A fourth paper was then reported later in 1985 that focused on the protonic doping of polyaniline.[190] Although this paper starts with a nearly identical statement to that given in his first paper, he does go on to correctly credit Jozefowicz with previously demonstrating the dependence of pH and humidity on the conductivity of polyaniline, which he states has since been confirmed by several groups. In terms of the data presented, it is essentially the same data previously reported in his first paper, although with greater discussion. In particular, he attempts to propose how the protonation could influence the chemical structure of the polymer, thus resulting in the dramatic increase in conductivity.

A final 1985 book chapter was then published that focused on the optical absorption of the oxidized polyaniline in both its base and salt form.[191] Interestingly, he states in the introduction that "Polyaniline has been rediscovered recently. . .," citing his first paper discussed earlier. The results presented are compared with the compound "nigrosine," which is given as the octamer of pernigraniline, although neither its synthesis, evidence of structure, nor any reference to the literature is given. In terms of the absorption data reported, it is difficult to make any real conclusions, and the data for the polyaniline salt does not seem to agree well with previously reported spectral data.[16,183]

This was then followed by a January 1986 paper that returned again to the protonic doping of polyaniline.[192] Interestingly, the structure of oxidized polyaniline is now described as containing relative ratios of the fully reduced and fully oxidized structures given in the previous papers, with the ratio dependent upon the conditions applied to the polymer. The five forms of polyaniline as previously reported by Green are then presented, followed by a discussion of the various possible protonated forms. Here, it is suggested that the imine nitrogens are preferentially protonated over the amine nitrogens, after which it is proposed that the most stable form corresponds to the emeraldine salt, which is then used as the basis of the remaining discussion. As such, this is now in line with the bulk of the previous literature that had been developed up to the point that MacDiarmid began his polyaniline studies.

The study then went on to further examine the effect of water and pH on the polymer conductivity. In the process, it highlights the difficulties with using H_2SO_4 solutions and emeraldine sulfate, particularly in control of water content. As such, emeraldine sulfate prepared via chemical oxidation was neutralized to the emeraldine base and then protonated with HCl for the reported measurements. As before, it is shown that the conductivity of pressed-pellet samples increases with decreasing pH, giving a maximum conductivity of circa 5 S cm^{-1}.[192] Furthermore, the samples produced from each acid treatment were analyzed by elemental analysis in order to investigate the extent of protonation at each pH, which was based on the relative chloride content of each sample. In this way, it was determined that the maximum conductivity corresponded to what was described as 50% doping of the emeraldine base, which would correspond to half the nitrogens protonated by HCl. Assuming that the imine nitrogens were preferentially protonated, this would thus correspond to complete protonation of those nitrogens.

Finally, while his previous papers had described protonic doping as a form of p-doping not requiring oxidation or reduction of the polymer,[187] it is now clarified that while the conductivity of typical conducting polymers depends only on the extent of oxidation or reduction of the polymer, the conductivity of polyaniline depends on two variables—the extent of oxidation and the extent of protonation.[192] This point is then emphasized by the fact that conductivity of the leucoemeraldine base is less than circa 10^{-10} S cm^{-1} and shows no measurable increase with protonation. It is also stated that other forms, such as protoemeraldine or nigraniline, would be expected to exhibit different conductivity/doping relationships.

A second 1986 paper then followed that focused on the electrochemistry of polyaniline samples prepared by both chemical and electrochemical oxidation.[193] Samples of emeraldine hydrochloride prepared by either method were found to give essentially identical cyclic voltammograms, consisting of two redox couples with anodic peaks at circa 2.0 and 0.75 V versus SCE. Observable degradation after a few cycles of the full voltammogram was observed, yet no significant degradation was found if limited to potential sweeps of −0.2–0.6 V. The voltammogram profile, cathodic potentials, and observed stability are all in good agreement with the previously reported studies of Kobayashi and coworkers.[182]

The voltammograms were then repeated at variable pH in order investigate any pH dependence. While the first anodic peak was found to be almost independent of the pH of the electrolyte solution, the second anodic peak was strongly pH dependent, with the anodic peak shifting to less positive potentials with increasing pH. This relationship was found to be linear over a pH range of –0.2–4.0. Further study of the first anodic peak at even lower pH values did result in a pH dependence over the range of –0.2 to –2.0, with the anodic potential shifting to higher potentials with decreasing pH. Various mechanistic processes were then proposed in an attempt explain the observed pH dependence for both redox couples.[193]

Two additional papers in early 1987 then focused on the study of protonic doping via EPR and Raman, which were carried out in collaboration with Arthur J. Epstein (1945–2019) at Ohio State University.[194,195] From the EPR measurements, it was found that the relative contribution from Pauli spins increases with increasing protonation, with the intensity of the EPR signal at 50% protonation circa 100 times that of the initial emeraldine base. To account for the high contribution from spins, it was proposed that the protonation of emeraldine base leads to the formation of a poly(semiquinone radical cation). This was envisioned to occur via the disproportionation of bipolaronic quinoid units with fully reduced units, thus resulting in isolated radical cation polarons and a polaron conduction band. Results from Raman studies were also believed to be consistent with this model.

A third paper with Epstein was then reported in the spring of 1987 that continued to build on this polaron model with further EPR studies.[196] The total magnetic susceptibility was viewed as a sum of contributions from core diamagnetism, Pauli susceptibility, and localized "Curie" susceptibility, where a sizable Pauli susceptibility was found to be proportional to the extent of protonation of emeraldine. Determination of the magnetic susceptibility as a function of both temperature and protonation level led to a model where localization of polarons and bipolarons at chain ends, at surfaces, and within the small granular metal particles could account for the temperature dependence of the observed Curie spins.

A final 1987 paper was then a collaboration between both Epstein and Jean-Luc Brédas (b. 1954), a computational chemist at the University of Namur, Belgium.[197] Here, polaron and bipolar lattice band-structure calculations were reported for the proton-doped emeraldine salt. It was concluded that the polaron-lattice band structure fully accounted for the observed optical transitions of the polymer. At the same time, the results were viewed to be in marked contrast to the electronic structure of other conducting polymers, in that only one single broad polaron band appeared deep in the gap, together with a very narrow band nearly degenerate with the conduction band edge.

While MarDiarmid's study of polyaniline might have begun with a shaky start in 1985, he was now contributing real advances to the understanding of conductivity in protonated emeraldine states of the polymer. He continued the study of polyaniline for the rest of his career and continued to make real contributions to the polyaniline literature. This can be illustrated by the fact that his Nobel lecture focused far more on polyaniline than the polyacetylene work generally recognized by the 2000 Nobel Prize.[198]

2.13 Conclusions

From the discussion in this chapter, it is quite clear that polyaniline has an impressively long history dating back to the first oxidation of aniline by Runge in 1834.[12] Of course, neither the structure nor the polymeric nature of Runge's products was yet known, which is not surprising considering that Jacob Berzelius (1779–1848) had only introduced the term *polymer* two years earlier in 1832 and the macromolecular model of polymers had to wait until its introduction by Hermann Staudinger (1881–1965) in the 1920s.[14] Still, the modern oxidative polymerization of polyaniline using chemical oxidants has changed little from the methods originally reported in 1834, thus allowing us to date this polymer to Runge's initial report. In addition, as this oxidative polymerization predates Eduard Simon's (1789–1856)[13] 1839 report on the production of polystyrene, polyaniline also has the distinction of being the oldest known fully synthetic organic polymer.[5,14]

Of course, the history of polyaniline also includes a number of other important milestones. Runge demonstrated its ability to be used as a cotton dye in 1834, making it the very first synthetic aniline dye, predating the more commonly credited *mauve* (aniline purple) and *magenta* (fuchsine) by more than 20 years.[14,55–58] Its interest as cotton dyes also led to it becoming the very first conjugated polymer to be patented, with patents as cotton dyes filed as early as 1860. These dyes were quickly commercialized under the names *emeraldine*, *azurine*, and *aniline black*, making polyaniline also the first conjugated polymer on the commercial market.[5–7,11] Furthermore, the name of the green dye, emeraldine, has been retained in modern nomenclature for the most stable and conductive form of polyaniline, which was also responsible for the green color of the original commercial dye.

Another significant milestone for polyaniline was its electropolymerization by Letheby in 1862,[5,18] making it the very first conjugated polymer to be produced by electrochemical methods. This paved the way for the later electropolymerization of polypyrroles and polythiophenes, while also remaining a common modern method for the production of polyaniline. Furthermore, it is noteworthy to recognize that all of the milestones given so far occurred before the start of the 20th century.

Of course, a complication with this early history is that the modern name polyaniline had not yet been introduced, with *aniline black* the most commonly used term to refer to the material prior to 1960. The fact that polyaniline was known as aniline black and other names for the first circa 125 years of its history has led many to completely underestimate the amount of early research carried out on these materials. Furthermore, the timeline of polyaniline's history is not sporadic isolated events, nor was it abandoned and "rediscovered" in the 1980s, as some would lead one to believe.[191] Rather, the discussion follows a nearly continuous development of these fascinating materials over a period of 150+ years.

Another important aspect of polyaniline that has been largely overlooked is that it was the very first doped conjugated polymer to have its conductivity measured, dating back to 1961.[159] While those initial values were quite low in comparison with later successes, this is actually where the interest in conjugated polymers as conducting materials started. Furthermore, conductivities as high as 100 S cm^{-1} had been accomplished by 1969,[173] by far the highest conductivity known for a doped conjugated polymer at the time and greater than the original values determined for doped polyacetylene films in 1977.

Overall, it is clear that there is much that we can learn from the history of this oldest member of the family of conjugated polymers. Furthermore, it is also clear that later polymers owe much to polyaniline, both in terms of the oxidative polymerization methods used for their preparation and the fact that it set the stage for interest in their electronic nature. As such, it is hoped that this chapter's discussion will provide plenty of food for thought, as well as new insight about the overall development of conjugated materials.

References and Notes

1. Feast, W. J.; Tsibouklis, J.; Pouwer, K. L.; Groenendaal, L. Meijer, E. W. Synthesis, processing and material properties of conjugated polymers. *Polymer* **1996**, *37*, 5017–5047.
2. MacDiarmid, A. G.; Epstein, A. J. Polyanilines: A novel class of conducting polymers. *Faraday Discuss. Chem. Soc.* **1989**, *88*, 317–332.
3. Geniès, E. M.; Boyle, A.; Lakowski, M.; Tsintavis, C. Polyaniline: A historical survey. *Synth. Met.* **1990**, *36*, 139–182.
4. Li, D.; Huang, J.; Kaner, R. B. Polyanilne nanofibers: A unique olymer nanostructure for versatile applications. *Acc. Chem. Res.* **2009**, *42*, 135–145.
5. Rasmussen, S. C. Conjugated and conducting organic polymers: The first 150 years. *ChemPlusChem* **2020**, *85*(7), 1412–1429.
6. Rasmussen, S. C. Early history of conjugated polymers: From their origins to the handbook of conducting polymers. In *Handbook of Conducting Polymers*, 4th ed. Reynolds, J. R.; Skotheim, T. A.; Thompson, B., Eds.; CRC Press: Boca Raton, FL, 2019, pp. 1–35.
7. Rasmussen, S. C. Early history of conductive organic polymers. In *Conductive Polymers: Electrical Interactions in Cell Biology and Medicine*. Zhang, Z.; Rouabhia, M.; Moulton, S., Eds.; CRC Press: Boca Raton, FL, 2017, pp. 1–21.

8. Rasmussen, S. C. Electrically conducting plastics: Revising the history of conjugated organic polymers. In *100+ Years of Plastics. Leo Baekeland and Beyond* Strom, E. T.; Rasmussen, S. C.; Eds.; ACS Symposium Series 1080, American Chemical Society: Washington, DC, 2011, pp. 147–163.

9. Rasmussen, S. C. The early history of polyaniline II: Elucidation of structure and redox states. *Substantia* **2022**, *6*(1), 107–119.

10. Rasmussen, S. C. The early history of polyaniline - revisited: Russian contributions of Fritzsche and Zinin. *Bull. Hist. Chem.* **2019**, *44*(2), 123–133.

11. Rasmussen, S. C. The early history of polyaniline: Discovery and origins. *Substantia* **2017**, *1*(2), 99–109.

12. Runge, F. F. Ueber einige Producte der Steinkohlendestillation. *Ann. Phys. Chem.* **1834**, *31*, 513–524.

13. Simon, E. Ueber den flüssigen Storax (Styrax liquidas). *Ann. Pharm.* **1839**, *31*, 265–277.

14. Rasmussen, S. C. Revisiting the early history of synthetic polymers: Critiques and new insights. *Ambix* **2018**, *65*, 356–372.

15. Constantini, P.; Belorgey, G.; Jozefowicz, M.; Buvet, R. Préparation, propriétés acides et formation de complexes anioniques de polyanilines oligomères. *C. R. Seances Acad. Sci.* **1964**, *258*, 6421–6424.

16. Jozefowicz, M.; Belorgey, G.; Yu, L.-T.; Buvet, R. Oxydation et réduction de polyanilines oligomères. *C. R. Seances Acad. Sci.* **1965**, *260*, 6367–6370.

17. Combarel, M.-F.; Belorgey, G.; Jozefowicz, M.; Yu, L.-T.; Buvet, R. Conductivité en courant continu des polyanilines oligomers: Influence de l'état acide-base sur la conductivité électronique. *C. R. Seances Acad. Sci.* **1966**, *262*, 459–462.

18. Letheby, H. On the production of a blue substance by the electrolysis of sulphate of aniline. *J. Chem. Soc.* **1862**, *15*, 161–163.

19. Heth, C. L.; Tallman, D. E.; Rasmussen, S. C. Electrochemical study of 3-(*N*-alkylamino)thiophenes: Experimental and theoretical insights into a unique mechanism of oxidative polymerization. *J. Phys. Chem. B* **2010**, *114*, 5275–5282.

20. Wei, D.; Kyarnstrom, C.; Lindfors, T.; Kronberg, L.; Sjoholm, R.; Ivaska, A. Electropolymerization mechanism of *N*-methylaniline. *Synth. Met.* **2006**, *156*, 541–548.

21. Hand, R. L.; Nelson, R. F. Anodic oxidation pathways of *N*-alkylanilines. *J. Am. Chem. Soc.* **1974**, *96*, 850–860.

22. Bacon, J.; Adams, R. N. Anodic oxidations of aromatic amines. III. Substituted anilines in aqueous media. *J. Am. Chem. Soc.* **1968**, *90*, 6596–6599.

23. D'Aprano, G.; Proynov, E.; Leboeuf, M.; Leclerc, M.; Salahub, D. R. Spin densities and polymerizabilities of aniline derivatives deduced from density functional calculations. *J. Am. Chem. Soc.* **1996**, *118*, 9736–9742.

24. Buncel, E. 1999 R.U. Lemieux Award Lecture Adventures with azo-, azoxy-, and hydrazoarenes: From the Wallach to the benzidine rearrangement. Molecular electronics. *Canadian J. Chem.* **2000**, *78*, 1251–1271.

25. Buncel, E.; Cheon, K.-S. Acid-catalysed disproportionation and benzidine rearrangement of phenylhydrazinopyridines: Reaction pathways, kinetics and mechanism. *J. Chem. Soc., Perkin Trans. 2*, **1998**, 1241–1247.

26. Shine, H. J.; Habdas, J.; Kwart, H.; Brechbiel, M.; Horgan, A. G.; Filippo, Jr., J. S. Benzidine rearrangements. 18. Mechanism of the acid-catalyzed disporportionation of 4,4'-diiodohydrazobenzene. Application of heavy-atom kinetic isotope effects. *J. Am. Chem. Soc.* **1983**, *105*, 2823–2827.

27. Dussauce, H. Aniline - Its history, properties, and preparation. *Sci. Am.* **1867**, *17*(21), 321–322.

28. Jackson, C. M. Synthetical experiments and alkaloid analogues: Liebig, Hofmann, and the origins of organic synthesis. *Hist. Stud. Nat. Sci.* **2014**, *44*, 319–363.

29. Unverdorben, O. Ueber das Verhalten der organifchen Körper in höheren Temperaturen. *Ann. Phys. Chem.* **1826**, *8*, 397–410.

30. Orna, M. V. Historic mineral pigments: Colorful benchmarks of ancient civilizations. In *Chemical Technology in Antiquity*, Rasmussen, S. C., Ed.; ACS Symposium Series 1211, American Chemical Society: Washington, DC, 2015, pp. 17–69.

31. McGovern, P. E.; Michel, R. H. Royal purple dye: The chemical reconstruction of the ancient Mediterranean industry. *Acc. Chem. Res.* **1990**, *23*, 152–158.

32. Koren, Z. C. Modern chemistry of the ancient chemical processing of organic dyes and pigments. In *Chemical Technology in Antiquity*, Rasmussen, S. C., Ed.; ACS Symposium Series 1211, American Chemical Society: Washington, DC, 2015, pp. 197–217.

33. Baeyer, A. Ueber die Verbindungen der Indigogruppe. *Ber. Dtsch. Chem. Ges.* **1883**, *16*, 2188–2204.

34. Although most authors give the name as *Krystallin*, Unverdorben clearly spelled it *Crystallin* in his original 1826 paper. The use of the more common alternate spelling dates as far back as 1840 and was likely due to the fact that "kristallin" is the German word for crystalline.

35. Runge, F. F. Ueber einige Produkte der Steinkohlendestillation. *Ann. Phys. Chem.*, **1834**, *31*, 65–78.

36. Fritzsche, J. Ueber das Anilin, ein neues Zersetzungsproduct des Indigo. *Bull. Sci. Acad. Imp. Sci. St. Petersb.* **1840**, *7*, 161–165.

37. Fritzsche, J. Ueber das Anilin, ein neues Zersetzungsprodukt des Indigo. *J. Prakt. Chem.* **1840**, *20*, 453–457.

38. Fritzsche, J. Ueber das Anilin, ein neues Zersetzungsprodukt des Indigo. *Ann. Chem. Pharm.* **1840**, *36*, 84–88.

39. As determined by modern means, the density of aniline is 1.0297 g/mL, with a boiling point of 184.13°C.

40. Such doubling of formulas was typical of the time period, usually the result of utilizing an atomic weight of 6 for carbon.

41. Erdmann, O. Nachschrift. *J. Prakt. Chem.*, **1840**, *20*, 457–459.

42. Liebig, J. Bemerkung zu vorstehender Notiz. *Ann. Chem. Pharm.* **1840**, *36*, 88–90.

43. Sinin, N. Beschreibung einiger neuer organischer Basen, dargestellt durch die Einwirkung des Schwefelwasserstoffs auf Verbindungen der Kohlenwasserstoffe mit Untersalpetersäure. *Bull. Sci. Acad. Imp. Sci. St. Petersb.* **1842**, *10*, 273–285.

44. Zinin, N. Beschreibung einiger neuer organischer Basen, dargestellt durch die Einwirkung des Schwefelwasserstoffs auf Verbindungen der Kohlenwasserstoffe mit Untersalpetersäure. *J. Prakt. Chem.* **1842**, *27*, 140–153.

45. Zinin, N. Organische Salzbasen, aus Nitronaphtalose und Nitrobenzid mittelst Schwefelwasserstoff entstehend. *Ann. Chem. Pharm.* **1842**, *44*, 283–287.

46. Fritzsche, J. Bemerkung zu vorstehender Abhandlung des Hrn. Zinin. *J. Prakt. Chem.* **1842**, *27*, 153.

47. Hofmann, A. W. Chemische Untersuchung der organischen Basen im Steinkohlen-Theeröl. *Ann. Chem. Pharm.* **1843**, *47*, 37–87.

48. Zinin, N. Ueber das Azobenzid und die Nitrobenzinsäure. *J. Prakt. Chem.* **1845**, *36*, 93–107.

49. Bechamp, A. De l'action des protosels de fer. Sur la nitronaphtaline et la nitrobenzine. Nouvelle méthode de formation des bases organiques artificielles de Zinin. *Ann. Chim. Phys.* **1854**, *42*, 186–196.

50. Kränzlein, G. Zum 100 jährigen Gedächtnis der Arbeiten von F. F. Runge. *Angew. Chem.* **1935**, *48*, 1–3.

51. Bussemas, H. H.; Harsch, G.; Ettre, L. S. Friedlieb Ferdinand Runge (1794–1867): "Self-grown pictures" as precursors of paper chromatography. *Chromatographia* **1994**, *38*, 243–254.

52. Anft, B.; Oesper, R. E. (Trans.). Friedlieb Ferdinand Runge: A forgotten chemist of the nineteenth century. *J. Chem. Ed.* **1955**, *32*, 566–574.

53. Schwenk, E. F. Friedlieb Runge and his capillary designs. *Bull. Hist. Chem.* **2005**, *30*, 30–34.

54. Noelting, E. *Scientific and industrial history of Aniline Black.* Wm. J. Matheson & Co.: New York, 1889, pp. 7–16.

55. Garfield, S. *Mauve: How One Man Invented a Color That Changed the World.* W. W. Norton: New York, 2002, pp. 6–8.

56. Orna, M. V. *The Chemical History of Color.* SpringerBriefs in Molecular Science: History of Chemistry. Springer: Heidelberg, 2013, pp. 69–75.

57. Holme, I. Sir William Henry Perkin: A review of his life, work and legacy. *Color. Technol.* **2006**, *122*, 235–251.

58. Johnston, W. T. The discovery of aniline and the origin of the term "aniline dye." *Biotech. Histochem.* **2008**, *83*, 83–87.

59. Runge, F. F. *Farbenchemie. Erster Theil: Die kunſt zu färben gegründet auf das chemische Berhalten der Baumwollenfaser zu den Salzen und Säuren.* E. S. Mittler: Berlin, 1834.

60. Runge, F. F. *Farbenchemie. Zweiter Theil: Die kunst zu druden gegründet auf das chemische Berhalten der Baumwollenfaser zu den Salzen und Säuren.* E. S. Mittler: Berlin, 1842.

61. Runge, F. F. *Farbenchemie. Dritter Theil: Die Kunst der Farbenbereitung.* E. S. Mittler: Berlin, 1850.

62. Poggendorff, J. C. *Biographisch-Literarisches Handwörterbuch zur Geschichte der Exacten Wissenschaften*, Vol. 1. J. A. Barth: Leipzig, 1863, pp. 808–809.

63. Butlerow, A. Carl Julius Fritzsche. *Ber. Dtsch. Chem. Ges.* **1872**, *5*, 132–136.

64. Harcourt, A. V. Anniversary meeting, March 30th, 1872. *J. Chem. Soc.* **1872**, *25*, 341–364.

65. Sheibley, F. E. Carl Julius Fritzsche and the discovery of anthranilic acid, 1841. *J. Chem. Educ.* **1943**, *20*, 115–117.

66. Lewis, D. E. *Early Russian Organic Chemists and Their Legacy*, Springer Briefs in Molecular Science: History of Chemistry, Springer: Heidelberg, 2012, p. 45.

67. Leicester, H. M. N. N. Zinin, an early Russian Chemist. *J. Chem. Educ.* **1940**, *17*, 303–306.

68. Fritzsche, J. Vorläufige Notiz über einige neue Körper aus der Indigoreihe. *Bull. Classe phys.-math. Acad. Sci. St.-Pétersbourg* **1843**, *1*, 103–108.

69. Fritzsche, J. Vorläufige Notiz über einige neue Körper aus der Indigoreihe. *J. Prakt. Chem.* **1843**, *28*, 198–204.

70. Travis, A. Perkin's mauve: Ancestor of the organic chemical industry. *Technology and Culture* **1990**, *31*, 51–82.

71. Garfield, S. *Mauve: How One Man Invented a Color That Changed the World*. W. W. Norton: New York, 2002, pp. 78–79.
72. Crace Calvert, F. *Lectures on Coal Tar Colours and on Recent Improvements and Progress in Dyeing & Calico Printing*. Henry Carey Baird: Philadelphia, 1863, p. 23.
73. Binney, E. W. Manchester Literary and Philosophical Society. Annual meeting, 1874. *Chem. News J. Phys. Sci.* **1875**, *31*, 56–57.
74. Crace-Calvert, F. *Dyeing and Calico Printing: Including an Account of the Most Recent Improvements in the Manufacture and Use of Aniline Colours*. Palmer & Howe: Manchester, UK, 1876, pp. ix–xiv.
75. Nicklas, C. C. *Splendid Hues: Colour, Dyes, Everyday Science, and Women's Fashion, 1840-1875*. Ph.D. dissertation, University of Brighton, November 2009, pp. 199–204.
76. Smith, R. A. Manchester Literary and Philosophical Society. Ordinary meeting, November 4th, 1873. *Chem. News J. Phys. Sci.* **1873**, *28*, 300.
77. Crace Calvert, F. On improvement and progress in dyeing and calico printing since 1851. *J. Soc. Arts* **1862**, *10*, 169–180.
78. Lightfoot, J. *The Chemical History and Progress of Aniline Black*. Lower House, Lancashire, UK, 1871, pp. 1–5.
79. Travis, A. S. From Manchester to Massachusetts via Mulhouse: The transatlantic voyage of aniline black. *Technology and Culture* **1994**, *35*, 70–99.
80. Crace Calvert, F. *Lectures on Coal-Tar Colours, and on Recent Improvements and Progress in Dyeing and Calico Printing*. Henry Carey Baird: Philadelphia, 1863, p. 63.
81. Noelting, E. *Scientific and Industrial History of Aniline Black*. Wm. J. Matheson & Co.: New York, 1889, pp. 37–42.
82. Travis, A. S. Artificial dyes in John Lightfoot's Broad Oak Laboratory. *Ambix* **1995**, *42*, 10–27.
83. Lightfoot, J. Black dye from aniline. *Chem. News J. Ind. Sci.* **1862**, *6*, 287.
84. Lightfoot, J. Improvement in Dyeing and Printing a Black Color on Fabrics with Aniline Compounds. U.S. Patent 38, 589, May 19, 1863.
85. Bernthsen, A. Heinrich Caro. *Ber. Dtsch. Chem. Ges.* **1912**, *45*, 1987–2042.
86. Darmstaedter, E. Caro. In *Das Buch der grossen Chemiker*, Vol. II, 2nd ed. Bugge, G., Ed.; Verlag Chemie: Berlin, 1930, pp. 298–309.
87. Travis, A. S. Heinrich Caro at Roberts, Dale & Co. *Ambix* **1991**, *38*, 113–134.
88. Bernthsen, A. Zum 70. Geburtstage von H. Caro. *Z. Angew. Chem.* **1911**, *24*, 1059–1064.
89. Wiedemann, M. Heinrich Caro. *Chem. Ind.* **1910**, *33*, 561.
90. Reinhardt, C.; Travis, A. S. *Heinrich Caro and the Creation of Modern Chemical Industry*. Kluwer Academic Publishers: Dordrecht, the Netherlands, 2000, pp. 55–61.
91. Hamlin, C. Letheby, Henry (1816–1876). In *Oxford Dictionary of National Biography*. Matthew, C.; Harrison, B., Eds.; Oxford University Press: Oxford, 2004, Vol. 33, pp. 505–506.
92. Archbold, W. A. J. Letheby, Henry (1816–1876). In *Dictionary of National Biography, 1885-1900*. Lee, S., Ed.; Smith, Elder & Co.: London, 1893, Vol. 33, p. 131.
93. Abel, F. Anniversary meeting, March 30th, 1876, *J. Chem. Soc.* **1876**, *29*, 617–640.
94. Dr. Letheby. *The Analyst* **1876**, *1*, 15.

95. Dr. Letheby. *Chem. News J. Ind. Sci.* **1876**, *33*, 146.

96. Grove, W. R. On voltaic series and the combination of gases by platinum. *Philos. Mag.*, 3rd ser. **1839**, *14*, 127–130.

97. Grove, W. Pile voltaïque d'une grande énergie électro-chimique. *C. R. Acad. Sci.* **1839**, *8*, 567–570.

98. Grove, W. R. On a new voltaic combination. *Philos. Mag.*, 3rd ser. **1839**, *14*, 388–390.

99. Heth, C. L. Energy on demand: A brief history of the development of the battery. *Substantia* **2019**, *3*(2) Suppl. 1, 73–82.

100. Feddersen, B. W.; von Oettingen, A. J., Eds. *J. C. Poggendorff's Biographisch-Literarisches Handwörterbuch zur Geschichte der Exacten Wissenschaften*, Vol. 3. J. A. Barth: Leipzig, 1898, p. 300.

101. Wurtz, C. A.; Gauthier, A. 11. Laboratoire de chimie de la Faculté de médecine. *Rapport sur l'École pratique des hautes études, 1876–1877*, **1876**, 40–41.

102. Coquillion, J.-J. Sur information du noir d'aniline, obtenu par l'éleclrolyse de ses sels. *C. R. Seances Acad. Sci.* **1875**, *81*, 408–409.

103. Bunsen, R. Ueber eine neue Construction der galvanischen Säule. *Ann. Chem. Pharm.* **1841**, *38*, 311–313.

104. Coquillion, J.-J. Sur la synthèse du noir d'aniline. *C. R. Seances Acad. Sci.* **1876**, *82*, 228–230.

105. Goppelsroeder, F. Sur l'électrotyse des corps de la série aromatique. *C. R. Seances Acad. Sci.* **1875**, *81*, 944–945.

106. Strahlmann, B. Friedrich Goppelsroeder. *Neue Deutsche Biographie* **1964**, *6*, 645–646.

107. Tamm, C. Friedrich Goppelsroeder. In *Historisches Lexikon der Schweiz (HLS)*, Version 28.11.2005, https://hls-dhs-dss.ch/de/articles/044500/2005-11-28/ (accessed December 30, 2021).

108. Fichter, F. Friedrich Goppelsroeder. *Verhandlungen der Naturforschenden Gesellschaft in Basel* **1920**, *31*, 133–152.

109. Fichter, F. Friedrich Goppelsroeder. *Verhandlungen der Schweizerische Naturforschende Gesellschaft* **1920**, *101*, A30–A31.

110. Ostwald, W. Zu Ehren Friedrich Goppelsroeder's. *Kolloid-Zeitschrift* **1912**, *10*, 1–3.

111. Goppelsroeder, F. Sur le noir d'aniline électrolytique. *C. R. Seances Acad. Sci.* **1876**, *82*, 331–333.

112. Goppelsroeder, F. Sur l'électrolyse des dérivés de l'aniline, du phénol, de la naphtylamine et de l'anthraquinone. *C. R. Seances Acad. Sci.* **1876**, *82*, 1199–1201.

113. Goppelsroeder, F. Analyse élémentaire du noir d'aniline électrolytique. *C. R. Seances Acad. Sci.* **1876**, *82*, 1392–1394.

114. Goppelsroeder, F. Sur une cuve au noir d'aniline et sur la transformation du noir d'aniline en une matière colorante rose fluorescente. *C. R. Seances Acad. Sci.* **1877**, *84*, 447–450.

115. Goppelsroeder, F. Zum Studium der Metamorphosen des Anilinschwarz. *Dinglers Polytech J.* **1877**, *224*, 439–448.

116. Rheineck, H. Ein Beitrag zur Kenntniss der Natur des Anilinschwarz. *Dinglers Polytech. J.* **1872**, *203*, 485–489.

117. *Verzeichnis der Beamten, Lehrer und Studirenden der königlich württembergischen Universität Tübingen in dem Sommerhalbjahr 1863.* University of Tübingen: Tübingen, Germany, p. 27.

118. Rheineck, H. Ueber das Verhalten des Allantoïns zu Natrium. *Ann. Chem. Pharm.* **1865**, *134*, 219-228.
119. Rheineck, H. Ein Colorimeter. *Dinglers Polytech. J.* **1871**, *201*, 433-435.
120. Rheineck, H. Ueber die Gührungserscheinungen. *Dinglers Polytech. J.* **1871**, *202*, 282-287.
121. Rheineck, H. Ueber eine maaßanalytische Beſtimmung des Eisens und Ferrocyans. *Dinglers Polytech. J.* **1871**, *202*, 154-159.
122. Rheineck, H. Ein Beitrag zur Kenntniß der Natur des Anilinschwarz. *Polytech. Centralblatt* **1872**, *38*, 667-670.
123. Rheineck, H. Beitrag zur Kenntniss der Natur des Anilinschwarz. *Chemisches Centralblatt* **1872**, *43*, 295-296.
124. Rheineck, H. berichtet über Versuche, die er betreffs der Bildung von Anilinschwarz auf dem Gewebe angestellt hat. Jahresbericht Leistungen. *Chem. Tech.* **1873**, *18*, 710-711.
125. Nietzki, R. Ueber Anilinschwarz. *Ber. Dtsch. Chem. Ges.* **1876**, *9*, 616-620.
126. Nietzki, R. Ueber Anilinschwarz. *Ber. Dtsch. Chem. Ges.* **1876**, *9*, 1168-1170.
127. Schwarz, H.-D. Nietzki, Rudolf Hugo. *Neue Deutsche Biographie* **1999**, *19*, 248.
128. Kurz, M.; Nietzki, Rudolf. *Historisches Lexikon der Schweiz (HLS)*, Version 03.06.2010. https://hls-dhsdss.ch/de/articles/045058/2010-06-03/ (accessed September 2021).
129. Noelting, E. Rudolf Nietski 1847-1917. *Helv. Chim. Acta* **1918**, *1*, 343-430.
130. Weinmeister, P. Ed., *J. C. Poggendorffs Biographisch-Literarisches Handwörterbuch*, Vol. 5, Verlag Chemie: Leipzig, 1925, p. 906.
131. Nietzki, R. Ueber Anilinschwarz. *Ber. Dtsch. Chem. Ges.* **1878**, *11*, 1093-1102.
132. Nietzki, R. Zur Darstellung der Chinone und Hydrochinone. *Ber. Dtsch. Chem. Ges.* **1877**, *10*, 1934-1935.
133. Nietzki, R. Zur Darstellung des Chinons. *Ber. Dtsch. Chem. Ges.* **1878**, *11*, 1102-1104.
134. Caro, H. Die Oxydation des Anilins. *Chemiker Zeitung* **1896**, *20*, 840.
135. Caro, H. Zur Kenntniss der Oxydation des Anilins. *Verh. Ges. Dtsch. Naturforsch. Aerzte* **1896**, *68*, 119-120.
136. Willstätter, R.; Moore, C. W. Über Anilinschwarz. I. *Ber. Dtsch. Chem. Ges.* **1907**, *40*, 2665-2689.
137. Berl, E. Richard Willstätter. *Chem. Eng. News* **1942**, *20*, 954.
138. Robinson, R. Richard Willstätter. 1872-1942. *Obit. Notices Fellows R. Soc.* **1953**, *8*, 609-634.
139. Robinson, R. Richard Willstätter. *J. Chem. Soc.* **1953**, 999-1026.
140. Adams, R. Richard Willstätter. *J. Am. Chem. Soc.* **1943**, *65*, 127-128.
141. Huisgen, R. Richard Willstätter. *J. Chem. Educ.* **1961**, *38*, 10-15.
142. Willstätter, R.; Dorogi, S. Über Anilinschwarz II. *Ber. Dtsch. Chem. Ges.* **1909**, *42*, 2147-2168.
143. Bucherer. H. T. Über den Mechanismus der Indamin- und Azinsynthese. Bemerkungen zu der Abhandlung R. Willstätters über Anilinschwarz. *Ber. Dtsch. Chem. Ges.* **1907**, *40*, 3412-3419.
144. Willstätter, R.; Dorogi, S. Über Anilinschwarz III. *Ber. Dtsch. Chem. Ges.* **1909**, *42*, 4118-4151.
145. Willstätter, R.; Cramer, C. Über Anilinschwarz IV. *Ber. Dtsch. Chem. Ges.* **1910**, *43*, 2976-2988.

146. Green, A. G.; Woodhead, A. E. Aniline-black and allied compounds. Part I. *J. Chem. Soc. Trans.* **1910**, *97*, 2388–2403.

147. Green, A. G. The chemical technology of aniline black. In *Seventh International Congress of Applied Chemistry, London, May 27th to June 2nd, 1909*. Partridge & Cooper, Ltd.: London, 1910, Sect. IVb, pp. 83–84.

148. Green, A. G. The chemical technology of aniline black. *J. Soc. Dyers Colour* **1909**, *25*, 188–192.

149. Saunders, K. H. Arthur George Green. *J. Soc. Dyers Colour* **1941**, *57*, 364–366.

150. Baddiley, J. Arthur George Green. *J. Chem. Soc.* **1946**, 842–852.

151. Saunders, K. H. The scientific and technical work of Professor A. G. Green, F.R.S. *J. Soc. Dyers Colour* **1944**, *60*, 81–93.

152. Friswell, R. J.; Green, A. G. On the relation of diaxobenaeneanilide to anzidoazobenzene. *J. Chem. Soc., Trans.* **1909**, *95*, 2202–2215.

153. Cross, E. J.; Speakman, J. B. Prof. F. M. Rowe, F.R.S. *Nature* **1947**, *159*, 53.

154. Willstätter, R.; Cramer, C. Über Anilinschwarz. V. *Ber. Dtsch. Chem. Ges.* **1911**, *44*, 2162–2171.

155. Green, A. G.; Woodhead, A. E. Anilinschwarz und seine Zwischenkorper. II. *Ber. Dtsch. Chem. Ges.* **1912**, *45*, 1955–1958.

156. Green, A. G.; Woodhead, A. E. Aniline-black and allied compounds. Part II. *J. Chem. Soc. Trans.* **1912**, *101*, 1117–1123.

157. Green, A. G.; Wolff, S. Anilinschwarz und seine Zwischenkorper III. *Ber. Dtsch. Chem. Ges.* **1913**, *46*, 33–49.

158. Green, A. G.; Johnson, W. Anilinschwarz und seine Zwischenkorper IV. *Ber. Dtsch. Chem. Ges.* **1913**, *46*, 3769–3779.

159. Parini, V. P.; Kazakova, Z. S.; Berlin, A. A. Polymers with conjugated bonds and with heteroatoms in the conjugated bond chain XIX. Some properties of aniline black. *Vysokomol. soedin.* **1961**, *3*, 1870–1873.

160. Westhaus, P. A.; Swamy, N. V. V. J. Herbert A. Pohl. *Physics Today* **1987**, *40*, 90–91.

161. Pohl, H. A.; Engelhardt, E. H. Synthesis and characterization of some highly conjugated semiconducting polymers. *J. Phys. Chem.* **1962**, *66*, 2085–2095.

162. Girault, J. Jozefowicz Marcel (Jozefonvicz dit). https://maitron.fr/spip.php?article136070, posted on December 17, 2010, last modification on December 18, 2010 (accessed April 10, 2022).

163. Foreign scientists awarded by the Russian Academy of Sciences the Honoris causa Doctorate Degree in Chemical Sciences in 1993–1994. *Russ. Chem. Bull.* **1994**, *43*, 1111–1113.

164. Jozefowicz, M.; Buvet, R. Préparation de poly-*p*-phénylènes oligomères. *C. R. Seances Acad. Sci.* **1961**, *253*, 1801–1803.

165. Jozefowicz, M.; Buvet, R. Caractérisation spectrophotométrique des mélanges de *p*-oligophénylènes. *C. R. Seances Acad. Sci.* **1962**, *254*, 284–286.

166. Liang-Tse, Y.; Petit, J.; Jozefowicz, M.; Belorgey, G.; Buvet, R. Conductivité en courant continu des sulfates acides d'éméraldine. *C. R. Seances Acad. Sci.* **1965**, *260*, 5026–5029.

167. Jozefowicz, M.; Yu. L. T. Relations entre propriétés chimiques et électrochimiques de semi-conducteurs macromoléculaires. *Rev. Gen. Electr.* **1966**, *75*, 1008–1013.

168. Yu, L. T.; Jozefowicz, M. Conductivité et constitution chimique pe semi-conducteurs macromoléculaires. *Rev. Gen. Electr.* **1966**, *75*, 1014–1018.

169. Yu, L. T.; Borredon, M. S.; Jozefowicz, M.; Belorgey, G.; Buvet, R. Etude experiment ale de la Conductivite en Courantvcontinu des Composes macromoleculaires. *J. Polym. Sci. Polym. Symp.* **1967**, *16*, 2931–2942.

170. De Surville, R.; Jozefowicz, M.; Yu, L. T.; Perichon, J.; Buvet, R. Electrochemical chains using protolytic organic semiconductors. *Electrochim. Acta* **1968**, *13*, 1451–1458.

171. Cristofini, F.; De Surville, R.; Jozefowicz, M.; L.-T. Yu; Buvet, R. Propriétés électrochemiques des sulfates de polyaniline. *C. R. Seances Acad. Sci., Ser. C* **1969**, *268*, 1346–1349.

172. Labarre, D.; Jozefowicz, M. Polymères conducteurs organiques fimogènes à base de polyanilines. *C. R. Seances Acad. Sci., Ser. C* **1969**, *269*, 964–966.

173. Jozefowicz, M.; Yu, L. T.; Perichon, J.; Buvet, R. Proprietes nouvelles des polymeres semiconducteurs. *J. Polym. Sci. Part C: Polym. Symp.* **1969**, *22*, 1187–1195.

174. Doriomedoff, M.; Hautiere-Cristofini, F.; de Surville, R.; Jozefowicz, M., Yu L. T. Buvet,R. Conductivite en courant continu des sulfates de polyanilines., *J. Chim. Phys. Phys.-Chim. Biol.* **1971**, *68*, 1055–1069.

175. Inzelt, G. *Conducting Polymers. A New Era in Electrochemistry.* Monographs in Electrochemistry, F. Scholz, Ed. Berlin: Springer-Verlag: Berlin, 2008, pp. 265–269.

176. Mohilner, D. M.; Adams, R. N.; Argersinger, W. J. Jr. Investigation of the kinetics and mechanism of the anodic oxidation of aniline in aqueous sulfuric acid solution at a platinum electrode. *J. Am. Chem. Soc.* **1962**, *84*, 3618–3622.

177. Adams Institute, University of Kansas. Ralph N. "Buzz" Adams. https:// adamsinstitute.ku.edu/ralph-n-adams (accessed May 17, 2024).

178. Diaz, A. F.; Logan, J. A. Electroactive polyaniline films. *J. Electroanal. Chem.* **1980**, *111*, 111–114.

179. Hernandez, R.; Diaz, A. F.; Waltman, R.; Bargon, J. Surface characteristics of thin films prepared by plasma and electrochemical polymerizations. *J. Phys. Chem.* **1984**, *88*, 3333–3337.

180. LaCroix, J.-C.; Diaz, A. F. Electrolyte effects on the switching reaction of polyaniline. *J. Electrochem. Soc.* **1988**, *135*, 1457–1263.

181. Noufi, R.; Nozik, A. J.; White, J.; Warren, L. F. Enhanced stability of photoelectrodes with electrogenerated polyaniline films. *J. Electrochem. Soc.* **1982**, *129*, 2261–2265.

182. Kobayashi, T.; Yoneyama, H.; Tamura, H. Polyaniline film-coated electrodes as electrochromic display devices. *J. Electroanal. Chem.* **1984**, *161*, 419–423.

183. Kobayashi, T.; Yoneyama, H.; Tamura, H. Electrochemical reactions concerned with electrochromism of polyaniline film-coated electrodes. *J. Electroanal. Chem.* **1984**, *177*, 281–291.

184. Kobayashi, T.; Yoneyama, H.; Tamura, H. Oxidative degradation pathway of polyaniline film electrodes. *J. Electroanal. Chem.* **1984**, *177*, 293–297.

185. Kaneko, M.; Nakamura, H. Photoresponse of a liquid junction polyaniline film. *J. Chem. Soc., Chem. Commun.* **1985**, 346–347.

186. DeBerry, D. W. Modification of the electrochemical and corrosion behavior of stainless steels with an electroactive coating. *J. Electrochem. Soc.* **1985**, *132*, 1022–1026.

187. MacDiarmid, A. G.; Chiang, J.-C.; Halpern, M.; Huang, W.-S.; Mu, S.-L.; Somasiri, N. L. D.; Wu, W.; Yaniger, S. I. "Polyaniline": Interconversion of metallic and insulating forms. *Mol. Cryst. Liq. Cryst.* **1985**, *121*, 173–180.

188. MacDiarmid, A. G.; Mu, S.-L.; Somasiri, N. L. D.; Wu, W. Electrochemical characteristics of "polyaniline" cathodes and anodes in aqueous electrolytes. *Mol. Cryst. Liq. Cryst.* **1985**, *121*, 187–190.

189. Salaneck, W. R.; Liedberg, B.; Inganäs, O.; Erlandsson R.; Lundström, I.; MacDiarmid, A. G.; Halpern, M.; Somasiri, N. L. D. Physical characterization of some polyaniline, $(\text{ØN})_x$. *Mol. Cryst. Liq. Cryst.* **1985**, *121*, 187–190.

190. MacDiarmid, A. G.; Chiang, J.-C.; Huang, W.; Humphrey, B. D.; Somasiri, N. L. D. Polyaniline: Protonic acid doping to the metallic regime. *Mol. Cryst. Liq. Cryst.* **1985**, *125*, 309–318.

191. Salaneck, W. R.; Lundström, I.; Liedberg, B.; Hasan, M. A.; Erlandsson, R.; Konradsson, P.; MacDiarmid, A. G.; Somasiri, N. L. D. Spectroscopic characterization of some polyanilines. In *Electronic Properties of Polymers and Related Compounds*. Kuzmany, H.; Mehring, M.; Roth, S.; Eds.; Springer Series in Solid-State Sciences 63, Springer: Heidelberg, 1985, pp. 218–222.

192. Chiang, J.-C.; MacDiarmid, A. G. "Polyaniline": Protonic acid doping of the emeraldine form to the metallic regime. *Synth. Met.* **1986**, *13*, 193–205.

193. Huang, W.-S.; Humphrey, B. D.; MacDiarmid, A. G. Polyaniline, a novel conducting polymer. Morphology and chemistry of its oxidation and reduction in aqueous electrolytes. *J. Chem. Soc., Faraday Trans. 1*, **1986**, *82*, 2385–2400.

194. MacDiarmid, A. G.; Chiang, J. C.; Richter, A. F.; Epstein, A. J. Polyaniline: A new concept in conducting polymers. *Synth. Met.* **1987**, *18*, 285–290.

195. Epstein, A. J.; Ginder, J. M.; Zuo, F.; Bigelow, R. W.; Woo, H.-S.; Tanner, D. B.; Richter, A. F.; Huang, W.-S.; MacDiarmid, A. G. Insulator-to-metal transition in polyaniline. *Synth. Met.* **1987**, *18*, 303–309.

196. Ginder, J. M.; Richter, A. F.; MacDiarmid, A. G.; Epstein, A. J. Insulator-to-metal transition in polyaniline. *Solid State Commun.* **1987**, *63*, 97–101.

197. Stafström, S.; Brédas, J. L.; Epstein, A. J.; Woo, H. S.; Tanner, D. B.; Huang, W. S.; MacDiarmid, A. G. Polaron lattice in highly conducting polyaniline: Theoretical and optical studies. *Phys. Rev. Lett.* **1987**, *59*, 1464–1467.

198. MacDiarmid, A. G. "Synthetic metals": A novel role for organic polymers (Nobel Lecture). *Angew. Chem. Int. Ed.* **2001**, *40*, 2581–2590.

Chapter 3
Polypyrrole

3.1 Introduction

Of the various parent conjugated polymers introduced in Chapter 1, polypyrrole (Figure 3.1) is notable for a number of significant milestones in the field of conjugated and conducting polymers.[1-5] Not only is it the second oldest of the known conjugated polymers, but it also provided the first real example of an organic polymer with significant conductivity, as initially reported in 1963 by Donald Weiss and coworkers in Australia.[6-8] Furthermore, it was the first conjugated polymer to be prepared as a plastic film.[9] Of course, the latter two milestones are typically accredited to polyacetylene, and the bulk of modern literature overlooks the first 64 years of reports on polypyrrole, instead pointing to the later work of Diaz and coworkers in 1979[10,11] when introducing the early studies of this polymer.[12-16] Of course, by placing the origin of conducting polypyrrole in 1979, such reports further reinforce the commonly held view that the history of conducting polymers begins with polyacetylene and that other members of the family of conjugated polymers followed afterward.

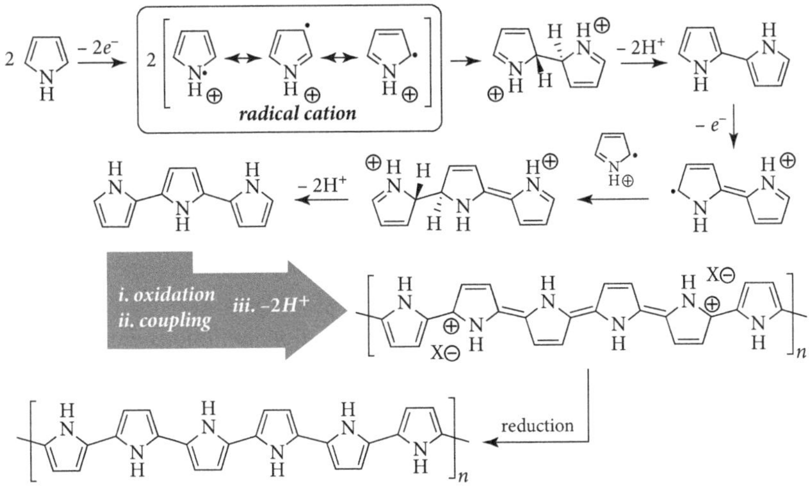

Figure 3.1 Modern oxidative polymerization mechanism of pyrrole.

The Origins and Early History of Conjugated Organic Polymers. Seth C. Rasmussen, Oxford University Press.
© Oxford University Press (2025). DOI: 10.1093/9780197638194.003.0003

Although none of the early researchers ever explicitly state it, there are too many indicators that point to the fact that the early development of polypyrrole seems to owe much to polyaniline (Chapter 2). Beginning with Runge as the common discoverer of both aniline and pyrrole, the two materials also began with very similar names (*aniline black* vs. *pyrrole black*), and other similar terms can be found in the histories of both polymers. Added to this is the fact that the two materials are also prepared in very similar ways. As with polyaniline, polypyrrole is most commonly synthesized via oxidative polymerization, which can be accomplished via the use of either chemical oxidizing agents or electrochemically via the application of anodic potentials.[12-16] Regardless of the mode of oxidation, the basic polymerization mechanism is the same and is commonly held to occur through the removal of an electron from the pyrrole π-system to form the corresponding radical cation (Figure 3.1).[16-19] While multiple resonance forms of the radical cation are possible, spin density studies support the localization of the unpaired electron primarily at the α-position of the pyrrole.[17,20] Due to their reactive nature, the radical cations quickly couple to form dimeric species. While such coupling occurs predominately through the α-positions to produce the corresponding α,α'-coupled dication, α,β'- and β,β'-defects are also possible. Deprotonation and aromatization then result in generation of the neutral α,α'-dimer. As the dimer is more easily oxidized than the monomeric pyrrole, it is immediately oxidized again to form a new radical cation, which can again undergo coupling with another radical cation.[17-20] Chain propagation then continues through sequential oxidation, coupling, and deprotonation steps.

As the resulting polymer undergoes oxidation at lower potentials than either the previous oligomers or the initial monomeric pyrrole, the final product of oxidative polymerization is the oxidized or p-doped form of polypyrrole, which thus contains counterions obtained from either the oxidizing agent or electrolyte in order to maintain charge neutrality. This p-doped material can then be converted to the neutral material via treatment with reducing agents or the application of cathodic potential. While functionalized polypyrroles can exhibit solubility in appropriate solvents, the unfunctionalized parent polymer is insoluble and nonprocessible. As such, it can be produced as insoluble powders or films depending on the exact polymerization conditions. Powders can then be compressed into pellets for characterization of material properties (i.e., conductivity, etc.).

The ease of producing conductive films via electropolymerization also made polypyrrole the first conducting polymer that could be simply and reproducibly generated as either modified electrodes or freestanding films without advanced synthetic techniques. As a result, electropolymerization quickly became the most common method for the generation of conjugated polymeric films, until it was ultimately supplanted by the current focus on soluble, processible materials. Of course, the susceptibility of pyrrole to oxidation made the generation of polypyrrole inevitable, and thus it is no surprise that its history dates back to quite early time periods. In order to generate such polypyrrole materials, however, a significant source of pyrrole first had to be developed.

3.2 A brief history of pyrrole

The history of pyrrole begins in 1834, with the investigation of distillates of coal tar by the German chemist F. Ferdinand Runge (1794–1867), who was first introduced in Chapter 2.[21–25] Runge separated the initial distillate into acidic and basic fractions, and, from the basic fraction, he isolated a red oil with the odor of beets.[21,22,24] Because of its red color, he named this oil *Pyrrol*, from a combination of the Greek *pyrros* ("the color of fire") and the Latin *oleum* ("oil").[21,23] Runge believed that pure pyrrole was a gas and developed a test for its detection in which a pine splint dipped in hydrochloric acid would develop a purple-red color when exposed to pyrrole vapor.[21,22,24] While this test was stated to be able to detect approximately 3 ppm pyrrole, Runge suggested that pyrrole was difficult to detect in coal tar oil itself because of the intense reactions of other components.[21,24] However, the pine splint test could easily detect pyrrole in the wash water of illuminating gas. He also noted that one could recognize the distinctive odor of pyrrole during the carbonization of bone and horn and that pyrrole could additionally be found in tobacco oil. As Runge had also isolated Kyanol (aniline) in the same 1834 study,[21] which he then treated with oxidizing agents to generate colored materials consistent with modern polyaniline, it is quite surprising that he did not do the same with pyrrole, although this may have been due to the minute quantities that he was able to isolate. Instead, the study of pyrrole's oxidation still had to wait another 81 years.

A little over a decade later, in 1849, the Scottish chemist Thomas Anderson (1819–1892, Figure 3.2) began investigating first the distillates of coal tar and then of oils produced from the destructive distillation of animal bones.[23,24,26] He found that the crude bone oil gave a positive response to Runge's pyrrole test, resulting in the characteristic dark reddish color when exposed to the oil vapor. Attempts to isolate Runge's pyrrole, however, were unsuccessful, even when starting with even increasing quantities of crude bone oil. Finally, he obtained 250 gallons of bone oil, weighing over a ton, and had the distillation carried out at a "manufactory," giving him a variety of casks numbered according to their corresponding fraction from the distillation.[27] After a series of large-scale acid and base extractions, he steam-distilled the acid solution to isolate a strong-smelling oil. The oil vapors gave a strong positive color test for pyrrole and upon standing acquired a reddish color, becoming nearly black after a few days.[23,27,28] Further tests, however, led Anderson to conclude that this oil still contained a mixture of several bases, one of which he suspected to be pyrrole.

Figure 3.2   Thomas Anderson (1819–1892).

Anderson subjected the pyrrole-concentrated oil to a systematic fractionation and, after several rectifications, obtained an oil that was perfectly colorless and transparent when freshly distilled. This oil soon acquired a brown color, though much more slowly than the previous mixture.[28] He washed this oil several times with dilute sulfuric acid, causing the quantity of oil to be diminished by about a third, but the presence of picoline or other bases appeared to have been removed. The remaining fraction was then carefully dried with potassium hydroxide and subjected to additional distillations. After 15 fractional distillations, the boiling point of the remaining material had narrowed to 274°F–280°F (ca. 134°C–138°C) to give a transparent and colorless oil, which slowly acquired a brown color when exposed to air and light. Anderson described it of having "*a strong fetid smell, quite distinct from that of picoline, and a hot pungent taste*," which gave a positive response to Runge's pyrrole test, resulting in a fine red color.[28]

Although he initially thought the pyrrole to now be pure, the reaction of the oil with potassium hydroxide disclosed the presence of a small quantity of some impurity. In order to further purify the pyrrole, Anderson boiled a mixture of pyrrole and potassium hydroxide in copper flasks (as the hydroxide degraded typical glass) over a period of a day or two (Figure 3.3). The flask was then fitted with a bent tube and heated to distill off all of the oil that could be obtained. The remaining molten pyrrole salt was then poured out on a copper plate. On cooling, it solidified into a hard white mass with a yellowish tinge, which had no odor when dry, but breathing on it gave a new odor somewhat like chloroform. When added to water this solid gradually dissolved, and a transparent and colorless oil collected on the surface of the solution, which was separated by distillation.[28]

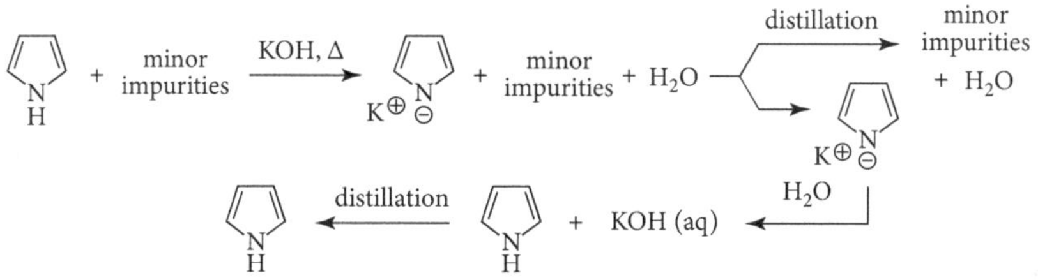

Figure 3.3 Anderson's final purification of pyrrole.

The final pyrrole sample exhibited a specific gravity of 1.077, boiled at 271°F (133°C), and gave a very powerful response to Runge's pyrrole test.[28] Combustion analysis gave a formula of C_8H_5N due to the fact that Anderson used a weight of 6 for carbon, rather than 12. Correction for the incorrect atomic weight would then result in the correct modern formula of C_4H_5N. Anderson had no doubt that this product isolated was Runge's pyrrole and so retained the name.[28]

Anderson reported his isolation and purification of pyrrole in 1857,[28] which was then followed by the first synthesis of pyrrole three years later. In a report from the University of Greifswald in 1860, the German chemist Hugo Schwanert (1828–1902) described the preparation of pyrrole via the dry distillation of ammonium mucate (Figure 3.4).[24,29] The resulting distillate separated into two layers consisting of an aqueous solution and an organic pyrrole fraction. The organic fraction was then separated, washed with water, dried with potassium hydroxide, and purified by a second distillation. Schwanert reported that large quantities of pyrrole could be easily obtained in this way and characterization of the product gave properties in excellent agreement with those previously reported by Anderson.[29]

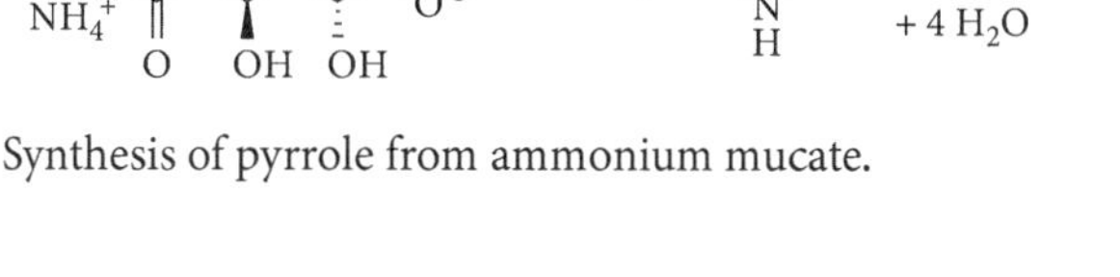

Figure 3.4 Synthesis of pyrrole from ammonium mucate.

The structural formula of pyrrole was finally proposed indirectly by Adolf von Baeyer (1835–1917, Figure 3.5) and A. Emmerling in 1870.[24,30,31] In fact, the structure in question was that of indole, which they correctly presented, stating that indole was composed of benzene and pyrrole in the same way that naphthalene was composed from two benzene rings (Figure 3.6). Of course, now that a suitable supply of pyrrole was available via Schwanert's synthetic methods, it was only a matter of time before someone began to study its oxidation.

Figure 3.5 Adolf von Baeyer (1835–1917).

Reproduced courtesy of the Edgar Fahs Smith Memorial Collection, University of Pennsylvania (CC0 1.0).

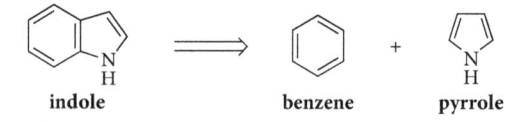

indole　　　　　　benzene　　　pyrrole

Figure 3.6 Structure of pyrrole proposed indirectly from the structure of indole.

The oxidation of pyrrole and its derivatives was reported as early as 1903, resulting in the production of various maleimide and maleic acid derivatives.[32–34] The bulk of these studies focused on functionalized pyrroles, however, and generally utilized conditions meant to provide limited oxidation.

Still, black products were reported in some cases, but these did not seem to be of interest and these materials were never characterized. As such, it was not until 1915 that the production and study of a black precipitate named "pyrrole black" was reported by the Italian chemist Angelo Angeli (1864–1931) at the University of Florence.[35-43]

3.3 Angeli and pyrrole black

Angelo Angeli (Figure 3.7) was born to parents Giovanni Angeli (1835–1906) and Caterina Carnelutti (1845–1907)[44] in the town of Tarcento, located in the Udine province of northeastern Italy (Figure 3.8), on August 20, 1864.[44-46] The son of a modest merchant, Angelo was forced to abandon his studies at the Technical School in Cividale del Friuli in order to help his father with

Figure 3.7 Angelo Angeli (1864–1931).
Reproduced courtesy of the "Ugo Schiff" Chemistry Department, University of Florence, Italy.

the family business in fabrics.[45] His early passion for chemistry, however, was maintained by his close relationship with his mother's brothers, Giuseppe and Giovanni Carnelutti, who were then studying at the Technische Hochschule in Vienna.[45,46] His uncles kept a small laboratory in their home in Tricesimo and allowed him to participate in chemical experiments there.[44,45]

Figure 3.8 Map of Italy highlighting the cities discussed in the current chapter.

At the insistence of his mother,[45] Angeli ultimately returned to his studies, enrolling first at the Technical Institute of Udine[44-46] and then the Vienna Institute of Technology.[44] Following a period of service in the military, Angeli then attended the University of Padua.[44-46] At Padua, Angeli began studying under the new chair of general chemistry, Giacomo Ciamician (1857–1922) (Figure 3.9).[45,46] Ciamician was so impressed with Angeli that when Ciamician moved to the University of Bologna in 1889, he asked Angeli to follow him as his private assistant. As a result, Angeli left Padua and completed his studies at Bologna in 1891,[44-46] after which he obtained the *libera docenza* in 1893.[45,46]

Due to his growing body of published works, Angeli finished first in an 1895 academic competition for the chair in docimastic chemistry (metallurgy and mineral assaying),[44-46] even though he did not have any specific qualifications in the field.[46] Unfortunately, however, the competition was canceled without filling the chair.[45,46]

Figure 3.9 Giacomo Ciamician (1857–1922).

Reproduced courtesy of the Edgar Fahs Smith Memorial Collection, University of Pennsylvania (CC0 1.0).

While at Bologna, Angeli had also come to know Adolf von Baeyer and had traveled to Monaco with him for a short time in 1894.[45] It has been said that von Baeyer recognized in Angeli a sure promise for Italian science and encouraged him to continue in his research.[44-46] After his disappointment over the decision concerning the chair of docimastic chemistry at Bologna, Angeli went on to win the chair of pharmaceutical chemistry at the University of Palermo in 1897.[45,46] However, he was unsure whether to actually accept the position, and it was ultimately von Baeyer that persuaded him, resulting in Angeli's move to Palermo in 1899.[45,46] The move seemed to agree with Angeli, and within a short time, he had gathered a circle of coworkers, friends, and devotees.[45]

After six years in the far south, Angeli was called to the Istituto di Studi Superiori[47] in Florence in order to take up the vacant chair of pharmaceutical chemistry resulting from the death of Augusto Piccini (1854–1905).[45,46] He was reluctant to leave his life in Palermo, but the declining health of his parents induced him to move back closer to family.[45] As his reputation continued to grow, he started to draw further offers from yet larger schools. In 1909, he was asked to move to Rome, but he declined and remained at Florence where, as in Palermo, he had found esteem among his collection of disciples, colleagues, and friends.[45,46] In 1916, Angeli then took the chair of organic chemistry at Florence, a position specifically created for him by the university.[44,46] He was called to move once again in 1922, when the University of Bologna wanted Angeli to fill the vacancy resulting from the death of his former mentor Giacomo Ciamician. Once again, Angeli declined, and he remained at the University of Florence for the rest of his life.[44-46]

Angeli has been characterized as shy and introverted,[44-46] finding it difficult to speak in public and limiting his participation in conferences or international meetings. It has even been said that he despised using the telephone.[44] Due to his nature, he was not one to seek honors and awards. Still, his various contributions were recognized, including being awarded the Military Cross for special services performed during World War I.[44,46] Starting in 1911, Angeli was also nominated for the Nobel Prize in Chemistry. Although he went on to be nominated a total of nine times before his death, he was never selected by the Nobel Committee.[44]

It was in the early hours of Sunday, May 31, 1931, that Angeli passed as the result of a fatal attack of pulmonary edema.[44-46] It was reported had he previously suffered from illness during the summer of 1929, and he is thought to have never completely recovered.[45] After working late on Saturday, Angeli was found in his bedroom of the modest hotel where he had lived since his arrival in Florence in 1905.[44,46]

Beginning as early as his time in Palermo, Angeli was investigating the chemistry of pyrrole and indole.[45] These efforts changed direction, however, starting in 1915. It was at this point that he began studying the treatment of pyrrole with hydrogen peroxide/acetic acid mixtures, which resulted in the formation of a black precipitate that he named "*nero di pirrolo*" (pyrrole black).[35-37] While Angeli said that the name was chosen for the sake of brevity, it is quite likely that he was influenced by the previous analogous names *carbon black* and *aniline black*. For his investigations, Angeli used synthetic pyrrole, and under typical conditions, one gram of pyrrole was dissolved in a sufficient amount of acetic acid, followed by the addition of 2–3 grams of a 50% H_2O_2 solution (this would roughly equate to 2.4–3.5 molar equivalents of H_2O_2 in relation to pyrrole). The resulting solution would quickly become greenish brown, ultimately turning black-brown over the space of a couple of days. The product could be isolated as a thin black powder, either via spontaneous precipitation, dilution of the final solution with water, or addition of aqueous sodium sulfate (Na_2SO_4).[36,37] The resulting precipitate was completely insoluble in everything but basic solutions. As such, purification methods included dissolution in a basic solution, followed by reprecipitation with either acetic acid or dilute sulfuric acid. The material was then isolated via filtration and dried at 120°C to give a fine, dark-brown-to-black powder.[36,37]

Angeli went on to find that pyrrole blacks could be obtained using a variety of oxidizing agents.[35] In addition to the original hydrogen peroxide conditions, Angeli also investigated oxidation via nitrous acid (HNO_2),[38] potassium dichromate ($K_2Cr_2O_7$)[39] or chromic acid ($H_2SO_4/Cr_2O_7^{2-}$),[42] lead oxide (PbO_2),[39] ferric chloride ($FeCl_3$),[39] light and oxygen,[41] ethylmagnesium iodide and oxygen,[41] potassium permanganate,[41] and various organic quinones.[42,43] The combined results obtained from the studies led Angeli to conclude:[4,39] "These facts are of special interest because it shows that the formation of pyrrole blacks is most likely preceded by a process of polymerization of the pyrrole molecule, which takes place more or less rapidly depending on the reagents that are used." In addition to these various polymerizations in solution, Angeli also investigated the potential of using pyrrole black as a fabric dye.[39] In these efforts, pyrrole was first treated with acetic acid to generate pyrrole acetate. A solution of pyrrole acetate was then used to impregnate cotton fabric, after which the treated cotton was passed through a bath of potassium dichromate. Fabric treated in this way was dyed black, the color of which was stable to both soap and light. Angeli stated that the result of this process was very similar to the color obtained from aniline black.

Of particular interest to Angeli was the strong resemblance of pyrrole black to melanins,[36,37,39,40,48] the natural family of pigments responsible for hair and skin color. The most common form of biological melanin is eumelanin, a brown-black pigment now commonly accepted to be a heterogeneous macromolecule comprised of 5,6-dihydroxyindole and 5,6-dihydroxyindole-2-carboxylic acid in various neutral and oxidized states (Figure 3.10).[49] The exact nature of the macromolecular structure, however, is still uncertain and two predominant structural models currently exist. The initial model was that of a traditional large heteropolymer, while more recent studies favor stacked oligomers of five to six indole units.

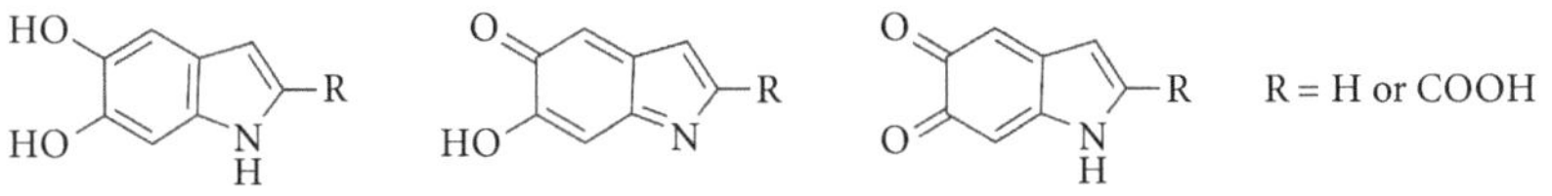

Figure 3.10 Monomeric units of eumelanin.

Chemical analysis of pyrrole black produced via hydrogen peroxide revealed a C:H:N atomic ratio of 5:4.87:1, very similar to that of melanin (C:H:N = 5:5:1), which further suggested some relationship between the two materials.[36,37] Unfortunately, further attempts to probe the structure of pyrrole black were significantly limited by its insoluble nature. Oxidative degradation, however, gave cleavage products consistent with pyrrole and indole derivatives,[36,37,39,41] which led Angeli to the conclusion that the pyrrole ring was retained in the structure of pyrrole black.

To further the scope of his study, Angeli found that the treatment of various functionalized pyrrole derivatives with oxidants could produce colored species but did not result in the production of the typical black precipitate.[38,39] The exception to this was β-nitrosopyrrole, in which both α-positions were free of functionalities and resulted in a black product Angeli called *nitrosopyrrole black* to distinguish it from normal pyrrole black.[38,40] Ultimately, Angeli's studies led him to propose that the structure of pyrrole black contained units consisting of direct carbon–carbon bonds between pyrroles as shown in Figure 3.11a.[39] Of particular interest is the fact that Angeli's proposed structure is remarkably similar to the currently accepted structure for oxidized sections of the polypyrrole backbone (Figure 3.2).

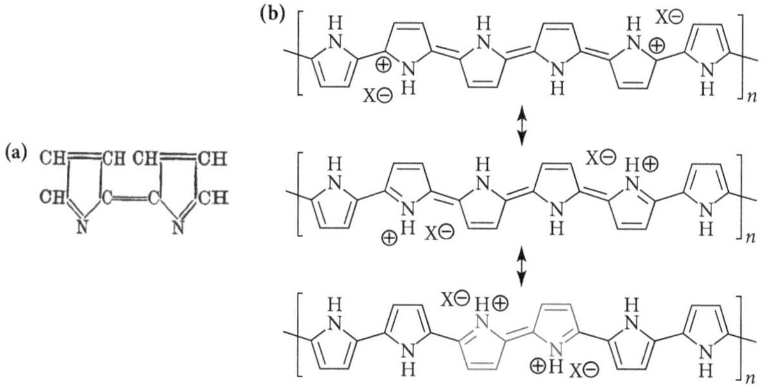

Figure 3.11 (a) Angeli's proposed basic unit for the structure of pyrrole black. (b) Modern resonance structures describing oxidized (p-doped) polypyrrole with Angeli's proposed unit highlighted in gray.

Part A reproduced from Angeli. Sopra i neri di pirrolo. *Rend. Accad. Lincei* **1918**, 27[I], 209–212.

After his primary research focus for nearly a decade, Angeli concluded his study of pyrrole black in the early 1920s in order to move onto other research topics, although his occasional coauthor, Antonio Pieroni (b. ca. 1885), continued publishing on the subject until 1926.[50–53] Pieroni's work during this period was carried out in the organic chemistry laboratory at Florence, and since Angeli had occupied the chair of organic chemistry since 1916, this was likely still at least indirectly under Angeli's supervision. Interestingly, Pieroni's 1922–1923 papers[50–52] seem to introduce the first use of the term *polypyrrole* (*polipirroli*), although this appears to be in reference to pyrrole oligomers (dimers and trimers), rather than true polymeric materials in the modern sense. A noteworthy example includes the pyrrole dimer (Figure 3.12a), which he states was successfully isolated by Angeli as an oxidation product of pyrrole.[52] It is important to remember that in determining composition via elemental analysis, oxygen content was determined by difference, rather than by a direct measurement. As such, the inclusion of the oxygen here was most likely the result of limitations in the accuracy of the combustion analysis. Pieroni points out that this species undergoes transformation to pyrrole black with greater ease than pyrrole itself.

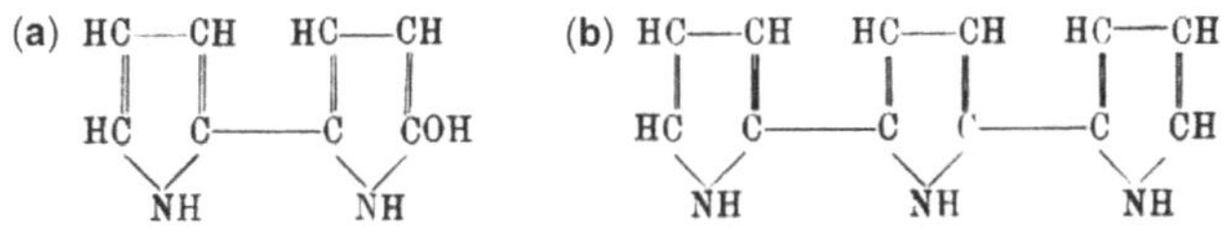

Figure 3.12 Pieroni's structures of the dimer (a) and trimer (b) of pyrrole.
Reproduced from A. Pieroni. Sopra i polipirroli. *Rend. Accad. Lincei* **1923**, *32*(II), 175–179.

In an attempt to extend this example, Pieroni aimed to synthesize the corresponding trimer (Figure 3.12b). Starting with diethyl succinate, he treated this with an excess of 2-pyrrolylmagnesium iodide (Figure 3.13). This initial product was then reacted with ammonium acetate in glacial acetic acid. However, even when limiting the heating of this step to just a few seconds, the desired trimer could not be isolated as it spontaneously polymerized. Analysis of the resulting pyrrole black gave an elemental composition consistent with a polymer of 2,2';5',2"-terpyrrole, although with some additional oxygen content.[52]

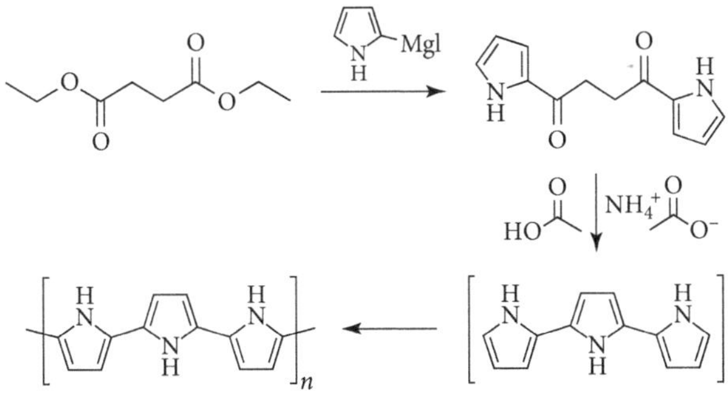

Figure 3.13 Pieroni's attempted synthesis of 2,2';5',2"-terpyrrole in 1923.

Angeli did return to the topic of pyrrole black with one final paper in 1930.[54] This last report consisted of a brief overview of the various types of pyrrole blacks and derivatives known to date, as well as updated discussion concerning the comparison of pyrrole black and melanines. At about the same time that Angeli had started to move away from his focus on pyrrole black, another student of Giacomo Ciamician, Riccardo Ciusa (1877–1965), began investigating the polymerization of heterocycles such as pyrrole, thiophene, and furan, all in an effort to generate heterocyclic analogs of graphite.[55-57]

3.4 Ciusa and heterocyclic analogs of graphite

Riccardo Ciusa was born April 27, 1877, in the city of Sassari, on the Italian island of Sardinia, located west of the Italian peninsula (Figure 3.8).[58,59] There, he attended the Sassari Technical School before enrolling at the University of Bologna in 1896. Three years later, Ciusa obtained a grant from a foundation associated with the "Carlo Alberto" Royal College in Turin to attend the University there.[58] He thus moved to the University of Turin in 1899, graduating *dottore*[60] in chemistry in 1902.[58,59] After graduation, he returned to the University of Sassari as an assistant of mineralogy under Professor Boeris.[58] In 1904, Cuisa obtained funding for two years of specialized training abroad, which allowed him to work in the laboratory of Johannes Thiele (1865–1918) in Strasbourg.[58,59]

Gaetano Minunni, professor of pharmaceutical chemistry at Sassari, then wrote to Giacomo Ciamician to recommend Ciusa for a position.[58] Minunni must have been convincing, as Ciusa moved to the University of Bologna as assistant of agricultural chemistry in 1906.[58,59] The following year, Ciusa became an assistant at the Istituto di Chimica Generale (Institute of General Chemistry) under the direction of Ciamician (Figure 3.14), replacing Joseph Bruni, who had left Bologna for another position.[58-59,61] Ciusa obtained the *libera docenza* in general chemistry in 1908, after which he taught the course on analytical chemistry at Bologna.[58,59] After the founding of the Royal School of Industrial Chemistry of Bologna in 1921, Ciusa began teaching the chemistry of dyes and coloring there, a relatively new subject at the time.[58]

From 1915 to 1918, Ciusa served as an infantry officer in the Italian Army as a part of World War I.[58,59] After his return to Bologna, he was named *aiuto* in 1919,[58] a promotion granted to a limited number of the best assistants. This position fell between that of a regular assistant and a professor, and it also included a small increase in salary.[59] Still, Ciusa continued as Ciamican's assistant until the professor's death in 1922. After Angeli declined to fill Ciamican's position,

Figure 3.14 Riccardo Ciusa (right, age 31) with Giacomo Ciamician at the University of Bologna in 1908.

Reproduced courtesy of Ned Heindel, Lehigh University.

Ciusa was appointed provisional director of the Istituto di Chimica Generale. In addition, he was charged with taking over the teaching of general chemistry.[58,59]

In November of 1922, Ciusa was a finalist for the position of professor of pharmaceutical chemistry at the University of Cagliari but stayed on at Bologna for two more years.[58] Following the founding of the new University of Bari in 1924, however, Ciusa was finally made professor of pharmaceutical chemistry there in November of the same year.[58,59] At Bari, his research focused on pyrrole and its derivatives, including natural pyrrole-based species such as chlorophyll, with an interest in plant chemistry and photochemistry later in his career.[61] Ciusa served as dean of faculty from 1932 to 1952[61] and retired as professor in 1952, after which he was appointed emeritus of organic chemistry in 1954.[58] In 1959, Ciusa moved to Rome where his health declined, suffering from frequent illnesses before finally passing on March 27, 1965.[58]

In his final year as Ciamican's assistant, Ciusa began investigating the possibility of generating heterocyclic analogs of graphite, in particular what he referred to as *"graphite of pyrrole"* and *"graphite of thiophene,"* and proposed that the presence of nitrogen or sulfur in natural graphite could be proof of the existence of such materials.[55] In order to accomplish the preparation of the proposed graphite of pyrrole, Ciusa utilized tetraiodopyrrole, which was known to eliminate ioidine upon heating, theoretically resulting in polymerization.[55–57] Initial attempts found that under either ambient atmosphere or vacuum, heating tetraiodopyrrole above 150°C resulted in the elimination of iodine to produce a black material that was insoluble in acids, bases, or ordinary solvents.[55] In his following papers, he provided a little more detail, specifying that the temperature applied to produce this material was 150°C–200°C (presumedly under ambient atmosphere)[56] or 150°C under vacuum.[57] Analysis of this material gave an elemental composition corresponding to $[C_4NHI]_n$ (Figure 3.15), which Ciusa viewed to be an intermediate in the formation of graphite.[55] He later proposed structures based on dimeric units as possible representations of this intermediate (Figure 3.16).[57]

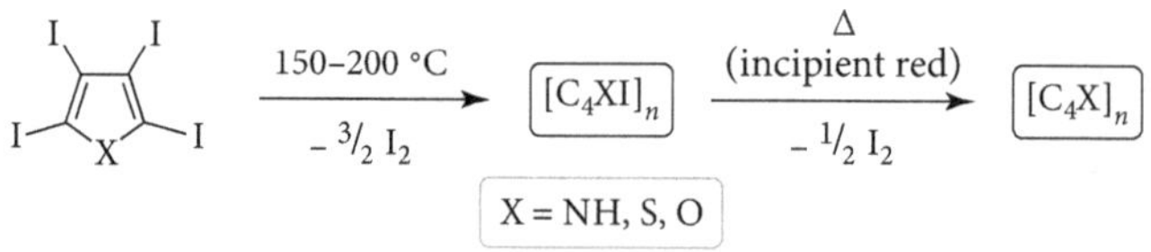

Figure 3.15 Thermal polymerization of tetraiodoheterocycles.

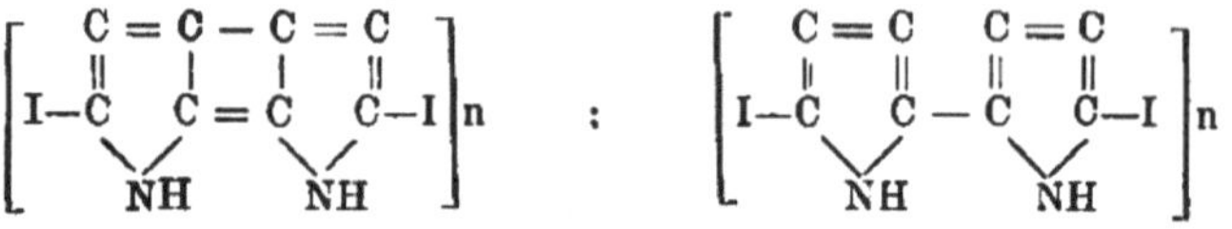

Figure 3.16 Ciusa's proposed structures for $[C_4NHI]_n$.

Reproduced from R. Ciusa. Sulle grafiti da pirrolo e da tiofene (Nota preliminare). Gazz. Chim. Ital. 1922, 52(II), 130–131.

Continuing these efforts, Ciusa heated this intermediate material a second time under a stream of inert gas at a temperature high enough to cause the material to glow red. This second, more rigorous heating appeared to liberate the final iodine atom to give a black material with an elemental composition of $[C_4NH]_n$ and an appearance similar to graphite flakes.[56] Ciusa went on to repeat this general process with the heterocyclic analogs thiophene and furan (Figure 3.15). Although the loss of the initial three iodides occurred at different temperatures for each of the three heterocycles, similar results were obtained overall.[56] He summarized this overall process as follows:[4,56] "It is remarkable that in the three cases studied so far, three atoms of iodine are eliminated at relatively low temperature and at the same time, while the fourth atom is eliminated in a later stage of the reaction and at a higher temperature." Finally, he applied an analogous approach to the thermal polymerization of hexaiodobenzene to produce a graphite material that he described as "very similar to ordinary graphite." Measurement of the resistivity of the synthesized graphite showed a low resistivity ($R = 0.3055$ ohms) as expected, although it was approximately six times more resistive than an authentic sample of graphite measured under the same conditions ($R = 5.566 \times 10^{-2}$ ohms).[57]

Unfortunately, Ciusa did not determine resistivities for the bulk of the heterocyclic graphite materials, with only the very high value of 2.068×10^5 ohms reported for the graphite of thiophene.[57] Nor did he ever go on to report any further significant characterization of these materials other than their graphite-like appearances and elemental compositions. Still, almost 40 years later and nearly halfway around the world, Ciusa's thermal polymerization studies became the foundation of efforts to produce conductive organic polymers in Melbourne, Australia, led by Donald Weiss (1924–2008) of the Council for Scientific and Industrial Research (CSIR) Division of Industrial Chemistry.[6-8]

3.5 Weiss and conducting polypyrrole

Donald Eric Weiss (Figure 3.17) was born in St. Kilda, a suburb of Melbourne, in Victoria, Australia, on October 4, 1924.[62-66] Known simply as Don,[62,64] Weiss was the only child of Herbert Vernon Weiss and Lillian Kate, née Le Lievre.[62,63] He was only three years old when his parents separated, after which

Figure 3.17 Donald E. Weiss (1924–2008).
Reproduced courtesy of Robert Weiss.

Weiss and his mother moved to Adelaide to live with her sisters.[62] In Adelaide, Weiss attended Mitcham Primary School before enrolling in Scotch College in 1937.[62-64,67] Weiss had little interest in his studies until introduced to chemistry. Fortunately, he had two first-class chemistry teachers at Scotch, John E. Smith and John Dow, who inspired his love of practical chemistry.[62,64] To support his growing chemical activities, he assembled a home laboratory and taught himself to use chemical glassware. In his last year at Scotch College, he won the Science Prize and shared both the Mathematics Prize and the Special English Prize.[62]

By the time he had finished his studies at Scotch College, Weiss had decided that he was going to be an industrial chemist. As such, he enrolled in the Diploma of Industrial Chemistry at the South Australian School of Mines and Industry (now part of the University of South Australia) in 1942.[62-64,66,67] During his first two years, however, the bulk of his courses were actually taken at the nearby University of Adelaide, with the two campuses separated by only a couple minutes walk. As a result, he finally transferred to a BSc course at Adelaide in 1944.[62,64,66,67] He failed a course in electrical engineering in his third year and had to repeat it via correspondence from Burnie, Tasmania, where he worked part-time for APPM (Australian Pulp and Paper Manufacturers)[68] during vacations.[62,63] He was ultimately awarded his BSc from the University of Adelaide in 1945[62,63,65,66] and worked full-time as a shift chemist for APPM from 1945 to 1946.[62,64-67] Wanting to be involved in research, Weiss then took a position the following year as a development chemist at Commonwealth Serum Laboratories (CSL, Melbourne), where he attempted to improve their new penicillin production process.[62-67]

Early in that same year, Weiss attended a meeting of the Royal Australian Chemical Institute, where he met Richard (Dick) Grenfell Thomas (1901–1974), the leader of the Minerals Utilization Section of the CSIR Division of Industrial Chemistry.[62,64,67] This chance meeting ultimately led to Weiss moving to CSIR on January 2, 1948, to work under David (Dirk) Zeidler (1918–1998) as a research officer in the Chemical Engineering Section of the Division of Industrial Chemistry at Fishermen's Bend (another Melbourne suburb).[62-67] In May of 1949, CSIR become CSIRO (Commonwealth Scientific and Industrial Research Organisation), and Weiss was later transferred to its Physical Chemistry Section in 1952.[62,66] Weiss was promoted to senior research officer in 1953 and then to principal research officer in 1955.[62] The Division of Industrial Chemistry was split into three divisions and three sections in October of 1958, with Weiss assigned to the Division of Physical Chemistry.[62,63,65] That same month Weiss submitted a collection of his research to the University of Adelaide in support of his candidature for the degree of Doctor of Science (DSc). Weiss was promoted again to senior principal research officer in 1959[61] and was awarded his DSc in 1960 for a thesis entitled "Adsorbents and Adsorption Processes."[62,63,66,69]

Weiss continued to rise through the ranks of CSIRO, again being promoted in 1965, now to chief research scientist (I). In 1971, Weiss was designated assistant chief of the Division of Applied Chemistry and simultaneously promoted to chief research scientist (II).[62,66] In February of 1974, the division was split to form the Divisions of Chemical Technology and Applied Organic Chemistry, with Weiss being appointed chief of the Division of Chemical Technology.[62] It was during these years that he made his most well-known contributions. This included the development of both the SIROTHERM ion exchange process for desalination and the SIROFLOC water clarification process in the 1970s,[62,64,66,70] as well as overseeing the research for a new reconstituted wood product, Scrimber, as an asbestos replacement for application in asbestos cement products.[62,66] Finally, he was appointed director of CSIRO's newly established Planning and Evaluation Advisory Unit in February of 1979.[62,66]

Weiss received numerous awards and recognitions during his career, particularly from the Royal Australian Chemical Institute (RACI).[62,63] He was elected a RACI fellow in 1957 and went on to win all four flagship medals of the RACI, beginning with the 1950 Rennie Memorial Medal. He was then awarded the H. G. Smith Memorial Medal in 1966 and the Leighton Memorial Medal in 1977. In 1980, the institute introduced its Applied Research Award, with the medal named each year after a notable applied chemist. Weiss won the award the following year, which was named the K. L. Sutherland Medal. In addition, he was awarded the Archibald D. Olle Prize of the New South Wales Branch of the institute in 1971. Finally, he was elected the vice president of the institute in 1982 and served as its president in 1983.[62,66] Outside of the institute, he was elected a fellow of the Australian Academy of Science in 1971 and a foundation fellow of the Australian Academy of Technological Sciences in 1976.[62,63,66] That same year he was also made an officer of the Most Excellent Order of the British Empire (OBE) for his contribution to science, followed by a CSIRO Medal in 1989.[63,64,66] Weiss retired from CSIRO in October of 1984, ending a successful 36 years with the organization.[62–64,70] In his later years, Weiss developed lung cancer, with his health steadily declining during the spring of 2008. A case of pneumonia finally brought this battle to a close, with Weiss passing away on July 30, 2008, in the eastern suburb of Blackburn.[62,64]

In the mid-1950s, Weiss had turned his attention to municipal water treatment as a challenging major problem worthy of long-term CSIRO attention.[62,70] Focusing on the issue of desalination, he began investigating the potential of developing easily regenerated adsorbents for ion exchange technology. Along these lines, he began to consider the possibility of an electrochemical approach using activated carbon electrodes. At the time, however, such carbon materials and their electronic properties were still poorly understood. As such, he felt that better insight was first needed before these materials could be applied to his proposed methods. As the German physical chemist Victor Garten had recently come to CSIRO to investigate the use of carbon black to reinforce rubber, they decided to work together on a study of the fundamental properties of activated carbon and carbon blacks.[62,70] This collaboration was fruitful, resulting in a number of publications on the study of these materials,[71-77] but Weiss ultimately came to the conclusion that more simple, synthetic polymers might be made with more direct interest than carbon to his proposed electrical process for water desalination.[70]

These new efforts to produce electrically conducting polymers then began in 1959 with the preparation of xanthene polymers.[70,78] The approach to these polymers was based on the well-known synthesis of fluorescein from resorcinol and phthalic anhydride (Figure 3.19). In an attempt to generate a linear polymer via further condensation, initial attempts reduced the ratio of resorcinol to phthalic anhydride to 1:1, resulting in a black tar that was soluble in alkali, but insoluble in ethanol. This approach was then further modified through the replacement of a portion of the phthalic anhydride with pyromellitic dianhydride in order to generate cross-links. While this did reduce solubility, the resulting material was still partially soluble. Lastly, the resorcinol was replaced with hydroquinone to give a completely insoluble product. An idealized structure of the xanthene product as proposed by Weiss is given in Figure 3.18.

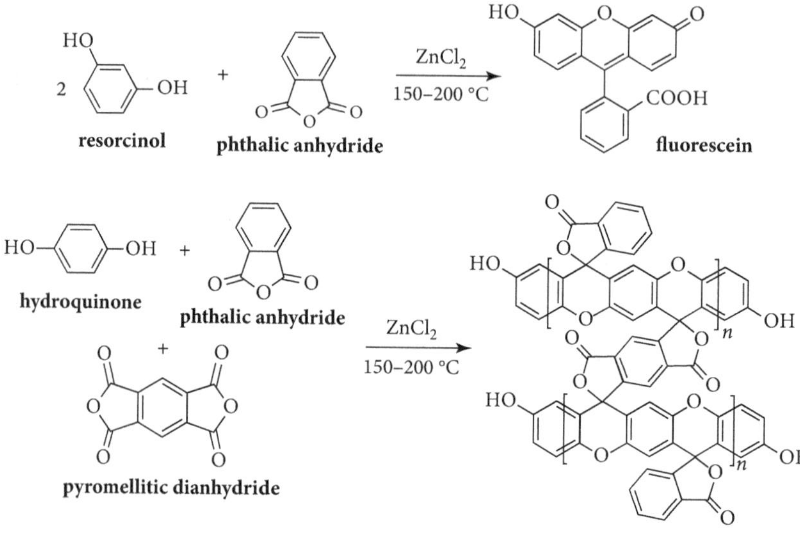

Figure 3.18 Synthesis of fluorescein and related xanthene polymers.

The electronic properties of the resulting materials were then investigated as pressed pellets under a nitrogen atmosphere at 25°C.[76] A linear relationship was found between the logarithm of the polymer resistance and the reciprocal of the absolute temperature. Furthermore, the lack of polarization observed during the resistance measurements supported the electronic nature of the conductivity. As such, these materials behaved as typical electronic semiconductors, although the measured resistivity was still quite high compared with activated carbon, even for the most optimized materials ($R_{min} = 7$–35×10^3 Ω cm). Lastly, it was found that the resistance increased in the presence of oxygen. Thus, while the conductivity of these materials was not sufficient for his desired applications, Weiss felt that this was a good proof of concept, stating:[78] "These considerations suggest that in the future, further knowledge of the organic chemistry of such polymers must result ultimately in the production of a wide variety of polymers with even lower resistances and individually synthesized for specific purposes." Weiss then came across the previously published papers of Ciusa (see Section 3.4),[55–57] which suggested that Ciusa's "graphite of pyrrole" might provide a new approach to conductive materials.[70] As the electrical properties of the pyrrole materials had never been reported, the CSIRO team began by reproducing the Ciusa's production of the pyrrole products. In the end, however, Weiss did not follow the exact methods reported by Ciusa. Rather, the conditions were modified such that the tetraiodopyrrole was heated at temperatures as low as 100°C –120°C in a rotating flask under a flow of nitrogen. Here, the nitrogen flow was used to both maintain an inert atmosphere and to transfer iodine vapor produced during the reaction away from the products.[6] Heating the samples at 100°C–120°C initiated the liberation of iodine vapor, although somewhat erratically, and was accompanied by the release of heat that caused the temperature to rise abruptly to about 180°C–200°C. Weiss proposed that this evolution of heat was possibly be due to chemisorption of iodine by the polymer. The isolated products were reported to be insoluble, black powders, which were described as "polypyrroles" consisting of "a three-dimensional network of pyrrole rings cross-linked in a nonplanar fashion by direct carbon to carbon bonds."[6]

Analysis of the products generated by the CSIRO team differed considerably from that found by Ciusa and did not correspond to either of the previously reported formulas of $[C_4NHI]_n$ or $[C_4NH]_n$. In addition, the new samples also showed the presence of considerable amounts of oxygen. Of course, it must be noted that Ciusa prepared his material under vacuum, rather than the nitrogen atmosphere used by Weiss, which could at least partially account for the differences. Weiss ultimately concluded that the oxygen arose from the adsorption of water and oxygen during the isolation and grinding of the product, a conclusion at least partially verified by the fact that the oxygen could be driven off with heat, with the polymer reabsorbing oxygen again upon standing.[6]

In addition to the oxygen, it was found that the materials contained substantial amounts of iodine, which was thought be due to a combination of unreacted iodine sites, described as "iodine of substitution," as well as "adsorbed molecular iodine."[6] In order to distinguish between the two, samples were treated with thiosulphate, which led to the conclusion that the bulk of the iodine (2.5–3.0 equivalents) was due to adsorbed molecular iodine. It was also found that this molecular iodine could be removed from the polymer by electrochemical reduction[6,7] or by heating the polymer at temperatures above 150°C under a high vacuum. The remainder of the iodine, determined by difference from the total value for iodine found by analysis, was assumed to be present as iodine of substitution. Based on the various descriptions given by the CSIRO team, a hypothetical structure for the polymeric material is given in Figure 3.19. It is critical to point out, however, that this structure corresponds only to the neutral polymer containing iodine of substitution, and it does not attempt to illustrate the effects of the adsorbed molecular iodine, which will be discussed further.

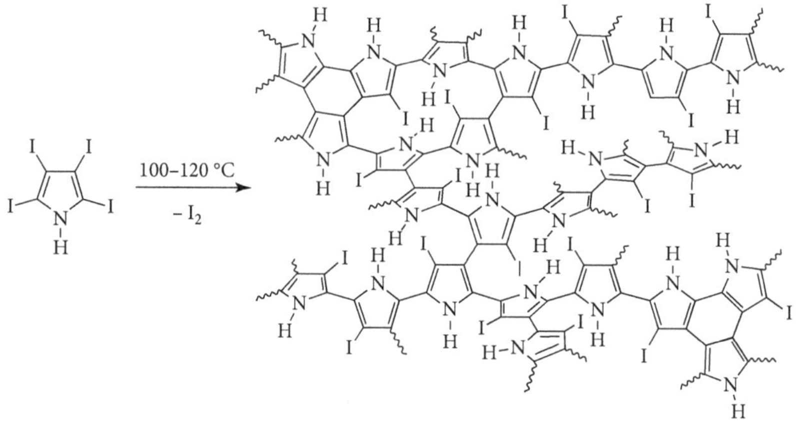

Figure 3.19 Hypothetical neutral structure of Weiss's polypyrrole.

The resistivity (R) of the polypyrrole products was then measured as pressed pellets under a stream of nitrogen to give values of 11–200 Ω cm at 25°C.[8] It should be noted that resistivity values as low as 1 Ω cm were also reported, but these were for samples that had been subjected to additional pyrolysis above 500°C, making the true identity of these samples more complicated. As with the previous xanthene polymers,[78] the polypyrroles prepared below 500°C exhibited a linearly correlation between the logarithm of the resistivity with the reciprocal of the absolute temperature over the temperature range of 25°C–150°C.[8]

These reported resistivity values correspond to conductivities ($1/R$) of 0.005–0.09 Ω^{-1} cm^{-1}, which were still lower than conductivities possible via carbon black. Still, this was a drastic improvement over the previous xanthene polymers.[78] Furthermore, at the time, this represented the highest reported conductivity for a nonpyrolyzed organic polymer[79–83] and remained so until reports of conductive polyaniline in 1966.[1] Weiss described the general nature of polypyrrole conductivity as follows:[8] "However it is apparent that the polymers are relatively good conductors of electricity. Since no polarization was observed during the measurement of the electrical resistance, even over substantial periods of time, it is assumed that the conductivity is of electronic origin." Of significant interest was the discovery that removal of the adsorbed molecular iodine from the polymer samples resulted in a significant increase in resistance.[6–8] For the most part, this was not dependent on the method of iodine removal, with either chemical/electrochemical reduction[7] or solvent extraction[6,8] giving similar results (Figure 3.20). The effect of removing iodine via heating under high vacuum gave conflicting results, but as this also involved the chemisorption of water and oxygen, the multiple processes involved made it more difficult to obtain a clear correlation between iodine and resistivity.[6]

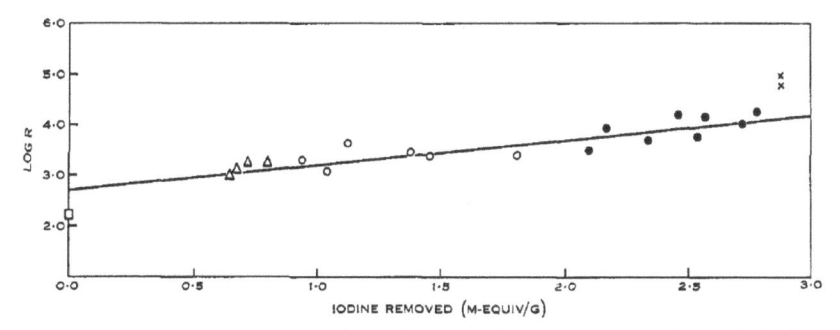

☐ Untreated polymer; Δ samples from blank experiments not subjected to polarization; ○ samples reduced at −0·60 V *versus* MSE. ● samples reduced at potentials of −1·50 V *versus* MSE and higher; × alkali-treated polymer.

Figure 3.20 The effect of removal of iodine on polymer resistance.

Used with permission of CSIRO Publishing, from B. A. Bolto, D. E. Weiss. Electronic conduction in polymers. II. The electrochemical reduction of polypyrrole at controlled potential. *Aust. J. Chem.* **1963**, *16*, 1076–1089; permission conveyed through Copyright Clearance Center, Inc.

Electron spin resonance (ESR) measurements were then performed, in which it was found that the spin concentration was greatest in samples with the largest proportions of adsorbed molecular iodine. Furthermore, the number of spins and the linewidth both decreased markedly as iodine was removed from the samples, leading to the conclusion that the paramagnetic species resulted from contributions arising from a charge-transfer interaction between polypyrrole and iodine.[8] As explained by the CSIRO team,[8] "Charge-transfer complexes of strength sufficient to cause partial ionization induce extrinsic [semiconductor] behaviour by changing the ratio of the number of electrons to the number of holes." Thus, it was understood that[6] "The presence of the oxidant iodine, and in its absence oxygen, facilitates oxidation of the polymer." Of course, such oxidative processes as illustrated in Figure 3.21 describe what would later be called p-doping of the polymer, which was ultimately determined to be the key to producing highly conductive organic polymers.[1-3] Weiss, however, admitted that the full role of the iodine oxidation to conductivity was not realized at the time.[70]

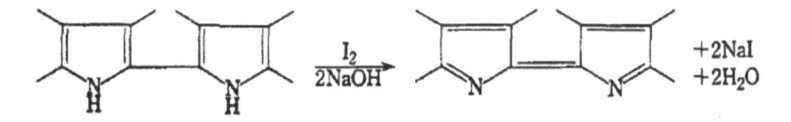

Figure 3.21 Iodine oxidation of polypyrrole under basic conditions

Used with permission of CSIRO Publishing, from B. A. Bolto, D. E. Weiss. Electronic conduction in polymers. II. The electrochemical reduction of polypyrrole at controlled potential. *Aust. J. Chem.* **1963**, *16*, 1076–1089; permission conveyed through Copyright Clearance Center, Inc.

The CSIRO team went on to study additional related materials, including conjugated polymers based on either pyridine or imidazole,[84] as well as materials from the polymerization of either hexachlorobenzene[85] or tetraiodophthalic anhydride.[86] As with the pyrrole-based materials, these all exhibited low resistivities, but no clear improvements over the previous polypyrroles. As a result, it was ultimately determined that the use of such organic conducting polymers for an electrical desalination process was impractical, and Weiss and the CSIRO team moved on to other projects by 1967.[70]

3.6 The University of Parma and pyrrole black revisited

About the same time that the CSIRO team was wrapping up its focus on the polypyrrole-iodine materials, a resurgence of the investigation of pyrrole black was happening back in Italy at the University of Parma due to the collective efforts of Luigi Chierici (d. 1967), Gian Piero Gardini (d. 2001), and Vittorio Bocchi.[9,87-92] These efforts were originally led by Chierici,[93] who was already studying pyrrole black as early as 1953,[94,95] prior to either of the other two arriving at Parma. However, it is Gardini of whom we have the most knowledge. After completing a degree in industrial chemistry from the University of Bologna in 1961, Gardini moved to the University of Parma, where he joined Chierici as a professor in the Institute of Pharmaceutical Chemistry in 1962.[87,96] The final member, Bocchi, then began working with Chierici and Gardini in about 1966.

The three researchers did not work collectively for long, however, as Chierici died prematurely on March 3, 1967.[90-93]

Chierici's death occurred during the establishment of Parma's Institute of Organic Chemistry, which was officially founded in 1968 with Giuseppe Casnati (1923–1992) as the institute's director.[93,97] As a result of this confluence of events, Gardini and Bocchi both moved to the Institute of Organic Chemistry, where they continued their collaborative efforts on pyrrole black under the direction of Casnati.[91-93] Gardini was appointed chair of Organic Chemistry in 1980 and then succeeded Casnati as director of the Institute of Organic Chemistry in 1984.[96,97] The institute then became the Department of Organic and Industrial Chemistry in 1992, with Gardini continuing as director of the new department.[96-98] He remained director of the department until 1998, after which he continued as the chair of Organic Chemistry until his death in 2001.[96,98]

The bulk of their collective efforts focused on continuing the previous studies of Angeli and Pieroni (Section 3.3) on pyrrole black generated from the treatment of pyrrole with H_2O_2. Here, the emphasis was on identifying the intermediates and byproducts formed during the polymerization in order to better understand the mechanisms involved.[88-92] While these efforts successfully identified new intermediates and byproducts, as well as elucidating the structures of previously reported species, Gardini concluded in 1973 that the mechanism still needed much more study in order to be suitably well defined.[35]

However, it was a larger collaboration on ESR studies of pyrrole blacks with Gennaro D'Ascola and coworkers from Parma's Institute of Physics that yielded the most significant results.[9,87] The initial 1966 study focused on the study of pyrrole blacks produced via Angeli's initial H_2O_2/acetic acid conditions.[87] Here, ESR was used to monitor the polymerization reaction, as well as to characterize the final polymer products. The measured ESR signal was found to be free from hyperfine structure, with a reported g factor of 2.0035 ± 0.0002. The maximum number of spins (1.5×10^{19} spins g^{-1}) was found during the ongoing progress of the polymerization, with the final product consisting of the lower value of 2.4×10^{18} spins g^{-1}. In the process, these results were also compared with the previous report of Weiss and coworkers,[8] in which it was concluded that the nature of the blacks prepared from H_2O_2 was different from that of the blacks reported by Weiss. As an example of this, the conductivity of the H_2O_2-generated pyrrole black was measured and reported to be "almost zero" ($<10^{-8}$ Ω^{-1} cm^{-1}).[87]

A second study was then reported in 1968, after Chierici's death.[9] Unlike the previously reported polymers from chemical oxidation, the polymeric material under study here was obtained via electrolysis and represents the earliest known example of an electropolymerized polypyrrole. A laminar black film was formed on the surface of a platinum electrode by applying a constant current of 100 mA to a solution of pyrrole in H_2SO_4, over a period of two hours. The film was then removed from the electrode, rinsed with distilled water, and dried under vacuum.[9,91] It should be pointed out that this is the first report of a conducting polymer in the form of a plastic, freestanding film, an innovation normally attributed to the production of polyactylene films nearly a decade later.[1-3,5] The composition of the films was then characterized via quantitative analysis to give results consistent with one SO_4^{2-} anion for every 5–6 pyrrole rings.

ESR measurements gave a signal free from hyperfine structure, with a reported g factor of 2.0026 ± 0.0001, very similar to that previously found for the H_2O_2-generated material.[87] Furthermore, the ESR characterization indicated the presence of antiferromagnetic interactions, strongly interacting polymer chains, and highly mobile electrons, although the spin density was found to not be very high.[9] X-ray analysis of the film indicated the material to be essentially amorphous, and conductivity measurements gave a room temperature value of 7.54 Ω^{-1} cm^{-1}, considerably higher than either the previous H_2O_2-generated material[87] or that reported by Weiss and coworkers for the thermally produced polypyrrole-iodine materials.[8] The measured conductivity was temperature dependent over the range 100–300 K and consistent with an intrinsic semiconductor.

Chierici's final papers, published posthumously in 1968[91] and 1970,[92] respectively, reported attempts to study the effect of polymerization conditions on the reaction rate and formation of byproducts, as well as attempts to gain insight into the structures of the various species produced by examining products generated via oxidative degradation of the polymers. While the bulk of this work focused on polymers generated from the H_2O_2 oxidation of pyrrole, the 1968 report concluded with a section on the electropolymerized materials, which were referred to as "black electrolytic" to differentiate them from the previous polymers.[91] It should be noted that the very similar description "electrolytic black" had been previously used by Goppelsroeder to refer to electropolymerized aniline black (polyaniline), again to differentiate it from aniline black produced via chemical oxidants.[99] The oxidative degradation of the electropolymerized material revealed some additional unidentified products in comparison with the traditional H_2O_2-generated materials, but the major product for all materials was pyrrole-2,5-dicarboxylic acid. This common degradation product led to the conclusion that all of the pyrrole blacks studied consisted of chains of α,α'-linked pyrroles (Figure 3.1).

Beginning in 1975, Gardini then spent time as a visiting scientist at the IBM Research Laboratory in San Jose, California. This first visit was followed up with additional stays in 1978 and 1981. It was during these visits that Gardini began working with Arthur Diaz (b. 1938), the results of which are discussed in the next section. Gardini and Bocchi then later returned to the study of polypyrrole at Parma with a series of papers over the time period 1986–1991.[100–102] These included materials generated via the chemical oxidation of pyrrole with $FeCl_3$ and other transition metal species, either as plastic films,[100,102] composite materials,[100] or insoluble powders.[101] The reported materials exhibited conductivities as high as 50–80 Ω^{-1} cm^{-1}.

3.7 Diaz and electropolymerized polypyrrole films

Arthur Fred Diaz was born in Southern California in 1938. After completing high school, he continued his education at San Diego State University, receiving a BS in Chemistry in 1960.[103,104] He then moved to the University of California, Los Angeles (UCLA), where he pursued graduate studies under Saul Winstein (1912–1969). Diaz completed his Ph.D. in 1965 with a dissertation entitled "Exchange and Substitution of Benzhydryl Derivatives."[103–105] Following his graduate studies, Diaz continued research activities at UCLA, first as a postdoctoral researcher and then a research associate, before eventually being hired by TRW Systems Group.[103] In 1969, however, he returned to academics as an assistant professor in the Department of Chemistry at the University of California, San Diego.[103,104] He then left in 1974 to spend a year as a program officer for the Educational Division of the National Science Foundation, after which he returned to California to join the IBM Almaden Research Laboratory in San Jose as manager of the Advanced Materials group. At IBM, he contributed to 18 US patents filed in the areas of display and printing technologies, and novel electrochromic materials, as well as receiving several IBM Innovation Awards in the areas of thin film coatings and materials for printing technologies.[103] After 20 years at IBM, he retired in 1995 and joined the Department of Chemical and Materials Engineering at San Jose State University.[103,104] At the same time, he also held an appointment at the Technical Institute of Baja California, where he helped develop a proposal to install first a bachelor's program in chemistry, and then subsequently an MS and Ph.D. program in chemistry.[103]

Upon arriving at IBM in 1975, Diaz was charged with developing a new project of significant impact. As IBM was interested in building capabilities in the field of electrochemistry, the application of such techniques was especially desirable.[103] This led to an interest in the preparation and study of modified electrodes, resulting in some cursory studies of monolayers and thin films. By 1978, conducting polymers from polyacetylene was a hot topic, and he thus considered the application of such conducting materials for modified electrodes but was unsure as to how to accomplish this with polyacetylene. As luck would have it, however, a potential solution then presented itself. It was during one of Gardini's visits to IBM that he mentioned the work on pyrrole black being carried out at Parma, particularly the most recent application of electropolymerization.[103] As Diaz acknowledged in a 1981 overview of his work on polypyrrole,[106] "I wish to thank my coworkers who have contributed significantly to this effort . . . in particular to G. P. Gardini who first suggested the potential of polypyrrole." The material's combination of intractability and conductivity intrigued Diaz, and he thus began investigating the generation of electropolymerized films of pyrrole.

After a matter of time, Diaz optimized the electropolymerization of pyrrole under controlled conditions, allowing the modification of electrode surfaces with strongly adhered films in a reasonable and repeatable manner.[10,103] The first of several papers was reported in 1979, with Gardini as a coauthor.[10] Diaz found that the use of aprotic solvents under oxygen-free conditions resulted in better material properties[10,11,107-109] in comparison with the previously applied aqueous conditions.[9,92] Under optimum conditions, the polypyrrole films were prepared galvanostatically on a platinum electrode from 0.06 M pyrrole in a 99:1 acetonitrile–water mixture with 0.1 M tetraethylammonium tetrafluoroborate (Et_4NBF_4) as the supporting electrolyte.[10,11,107-111] The two-electrode cell used in these cases is shown in Figure 3.22.[107] Similar, but somewhat modified, results were also found when using tetramethylammonium hexafluorophosphate (Me_4NPF_6) or tetraethylammonium perchlorate (Et_4NClO_4) as the supporting electrolyte.[109,111] It was found that the adherence of the film to the electrode could be controlled by the water content of the monomer solution. While the use of strictly anhydrous solvents produced poorly adhering, nonuniform films, adherence improved with the introduction of small amounts of water.[107] Overall, the films were found to be quite stable and could be cycled repeatedly without any evidence of decomposition.[111] In addition, it was shown that these films could be further modified via postpolymerization derivatization, in which the surface of the film was nitrated via treatment with HNO_3.[10,107]

Elemental analysis of the material indicated that it was comprised primarily of α,α'-coupled pyrrole units, plus BF_4^- anions, in a ratio of near 4:1, with a typical composition of $C_{4.0}N_{0.87}H_{3.44}(BF_4)_{0.25-0.30}$.[10,11,106,107,110,111] The exact composition of the films, however, was dependent on the specific conditions of the electropolymeration.[11,107] It was believed that this represented a polypyrrole backbone that carried a partial positive charge that was balanced by the BF_4^- ions.[11,106,107,110,111] Raman and reflective IR analysis showed bands characteristic of pyrroles, thus supporting the presence of pyrrole rings in this proposed structure.[11,105] Consistent with the electropolymerized films previously reported by the Parma group,[9] the films exhibited very little structural order, with X-ray diffraction showing no discernible peaks and electron diffraction showing only diffuse rings corresponding to a lattice spacing of 3.4 Å.[11,107]

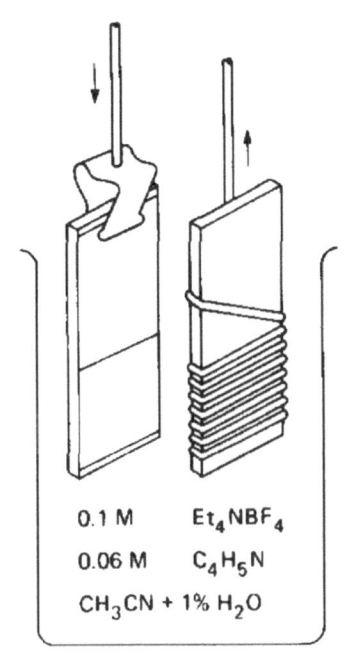

0.1 M Et_4NBF_4

0.06 M C_4H_5N

CH_3CN + 1% H_2O

Figure 3.22 The single compartment, two-electrode electropolymerization cell utilized by Diaz. The working electrode consisted of a platinum evaporated onto a glass slide, while the counter electrode consisted of gold wire tightly wrapped around a second glass slide.

Reprinted from K. K. Kanazawa, A. F. Diaz, W. D. Gill, P. M. Grant, G. B. Street, G. P. Gardini, J. F. Kwak. Polypyrrole: An electrochemically synthesized conducting organic polymer. *Synth. Met.* **1980,** *1,* 329–336, Copyright 1980, with permission from Elsevier.

It is pointed out, however, that the diffraction rings are arced under certain circumstances, suggesting some preferred orientation of the polymer chains.[11]

The electric properties of thicker freestanding films (5–50 μm) were determined via four-point probe to give room temperature conductivities of 10–200 Ω^{-1} cm^{-1}, with the possibility of consistently obtaining the upper range of these values.[10,11,106,107,110] This was a considerable improvement over the values of circa 8 Ω^{-1} cm^{-1} previously reported by the Parma group,[9] which Diaz attributed at least partially to higher quality films as the result of limited film thicknesses (as thin as 5–10 μm)[10,106] and the fact that he grew his films quite slowly.[103] In comparison, his impression was that the Parma films were quite thick.[103] It is now fairly well understood that the structural order of electropolymerized films decreases with both increased rates of deposition and increased film thickness, and enhanced order contributes to enhanced conductivity. At the same time, however, characterization did not provide strong evidence of increased crystallinity. Based on the determined composition of both materials, though, it does appear that the IBM materials were more highly doped than the previous materials from Parma. As with the previous materials reported by either Weiss[8] or the Parma group,[9] the conductivity of the new electropolymerized films revealed a temperature dependence consistent with a classical semiconductor.[11,107] Additional electrochemical evidence that the films were insulating in their neutral states were also reported.[109,110]

In 1981, Diaz then reported a study in which the relationship between conjugation length and the corresponding electronic and optical properties was detailed.[112] Thus, the series pyrrole, 2,2'-dipyrrole, and 2,2';5',2"-terpyrrole were analyzed via electrochemical and spectroscopic methods, during which it was found that the redox potentials scale linearly with the lowest energy electronic absorption, and that both characteristics scale linearly with the number of pyrrole units. Of particular interest was the fact that the measured values for polypyrrole were also in agreement with this linear relationship. These relationships were then coupled with additional spectroelectochemical studies of pyrrole's oxidative polymerization in order to present a detailed polymerization mechanism in 1983.[113] Although details have been modified over time, this initial mechanism was very similar to that given in Figure 3.1.

Diaz also extended the successful electropolymerization methods to functionalized species such as N-methyl-, N-ethyl-, N-propyl-, N-butyl-, and N-phenyl-pyrrole.[11,106,110,111] While these also produced polymer films, the functionalized pyrroles gave materials with significantly reduced conductivities. For example, both poly(N-methylpyrrole) and poly(N-phenylpyrrole) exhibited room temperature values of circa 1×10^{-3} Ω^{-1} cm^{-1}.[11,110] Interestingly, it was found that treating poly(N-methylpyrrole) films with bromine vapor caused the conductivity to increase by a factor of 20.[108] Pyrroles containing α-functionalities were also investigated, but these did not lead to any polymeric products, further supporting the view that polymerization occurred via α,α'-coupling of the pyrrole rings.[114]

By 1983, Diaz had moved on to the study of electropolymerized polythiophenes (see Chapter 6).[115] By this time, however, the electropolymerization methods developed by Diaz had become the default method for the preparation of conducting polypyrroles. More critically, these still remain the most common methods used for the modern preparation of these materials.

3.8 Conclusions

The discussion in this chapter has hopefully illustrated the long history of polypyrrole dating back to the earliest work of Angeli in 1915. While most of these reports referred to these materials using Angeli's original term *pyrrole black*, it is quite clear from the methods and characterization reported that these materials were all polypyrrole. In addition, the electrical conductivity of these materials has been known since the work of Weiss and coworkers in 1963, with the magnitude of its conductivity increasing significantly with each sequential improvement in the production of the polymer. Furthermore, the research carried out by each of these groups was not carried out in isolation, with the bulk of these studies citing the results of the previously reported materials. Lastly, it is important to note that with the exception of the work of Diaz, all of the work presented here occurred before the collaborative polyacetylene work of Heeger, MacDiarmid, and Shirakawa.[1–3] As such, it is hard to support the common claims that the discovery of conductive polyacetylene led to the development of conductive polypyrroles. While it is true that the reports of polyacetylene did initiate Diaz's interest in conducting polymers, the development of conductive polypyrrole was a direct extension of the previous work at Parma.

Of course, it is important to point out that the work of Diaz and coworkers presented here has significance beyond the simple fact that it represented the highest reported conductivity of polypyrrole in the early years of the growing field of conjugated organic polymers. While the electropolymerization of a conjugated polymer dates back to the work of Henry Letheby (1816–1876) on polyaniline in 1862[116] (see Chapter 2), and Diaz was not even the first to electropolymerize polypyrrole, he was the one that not only optimized this method in terms of quality and reproducibility but then also demonstrated that this could be broadened to a wide family of conjugated systems.[106] As a result, electropolymerization quickly became the most common method for the generation of conjugated polymeric films, until it was ultimately supplanted by the current focus on soluble, processible materials.

References and Notes

1. Rasmussen, S. C. Conjugated and conducting organic polymers: The first 150 years. *ChemPlusChem* **2020**, *85*(7), 1412–1429.
2. Rasmussen, S. C. Early history of conjugated polymers: From their origins to the handbook of conducting polymers. In *Handbook of Conducting Polymers*, 4th ed. Reynolds, J. R.; Skotheim, T. A.; Thompson, B.; Eds.; CRC Press: Boca Raton, FL, 2019, pp. 1–35.
3. Rasmussen, S. C. Early history of conductive organic polymers. In *Conductive Polymers: Electrical Interactions in Cell Biology and Medicine*. Zhang, Z.; Rouabhia, M.; Moulton, S.; Eds.; CRC Press: Boca Raton, FL, 2017, pp. 1–21.
4. Rasmussen, S. C. Early history of polypyrrole: The first conducting organic polymer. *Bull. Hist. Chem.* **2015**, *40*, 45–55.
5. Rasmussen, S. C. Electrically conducting plastics: Revising the history of conjugated organic polymers. In *100+ Years of Plastics. Leo Baekeland and Beyond*; E. T. Strom; S. C. Rasmussen; Eds.; ACS Symposium Series 1080, American Chemical Society: Washington, DC, 2011, pp. 147–163.

6. McNeill, R.; Siudak, R.; Wardlaw, J. H.; Weiss, D. E. Electronic conduction in polymers. *Aust. J. Chem.* **1963**, *16*, 1056–1075.

7. Bolto, B. A.; Weiss, D. E. Electronic conduction in polymers. II. The electrochemical reduction of polypyrrole at controlled potential. *Aust. J. Chem.* **1963**, *16*, 1076–1089.

8. Bolto, B. A.; McNeill, R.; Weiss, D. E. Electronic conduction in polymers. III. Electronic properties of polypyrrole. *Aust. J. Chem.* **1963**, *16*, 1090–1103.

9. Dall'Olio, A.; Dascola, G.; Varacca, V.; Bocchi, V. Resonance paramagnètique èlectronique et conductiviè d'un noir d'oxypyrrol èlectrolytique. *C. R. Acad. Sci., Ser. C* **1968**, *267*, 433–435.

10. Diaz, A. F.; Kanazawa, K. K.; Gardini, G. P. Electrochemical polymerization of pyrrole. *J. Chem. Soc., Chem. Commun.* **1979**, 635–636.

11. Kanazawa, K. K.; Diaz, A. F.; Geiss, R. H.; Gill, W. D.; Kwak, J. F.; Logan, J. A.; Rabolt, J. F.; Street, G. B. "Organic Metals": Polypyrrole, a stable synthetic "metallic" polymer. *J. Chem. Soc., Chem. Commun.* **1979**, 854–855.

12. Wynne K. J.; Street, G. B. Conducting polymers. A short review. *Ind. Eng. Chem. Prod. Res. Dev.* **1982**, *21*, 23–28.

13. Curran, D.; Grimshaw, J.; Perera, S. D. Poly(pyrrole) as a support for electrocatalytic materials. *Chem. Soc. Rev.* **1991**, *20*, 391–404.

14. Toshima, N.; Hara, S. Direct synthesis of conducting polymers from simple monomers. *Prog. Polym. Sci.* **1995**, *20*, 155–183.

15. Badgujar, A. G.; Bambole, V. A.; Mahanwar, P. A. Polypyrrole: Synthesis, modifications and applications. *Paintindia* **2011**, *61*, 53–66.

16. Pang, A. L.; Arsad, A.; Ahmadipour, M. Synthesis and factor affecting on the conductivity of polypyrrole: A short review. *Polym. Adv. Technol.* **2021**, https://doi.org/10.1002/pat.5201

17. Waltman, R. J.; Bargon, J. Reactivity/structure correlations for the electropolymerization of pyrrole: An INDO/CNDO study of the reactive sites of oligomeric radical cations. *Tetrahedron* **1984**, *40*, 3963–3970.

18. Waltman, R. J.; Bargon, J. Electrically conducting polymers: A review of the electropolymerization reaction, of the effects of chemical structure on polymer film properties, and of applications towards technology. *Can. J. Chem.* **1986**, *64*, 76–95.

19. John, R.; Wallace, G. G. The use of microelectrodes to probe the electropolymerization mechanism of heterocyclic conducting polymers. *J. Electroanal. Chem.* **1991**, *306*, 157–167.

20. Smith, J. R.; Cox, P. A.; Campbell, S. A.; Ratcliffe, N. M. Application of density functional theory in the synthesis of electroactive polymers. *J. Chem. Soc. Faraday Trans.* **1995**, *91*, 2331–2338.

21. Runge, F. F. Ueber einige Produkte der Steinkohlendestillation. *Ann. Phys. Chem.* **1834**, *31*, 65–78.

22. Kränzlein, G. Zum 100jährigen Gedächtnis der Arbeiten von F. F. Runge. *Angew. Chem.* **1935**, *48*, 1–3.

23. Anft, B. Friedlieb Ferdinand Runge: A forgotten chemist of the nineteenth century. *J. Chem. Educ.* **1955**, *32*, 566–574.

24. Anderson, H. J. Pyrrole. From Dippel to Du Pont. *J. Chem. Educ.* **1995**, *72*, 875–878.

25. Schwenk, E. F. Friedlieb Runge and his capillary designs. *Bull. Hist. Chem.* **2005**, *30*, 30–34.

26. Anderson, T. On the constitution and properties of picoline, a new organic base from coal-tar. *Trans. Royal Soc. Edin.* **1849**, *16*, 123–136.

27. Anderson, T. On the products of the destructive distillation of animal substances. Part II. *Trans. Royal Soc. Edin.* **1853**, *20*, 247–260.

28. Anderson, T. On the products of the destructive distillation of animal matters. Part IV. *Trans. Royal Soc. Edin.* **1857**, *21*, 571–595.

29. Schwanert, H. Ueber einige Zersetzungsproducte der Schleimsäure. *Liebigs Ann. Chem.* **1860**, *116*, 257–287.

30. Partington, J. R. *A History of Chemistry*. Martino Publishing: Mansfield Center, CT, 1998, Vol. 4, p. 782.

31. Baeyer A.; Emmerling, A. Reduction des Isatins zu Indigblau. *Ber.* **1870**, *3*, 514–517.

32. Plancher, G.; Cattadori, F. Sull'ossidazione del dimetilpirrolo asimmetrico. *Atti Accad. Naz. Lincei, Rend.* **1903**, *12*, 10–13.

33. Plancher, G.; Cattadori, F. Sull'ossidazione del pirrolo ad immide maleica. *Atti Accad. Naz. Lincei, Rend.* **1904**, *13*, 489–492.

34. Ciamician, G. Ueber die Entwickelung der Chemie des Pyrrols im letzten Vierteljahrhundert. *Chem. Ber.* **1904**, *37*, 4200–4255.

35. Gardini, G. P. The oxidation of monocyclic pyrroles. *Adv. Heterocyclic Chem.* **1973**, *15*, 67–98.

36. Angeli, A. Sopra il nero del pirrolo. Nota preliminare. *Rend. Accad. Lincei* **1915**, *24*, 3–6; *Gazz. Chim. Ital.* **1916**, *46*(II), 279–283.

37. Angeli A.; Alessandri, L. Sopra il nero pirrolo II. Nota. *Rend. Accad. Lincei* **1916**, *25*(I), 761–774; *Gazz. Chim. Ital.* **1916**, *46*(II), 283–300.

38. Angeli A.; Cusmano, G. Sopra i neri di nitropirrolo. *Nota. Rend. Accad. Lincei* **1917**, *26*(I), 273–278; *Gazz. Chim. Ital.* **1917**, *47*(I), 207–213.

39. Angeli, A. Sopra i neri di pirrolo. Nota. *Rend. Accad. Lincei* **1918**, *27*(I), 209–212; *Gazz. Chim. Ital.* **1918**, *48*(II), 21–25.

40. Angeli, A. I neri di pirrolo e le melanine. *Rend. Accad. Lincei* **1918**, *27*(I), 417–421; *Gazz. Chim. Ital.* **1918**, *48*(II), 67–72.

41. Angeli, A.; Pieroni, A. Sopra un nuovo modo di formazione del nero di pirrolo. Nota. *Rend. Accad. Lincei* **1918**, *27*(II), 300–304; *Gazz. Chim. Ital.* **1919**, *49*(I), 154–158.

42. Angeli A.; Lutri, C. Nuove ricerche sopra i neri di pirrolo. Nota. *Rend. Accad. Lincei* **1920**, *29*(I), 14–22; *Gazz. Chim. Ital.* **1920**, *50*(I), 128–139.

43. Angeli A.; Lutri, C. Ricerche sopra i neri di pirrolo. (VII). *Rend. Accad. Lincei* **1920**, *29*(I), 420–430; *Gazz. Chim. Ital.* **1921**, *51*(I), 31–34.

44. Fontani, M.; Orna, M. V. The shy angel who missed the Nobel Prize. *Chimica Oggi* **2012**, *30*, 58–60.

45. Cambi, L. A la memoria di Angelo Angeli. *Gazz. Chim. Ital.* **1933**, *63*, 527–560.

46. Gelsomini, N.; Costa, M. G.; Manzelli, P.; Fiorentini, C. La chimica a Firenze fra 800 e 900. Relazione presentata al IV Convegno Nazionale di "Storia e Fondamenti della Chimica" (Venezia, November 7–9, 1991); http://petrina.sm.chim.unifi.it/~castel/ugo_schiff/chi_storia.html (accessed March 27, 2014).

47. The Istituto di Studi Superiori Pratici e di Perfezionamento was officially made the University of Florence by the Italian Parliament in 1923.

48. Angeli, A.; Pieroni, A. A proposito di un lavoro del prof. E. Salkotvski sopra le melanine. *Rend. Accad. Lincei* **1921**, *30*(I), 241–245.

49. Meredith, P.; Sarna, T. The physical and chemical properties of eumelanin. *Pigment Cell Res.* **2006**, *19*, 572–594.

50. Pieroni, A.; Moggi, A. Sopra la costituzione di alcuni polipirroli. *Rend. Accad. Lincei* **1922**, *31*(I), 381–385.

51. Pieroni, A.; Moggi, A. Sopra alcuni polipirroli. *Gazz. Chim. Ital.* **1923**, *53*, 120–135.

52. Pieroni, A. Sopra i polipirroli. *Rend. Accad. Lincei* **1923**, *32*(II), 175–179.

53. Pieroni, A.; Veremeenco, P. Prodotti di ossidazione di vari composti pirrolo. *Gazz. Chim. Ital.* **1926**, *56*, 455–479.

54. Angeli, A. Sopra i neri di pirrolo. *Rend. Accad. Lincei* **1930**, *11*, 439–442.

55. Ciusa, R. Sulla scomposizione dello iodolo. *Rend. Accad. Lincei* **1921**, *30*(II), 468–469.

56. Ciusa, R. Sulle grafiti da pirrolo e da tiofene (Nota preliminare). *Gazz. Chim. Ital.* **1922**, *52*(II), 130–131.

57. Ciusa, R. Su alcune sostanze analoghe alla grafite. *Gazz. Chim. Ital.* **1925**, *55*, 385–389.

58. Musajo, L. Commemorazione ufficiale tenuta nell' Aula Magna dell'Universita di Bari il 27-11-1965. In *Riccardo Ciusa, 1877–1965.* Istituto di Chimica Farmaceutica e Tossicologica. Carelli, V., Ed.; Dell'Universita di Bari: Bari, 1968, pp. 15–28.

59. Nebbia, G. Personal communication, 2014.

60. Prior to the mid-1980s, *dottore* was the terminal degree in Italy, consisting of a total of four to five years of study and a thesis. In many respects, it would be similar to a modern master's degree.

61. *80 anni della Facolta di Farmacia di Bari, 1932–2012, Giornata Pugliese su Farmaco e Prodotti per la Salute.* Università di Bari: Bari, Italy, 2012.

62. Spurling, T. H. Donald Eric Weiss 1924–2008. *Hist. Rec. Aust. Sci.* **2011**, *22*, 152–170.

63. Weiss, D. E. Curriculum vitae, 2001.

64. Weiss, R. Tribute to Don Weiss by his son Robert. Tribute presented at the funeral of Don Weiss, August 4, 2008.

65. Walker, R. Weiss, Donald Eric (1924–2008). *Encyclopedia of Australian Science*, http://www.eoas.info/biogs/P004056b.htm (accessed April 26, 2021).

66. Obituaries: Don Weiss. *Australian Academy of Science Newsletter* **2008**, *73*, 11.

67. Mackey, B. Don Weiss a dominant figure in chemistry. *ATSE Focus* **2008**, *153*, 40.

68. Sources differ on the full name of APPM, with various sources giving it as Australian Pulp and Paper Manufacturers, Australasian Pulp and Paper Manufacturers, or Associated Pulp and Paper Mills.

69. Weiss, D. E. Adsorbents and adsorption processes. D.Sc. thesis, University of Adelaide, Australia, 1960.

70. Weiss, D. E. Notes on my professional career [Personal document]. 2001.

71. Garten, V. A.; Weiss, D. E. Quinonehydroquinone character of activated carbon and carbon black. *Aust. J. Chem.* **1955**, *8*, 68–95.

72. Garten, V. A.; Eppinger, K.; Weiss, D. E. Studies on the abrasion and wear of rubber. 1. The chemistry of black carbon and its effect on abrasion as determined by the National Bureau of Standards method. *Aust. J. Appl. Sci.* **1956**, *7*, 148–149.

73. Garten, V. A.; Weiss, D. E. Ion- and electron-exchange properties of activated carbon in relation to its behaviour as a catalyst and adsorbent. *Rev. Pure Appl. Chem.* **1957**, *7*, 69–112.

74. Garten, V. A.; Weiss, D. E.; Willis, J. B. A new interpretation of the acidic and basic structures in carbons. 1. Lactone groups of the ordinary and fluoroescein types in carbons. *Aust. J. Chem.* **1957**, *10*, 295–308.

75. Garten, V. A.; Weiss, D. E. A new interpretation of the acidic and basic structures in carbons. 2. The chromene-carbonium ion couple in carbon. *Aust. J. Chem.* **1957**, *10*, 309–328.

76. Garten, V. A.; Weiss, D. E. Copolymerisation of carbon black with elastomers and sulphur. *Ind. Chemist* **1959**, *35*, 525–530.

77. Garten, V. A.; Weiss, D. E. Functional groups in activated carbon and carbon black with ion and electron-exchange properties. In *Proceedings of the American Carbon Society Carbon Conference 1957, 3rd Biennial Carbon Conference held at the State University of New York, Buffalo.* The American Carbon Society: New York, 1959, pp. 295–313.

78. McNeill R.; Weiss, D. E. A xanthene polymer with semiconducting properties. *Aust. J. Chem.* **1959**, *12*, 643–656.

79. Pohl, H. A. Semiconduction in polymers. In *Organic Semiconductors. Proceedings of an Inter-Industry Conference.* Brophy, J. J.; Buttrey, J. W.; Eds.; The Macmillan Company: New York, 1962, pp. 134–141.

80. Pohl, H. A.; Bornmann, J. A.; Itoh, W. Studies of some semiconducting polymers. In *Organic Semiconductors. Proceedings of an Inter-industry Conference.* Brophy, J. J.; Buttrey, J. W.; Eds.; The Macmillan Company: New York, 1962, pp. 142–158.

81. Okamoto, K.; Brenner, W. *Organic Semiconductors.* Reinhold Publishing Corporation: New York, 1964, pp. 125–158.

82. Weiss, D. E.; Bolto, B. A. Organic polymers that conduct electricity. In *Physics and Chemistry of the Organic Solid State.* Fox, D.; Labes, M. M.; Arnold Weissberger, A., Eds.; Interscience Publishers: New York, 1965; Vol. II, Chapter 2, pp. 67–120.

83. Gutmann, F.; Lyons, L. E. *Organic Semiconductors.* John Wiley & Sons: New York, 1967, pp. 471–484.

84. McNeill, R.; Weiss, D. E.; Willis, D. Electronic conduction in polymers IV. Polymers from imidazole and pyridine. *Aust. J. Chem.* **1965**, *18*, 477–486.

85. Bolto, B. A.; Weiss, D. E.; Willis, D. Electronic conduction in polymers V. Aromatic semiconducting polymers. *Aust. J. Chem.* **1965**, *18*, 487–491.

86. Macpherson, A. S.; Siudak, R.; Weiss, D. E.; Willis, D. Electronic conduction in polymers VI. The chemical consequences of space charge effects in conducting polymers. *Aust. J. Chem.* **1965**, *18*, 493–505.

87. Dascola, G.; Giori, D. C.; Varacca, V.; Chierici, L. Rèsonance paramagnètique èlectronique des radicaux libres crèès lors de la formation des noirs d'oxypyrrol. Note. *C. R. Acad. Sci., Ser. C* **1966**, *262*, 1617–1619.

88. Chierici, L.; Gardini, G. P. Structure of the product $C_8H_{10}N_2O$ of oxidation of pyrrole. *Tetrahedron* **1966**, *22*(1), 53–56.

89. Bocchi, V.; Chierici, L.; Gardini, G. P. Structure of the oxidation product of pyrrole. *Tetrahedron* **1967**, *23*(2), 737–740.

90. Bocchi, V.; Chierici, L.; Gardini, G. P. Su un prodotto semplice di ossidazione del pirrolo. *Chim. Ind. (Milan, Italy)* **1967**, *49*(12), 1346–1347.

91. Chierici, L.; Artusi, G. C.; Bocchi, V. Sui neri di ossipirrolo. *Ann. Chim.* **1968**, *58*(8–9), 903–913.

92. Bocchi, V.; Chierici, L.; Gardini, G. P.; Mondelli, R. Pyrrole oxidation with hydrogen peroxide. *Tetrahedron* **1970**, *26*(17), 4073–4082.

93. Pochini, A. *La Chimica Universitaria a Parma.* University of Parma, 2015, https://scvsa.unipr.it/sites/st22/files/storia_chimica_unipr.pdf (accessed May 4, 2021).

94. Chierici, L.; Cella, A. Contibuto alla conoscenza dei polipirroli. *Ann. Chim.* **1953**, *43*, 141–147.

95. Chierici, L.; Serventi, G. α,α′-tripirrili e neri di ossipirrolo. *Gazz. Chim. Ital.* **1960**, *90*, 23–31.

96. Lions Club of Parma Farnese, Il Lions Club Parma Farnese presenta il Socio Gian Piero Gardini. http://www2.unipr.it/~gardini/home.htm (accessed May 4, 2021).

97. Department of Organic and Industrial Chemistry, University of Parma Faculty of Natural Sciences. Activity report 2008. http://www.unipr.it/arpa/chimorg/ (accessed July 10, 2014).

98. Università degli Studi di Parma. 8 giugno: assegnazione Premio tesi di dottorato in memoria del Prof. Gardini. http://www.unipr.it/notizie/8-giugno-assegnazione-premio-tesi-di-dottorato-memoria-del-prof-gardini (accessed July 16, 2014).

99. Goppelsroeder, F. Sur le noir d'aniline électrolytique. *C. R. Seances Acad. Sci.* **1876**, *82*, 331–333.

100. Bocchi, V.; Gardini, G. P. Chemical synthesis of conducting polypyrrole and some composites. *J. Chem. Soc., Chem. Commun.* **1986**, 148.

101. Rapi, S.; Bocchi, V.; Gardini, G. P. Conducting polypyrrole by chemical synthesis in water. *Synth. Met.* **1988**, *24*, 217–221.

102. Bocchi, V.; Gardini, G. P.; Zanella, G. Conductive polypyrrole by synthesis in strong alkaline medium. *Synth. Met.* **1988**, *24*, 217–221.

103. Diaz, A. F. personal communication, 2014.

104. Diaz, A. F. Curriculum vitae, 2006.

105. Diaz, A. F. Exchange and substitution of benzhydryl derivatives. Ph.D. dissertation, University of California, Los Angeles, 1965.

106. Diaz, A. Electrochemical preparation and characterization of conducting polymers. *Chemica Scripta* **1981**, *17*, 145–148.

107. Kanazawa, K. K.; Diaz, A. F.; Gill, W. D.; Grant, P. M.; Street, G. B.; Gardini, G. P.; Kwak, J. F. Poly-pyrrole: An electrochemically synthesized conducting organic polymer. *Synth. Met.* **1980**, *1*, 329–336.

108. Diaz, A. F.; Lee, W.-Y; Logan A.; Green, D. C. Chemical modification of a polypyrrole electrode surface. *J. Electroanal. Chem.* **1980**, *108*, 377–380.

109. Diaz, A. F.; Castillo, J. I. A polymer electrode with variable conductivity: Polypyrrole. *J. Chem. Soc., Chem. Commun.* **1980**, 397–398.

110. Diaz, A. F.; Castillo, J. I.; Logan, J. A.; Lee, W.-Y. Electrochemistry of conducting polypyrrole films. *J. Electroanal. Chem.* **1981**, *129*, 115–132.

111. Diaz, A. F.; Kanazawa, K. K.; Castillo, J. I.; Logan, J. A. Electrosynthesis and study of conducting polymeric films. In *Conductive Polymers.* Seymour, R. B., Ed.; Polymer Science and Technology, Vol. 15, Plenum Press: New York, 1981, pp. 149–153.

112. Diaz, A. F.; Crowley, J.; Bargon, J.; Gardini, G. P.; Torrance, J. B. Electrooxidation of aromatic oligomers and conducting polymers. *J. Electroanal. Chem.* **1981**, *121*, 355–361.

113. Genies, E. M.; Bidan, G.; Diaz, A. F. Spectroelectrochemical study of polypyrrole films. *J. Electroanal. Chem.* **1983**, *149*, 101–113.

114. Diaz, A. F.; Martinez, A.; Kanazawa, K. K. Electrochemistry of some substituted pyrroles. *J. Electroanal. Chem.* **1981**, *130*, 181–187.

115. Waltman, R. J.; Bargon, J.; Diaz, A. F. Electrochemical studies of some conducting polythiophene films. *J. Phys. Chem.* **1983**, *87*, 1459–1463.

116. Letheby, H. On the production of a blue substance by the electrolysis of sulphate of aniline. *J. Chem. Soc.* **1862**, *15*, 161–163.

Chapter 4
Polyphenylene and poly(phenylene vinylene)

4.1 Introduction

Among the primary parent conjugated polymers introduced in Chapter 1, polyphenylene, or poly(*para*-phenylene) (Figure 4.1), is often given little discussion, if not outright ignored. After all, the low solubility of the unfunctionalized parent prohibited the generation of higher molecular weight samples, which when coupled with the inherent steric interactions between neighboring benzene units resulted in materials with low effective conjugation.[1-3] As a result, these polymers exhibit lower conductivities in comparison with other conjugated materials and were initially of interest primarily for their high thermally stability.[1,3]

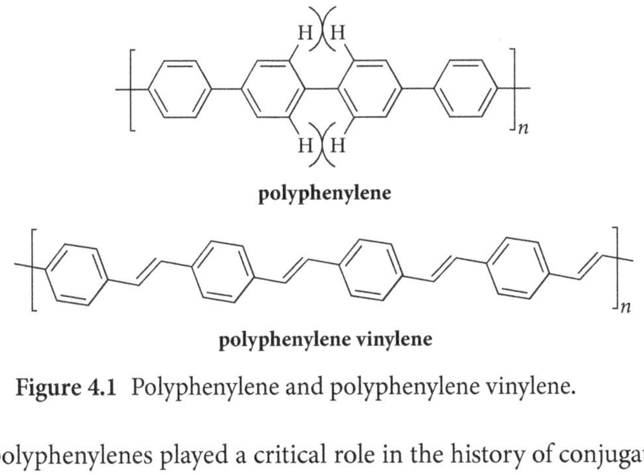

polyphenylene

polyphenylene vinylene

Figure 4.1 Polyphenylene and polyphenylene vinylene.

Still, polyphenylenes played a critical role in the history of conjugated materials, most importantly as the first example of these materials not to be produced via simple oxidative polymerization, but through the introduction of various metal-mediated cross-coupling methods.[1,2] At the same time, the synthesis of polyphenylenes also introduced the first use of dihalo-functionalized monomeric species, which have since become critical building blocks for the production of conjugated materials. Of course, the ongoing development of such metal-mediated methods, coupled with the application of dihaloaryl species,

The Origins and Early History of Conjugated Organic Polymers. Seth C. Rasmussen, Oxford University Press. © Oxford University Press (2025). DOI: 10.1093/9780197638194.003.0004

ultimately became the preferred methodology for the production of conjugated polymers, particularly for the more complex copolymeric materials that dominate the field today. As such, one can trace the origins of this overall synthetic approach back to polyphenylenes.

In addition, efforts to improve on the basic polyphenylene structure then led to the production of the related material poly(phenylene vinylene) (PPV, Figure 4.2). This material, which can be traced back to 1960, can be viewed as a hybrid of polyphenylene and polyacetylene and effectively reduces the steric issues inherent in the adjacent benzene units of the original parent polymer.[4-6] Another advancement on the initial polyphenylene backbone involved the incorporation of alkyl bridgeheads between adjacent phenylene units to give the fused-ring analog polyfluorene (Figure 4.2) in the mid-1980s.[7] Both PPV and polyfluorene then went on to become important conjugated polymers for fluorescent applications.

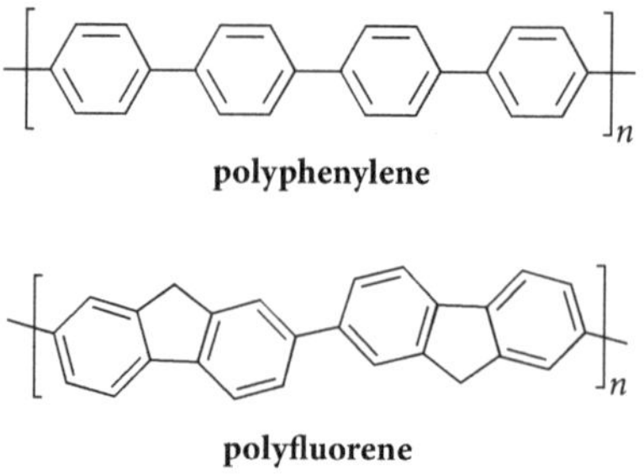

polyphenylene

polyfluorene

Figure 4.2 Polyfluorene as a fused polyphenylene.

Still, this historical evolution first required a steady supply of benzene. Perhaps more critically in this case was also a knowledge of the production of various necessary derivatives. Therefore, as with the previous chapters, we will begin with an overview of the history of benzene and its related chemistry.

4.2 A brief history of benzene

It was in 1825 that Michael Faraday (1791–1867, Figure 4.3) first isolated benzene during the examination of an oil that separated out in vessels of illuminated gas produced by the Portable Gas Company, a company that was marketing a gas prepared by heating whale oil. When the gas was compressed in the vessels, a fluid was deposited, from which Faraday separated two new "carburets of hydrogen" via distillation.[8,9] The first of these he gave the name *bicarburet of hydrogen*, which corresponds to our modern benzene.[9] Faraday describes the product as a "colourless transparent liquid, having an odour resembling that of oil gas, and partaking also of that of almonds."[8] Its specific gravity was found to be 0.85 at 60°F, and the liquid crystallized when cooled to 32°F.[10] Its boiling point in contact with glass was determined to be 186°F (85.6°C). Faraday determined it to be a binary compound of carbon and hydrogen, with a weight ratio of 12:1. However, as the atomic weight commonly used for carbon in England at the time was 6, he viewed its empirical formula to be C_2H. Finally, he went on to show that this product reacted with chlorine, nitric acid, and sulfuric acid, thus providing the start of functional derivatives.

Figure 4.3 Michael Faraday (1791–1867), circa 1820s.

Nine years later, in 1834, the generation of the same species was reported by the German chemist Eilhard Mitscherlich (1794–1863, Figure 4.4) via the distillation of a mixture of benzoic acid and slaked lime ($Ca(OH)_2$).[11] As the name *benzoin* had already been used, he decided on the name *benzin* for the product. Interestingly, however, the journal changed the name to *benzol*, as they thought the "-ol" suffix better described its properties and origin. Mitscherlich characterized the product as a clear, colorless liquid of a peculiar odor, with of specific gravity of 0.85. He found that it boiled at 86°C and became solid when placed in ice, with the resulting crystalline mass melting at 7°C. Benzin (or benzol) was found to be readily soluble in alcohol and ether, but only sparingly soluble in water. Analysis revealed a carbon and hydrogen weight ratio of circa 12:1, and he concluded that the product was identical to Faraday's bicarburet of hydrogen.

Figure 4.4 Eilhard Mitscherlich (1794–1863).

The next significant advancement was then in 1842, when benzene was isolated from coal tar first by John Leigh and then again three years later by August Wilhelm Hofmann (1818–1892).[12] Leigh is stated to have exhibited considerable quantities of benzene, nitrobenzene, and dinitrobenzene to the chemical section of the British Association for the Advancement of Science at the Manchester meeting in June 1842,[13] but so little information is given in the associated printed communication that it is not possible to identify the bodies in question.[14] In addition, benzene is not specifically mentioned in the communication.

While Hofmann's related report in 1845 primarily focused on a new method for the detection of benzene via its conversion to aniline, he applied this method to a variety of chemical species in order to determine benzene content, including distillates of coal tar.[15] In his isolation of benzene from coal tar, Hofmann began with the light hydrocarbons that had initially passed over during the rectification of the crude distillation product of coal tar (i.e., the second distillation).[15] He then redistilled this fraction, this time collecting everything that boiled up to 118°C. This new fraction was then distilled again to collect the material that boiled up to 105°C, at which point the collected fraction was analyzed to show that it contained a large quantity of benzene.

Charles Blachford Mansfield (1819–1855, Figure 4.5), Hofmann's student at the Royal College,[12,16] then improved the process of fractional distillation to the point where good quality benzene could be produced, patenting the isolation of benzene from the distillation of coal tar in 1847.[17] The details of his separation methods were then published in 1849.[18] Mansfield began his studies with light tar oil, which was the more volatile oil first isolated upon the initial distillation of coal tar. The light tar oil was then shaken with sulfuric acid, after which the oil was isolated and washed sequentially with water, a dilute potash

Figure 4.5 Charles Blachford Mansfield (1819–1855).

solution, and then water again. This washed oil was then fractionally distilled, collecting fractions by 5°C increments up to 200°C. The distillates obtained were then treated to additional fractional distillations, such that after 10 such distillations, a series of five products of significant purity were isolated. Of the isolated products, the fraction that boiled between 80°C–85°C was determined to be benzene.

During the various rectifications used for its purification, it was found that fractions that boiled between 80°C and 90°C crystallized at temperatures of –5°C, while the fractions that boiled at 75°C–80°C or 90°C–95°C remained liquid. As such, it was found that the number of fractional distillations could be significantly reduced by isolating the benzene via crystallization. Thus, after only the fifth distillation, the liquid was cooled to between –10°C and –12°C, resulting in a mass of solidified benzene that was separated from the remaining liquid through the use of a Beart's coffee machine (an early type of French press, Figure 4.6). Compression of the mixture thus allowed separation of the liquid and resulted in a compressed mass of benzene crystals. The crystalline mass was then placed in a funnel and allowed to warm to isolate the liquid benzene. By using this mix of distillation and crystallization, a pint of "almost pure benzene" could be isolated from two gallons of light tar oil. The properties of the isolated benzene agreed well with the previous results of both Faraday[8] and Mitscherlich.[11]

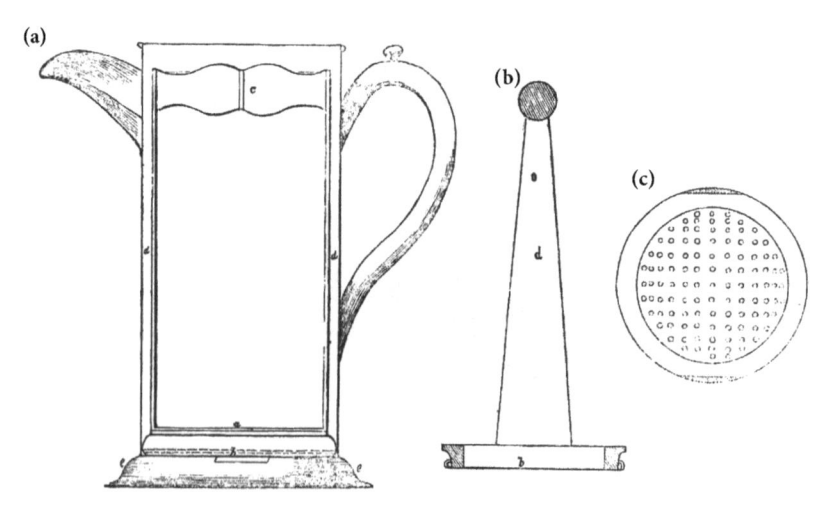

Figure 4.6 Beart's coffee machine: (a) vessel, (b) piston side view, (c) piston bottom view.

Reproduced from W. A. Robertson. Beart's patent coffee pot and filtering apparatus. *Mechanics Magazine* **1839**, *31*, 273–279.

Mansfield then further optimized this process by distilling the original light coal oil in a metal refluxing apparatus, collecting only the material that boiled below 100°C. The collected fraction was then distilled a second time, this time only collecting that which boiled below 90°C. This was then washed with either sulfuric or nitric acid, after which it was distilled a third time, again only collecting that which boiled below 90°C. The new fraction was then washed thoroughly with water and an alkaline solution, after which it was cooled to −20°C in order to separate the benzene crystals as described earlier.

The distillation of coal tar via Mansfield's methods then became the primary source of benzene, a critical reagent for the birth of the synthetic dye industry. Unfortunately, Mansfield did not live to see the fruits of his innovation, as he died only a few years later in 1855. While preparing benzene samples to send to the Paris Exhibition, a still overflowed, and Mansfield attempted to save the premises by carrying the blazing still into the street. In the process, he was so injured that nine days later he died at Middlesex Hospital.[16]

The empirical formula for benzene had been known dating back to Faraday's discovery, and by the 1850s, the atomic weight of 12 was replacing the previous weight of 6, thus giving CH for its newly corrected formula. Still, its molecular formula and structure were challenging to determine. Some, such as Archibald Scott Couper (1831–1892) and Josef Loschmidt (1821–1895), suggested possible structures, but too little evidence was available at the time to gain much support for any particular structure. Things began to become clearer in 1865, when the German chemist Friedrich August Kekulé (1829–1896, Figure 4.7) published a paper in the *Bulletin de la Société Chimique de Paris* suggesting that the structure contained a ring of six carbons with alternating single and double bonds.[19] The paper was then republished in German later that same year.[20]

This initial paper focuses on showing how his structure theory could be applied to explain structural isomerism and homology in benzene and its derivatives (i.e., adding methyl to phenyl gives the first homologue, toluene; two methyls give the second homologue, xylene, which is a structural isomer of ethylbenzene). However, what was later regarded as the two most essential and revolutionary aspects of the benzene theory, the cyclohexatriene structure itself and the positional isomerism of polysubstitution products, received very little emphasis in this first paper.[21] Nevertheless, Kekulé's cyclic structure explained a number of points about benzene and its derivatives commonly held by chemists at the time. This included the view that its minimum carbon content was six atoms, that there was only one known isomer of benzene, and that there was only a single known isomer of each monosubtituted benzene. Furthermore, as is often pointed out, it also explained the observation that there seemed to exist three isomers for doubly substituted benzene derivatives (now understood to correspond to the *ortho*, *meta*, and *para* isomers).[21,22]

Figure 4.7 Friedrich August Kekulé (1829–1896).

Reproduced courtesy of the Edgar Fahs Smith Memorial Collection, University of Pennsylvania (CC0 1.0).

Kekulé later said that he had discovered the ring shape of the benzene molecule after having a dream of a snake seizing its own tail, a popular story that is often repeated and discussed in the literature. As told by Kekulé in 1890 (more than 25 years later):[22]

During my residence in Ghent, in Belgium, I lived in an elegant bache- lor apartment on the main street. However, my study was situated along a narrow alley and had no light during the day. For a chemist who spends the day in the laboratory this was not a disadvantage. There I was sitting, working on my textbook, but it was not going well; my mind was on other matters. I turned my chair toward the fireplace and sank into half-sleep [Halb- schlaf]. Again the atoms fluttered before my eyes. This time smaller groups remained modestly in the background. My mental eye, sharpened by repeated visions of a similar kind, now distinguished larger forms [Gebilde] in a vari- ety of combinations. Long chains [Reihen], often fitted together more densely [dichter zusammengefügt]; everything in motion, twisting and turning like snakes. But look, what was that?! One of the snakes had seized its own tail, and the figure whirled mockingly before my eyes. I awoke as by a stroke of lightning, and this time, too, I spent the rest of the night working out the consequences of the hypothesis.

Although the authenticity of this event has occasionally been questioned, the historian Alan Rocke has pointed to the fact that "most historians are content to accept the story at face value."[21] As explained by him, the amount of detail included suggests that they probably happened pretty much as he describes it and there are not any persuasive grounds to accuse him of deliberate historical falsification. In addition, it would appear that Kekulé related the benzene story many times to friends and family before its first publication.

The cyclic, planar nature of benzene was finally confirmed by the British crystallographer Kathleen Lonsdale (1903–1971) in 1929 via X-ray diffraction of the hexamethyl derivative.[23] Prior to this point, attempts to study the structure of either various derivatives or crystalline benzene at low temperature had failed to distinguish between the two possible cyclic structures—either a planar or a puckered, diamond-type structure. Using samples of hexamethylbenzene crystallized from benzene, Lonsdale found that the crystals gave excellent reflections, with over 100 reflections observed, from which a unit cell was determined. Further analysis of the obtained diffraction data led to the conclusion that the

> benzene ring is similar in shape and size to the six-carbon ring in graphite, the aromatic carbon atoms having a diameter of 1.42 Å. Three of the valencies of aromatic carbon are co-planar, the ring itself and all the sidechain carbon atoms lying in one plane.[23]

In addition to the separation of benzene via the distillation of coal tar as described earlier, the production of benzene via the decomposition of petroleum at high temperatures had been studied since the late 19th century, which led to various cracking methods for the production of both gasoline and benzene.[24] Further advances in the 1940s occurred after Vladimir Haensel (1914–2002), a research chemist working for the Universal Oil Products Company (UOP), developed a catalytic reforming process using platinum on alumina as a dual-functional catalyst.[25] Haensel's process was then patented and commercialized by UOP in 1949 for producing a high-octane gasoline from low-octane naphthas,[26] and this commercialized process become known as the Platforming process.[25]

In catalytic reforming processes of this type, a mixture of hydrocarbons with boiling points between 60°C and 200°C is blended with hydrogen gas and then exposed to a bifunctional catalyst at temperatures up to 875°C and pressures up to 70 atm.[26] Through this process, aliphatic hydrocarbons form rings and lose hydrogen to become aromatic hydrocarbons.[25] The aromatic products of the reaction are then separated from the reaction mixture by extraction with any one of a number of solvents, and benzene is then separated from the other aromatics by distillation. Recovery of the aromatics by these steps is commonly referred to as BTX (benzene, toluene and xylene isomers). Such catalytic reforming processes then became a primary source for commercial benzene and provided greater access to the important aromatic feedstock.

4.3 Goldfinger, Kern, and the polycondensation of 1,4-dihalobenzenes

The condensation of halobenzenes dates back to the mid-19th century, with biphenyl first prepared by Rudolph Fittig (1835–1910) in 1862 via the condensation of bromobenzene.[27] Extension of such reactions in efforts to produce larger chains then followed with oligomers of up to 7 phenyl units from 1,4-dibromobenzene reported by Busch and Weber in 1936.[28] However, the first true *polyphenyl* (polyphenylene by modern nomenclature) is generally credited to George Goldfinger in 1949,[29] and it is thus with him that we will begin our discussion of polyphenylene materials.

George Goldfinger was born in Budapest, Hungary, on December 24, 1911.[30,31] He received a D.Chem. in Paris in 1937, after which he worked as a chemist and technical advisor in Romania from 1937 to 1940.[31] He then moved to the United States, where he worked for the U.S. Forest Service from 1940 to 1941.[31] Following time as a research fellow at the Polytechnical Institute of Brooklyn from 1941 to 1943, he then worked as a research chemist for G. L. Cabot in Boston, before joining the University of Buffalo as an associate professor of chemistry in 1945.[31] At the University of Buffalo, he taught chemistry from 1945 to 1956,[30] after which he moved again, this time to the Armour Research Foundation in Chicago.[32] He later taught textile chemistry at North Carolina State University, passing away in Raleigh, North Carolina, on November 1, 1984.[30]

It was in February of 1949, while at the University of Buffalo, that Goldfinger reported his initial efforts carried out during 1948 to prepare and study polyphenylene.[29] As shown in Figure 4.8, a solution of *para*-dichlorobenzene in dioxane, toluene, or xylene was treated with 2.4 equivalents of sodium metal and then heated at reflux for 24–48 hours. The excess sodium was then reacted with a mixture of water and alcohol, after which sequential Soxhlet extractions with first water and then alcohol were used to remove inorganic species. Further extraction with benzene, toluene, or xylene gave a dark brown solution, with an equal amount of insoluble material remaining. The product could be precipitated from the dark brown solution by addition of methanol to give a tan powder, or the solvent could be simply evaporated to give a brown, glassy material.

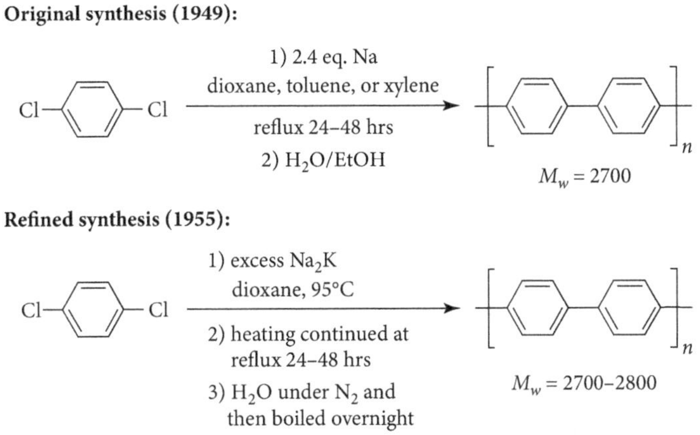

Original synthesis (1949):

Cl—⟨benzene⟩—Cl

1) 2.4 eq. Na
dioxane, toluene, or xylene
reflux 24–48 hrs
2) H_2O/EtOH

$[⟨benzene⟩—⟨benzene⟩]_n$

$M_w = 2700$

Refined synthesis (1955):

Cl—⟨benzene⟩—Cl

1) excess Na_2K
dioxane, 95°C
2) heating continued at
reflux 24–48 hrs
3) H_2O under N_2 and
then boiled overnight

$[⟨benzene⟩—⟨benzene⟩]_n$

$M_w = 2700$–2800

Figure 4.8 Goldfinger's preparation of polyphenylene via the Wurtz–Fittig reaction.

The product was believed to be a polymer consisting of a mixture of phenyl and quinoid linked via the *para* positions. The presence of quinoid content was stated to be supported by the position of the material's absorption maximum, although this value was not reported. Cryoscopic molecular weight determinations gave values of 2700 ± 100,[29] which would correspond to about 35 repeat units. Chlorine analysis of the product suggested a similar degree of polymerization of approximately 34 repeat units. Based on the polymer's relatively high viscosity, the polymer was believed to be rigid and planar. A second study coauthored with his graduate student Gerald Alonzo Edwards (1921–2005) followed in 1951,[33] which found that the polymer viscosity increased with an applied magnetic field.

A refined preparation of polyphenyl was then reported by Edwards and Goldfinger in 1955.[34] As before, the polymer was prepared from *para*-dichlorobenzene via the Wurtz–Fittig reaction. For these refined methods, however, this reagent was rigorously purified to give a highly crystalline material. The dichlorobenzene was then added to dioxane and heated to 95°C with stirring. The heating was then discontinued, and the liquid alloy Na_2K was added dropwise via a dropping funnel. The resulting reaction caused the mixture to reflux and induced the formation of a brown solid. Once the reaction had slowed to the point that reflux was no longer maintained, the reaction was again heated at reflux for an additional 24–48 hours. Nitrogen was passed through the reaction flask, and the excess alkali metal reacted with water, after which the mixture was heated to boiling overnight. A dark brown putty-like substance was then collected, washed with water, and dried overnight in a vacuum oven at 85°C to give a crude solid in a yield of 80%.

The benzene-soluble portion was then isolated by boiling the crude product in benzene for about 24 hours, followed by Soxhlet extraction for 48 hours. About twice the volume of methanol was then added to the combined benzene extracts, resulting in a "mud-colored" suspension, from which a tan-brown powder was collected by filtration. The collected powder was washed with a benzene-methanol solution, followed by pure methanol, and then dried to give a benzene-soluble product in circa 25% yield.[34]

This material was a tan-brown powder that did not melt at temperatures up to 550°C. It was soluble in pyridine, chloroform, aromatic hydrocarbons, and molten camphor but was insoluble in aliphatic hydrocarbons and more polar organic solvents. As with the initial 1949 report,[29] cryoscopic molecular weight determinations gave values of 2700–2800, with an error of ±100–500, depending on sample concentration.[34] Chlorine analysis again suggested a degree of polymerization of circa 34 repeat units. Absorption spectroscopy revealed an absorption maximum at around 3000 Å (300 nm), with a tailing out to 4900 Å (490 nm). As before, it was concluded that this absorption suggested a predominantly quinoid structure. Although X-ray diffraction resulted in diffuse rings characteristic of an amorphous solid, some evidence of orientation was indicated by the birefringence in slowly dried samples. A final, extremely brief letter then detailed the electrophoresis of the polymer solutions, with polymer solutions exhibiting increased conductivity in comparison with the solvent solutions alone.[32]

Shortly after Goldfinger's initial report in 1949, related work on the methyl-functionalized analog of polyphenylene via Ullmann coupling was reported by Werner Kern (1906–1985) in 1951.[35] Kern was born on February 9, 1906, in Tiengen, Germany, very close to the Swiss border.[36] He studied chemistry, physics, and mathematics at Heidelberg and Freiburg, before beginning his doctoral studies on polyoxymethylene under Hermann Staudinger (1881–1965) in 1928. He obtained his Ph.D. in 1930, after which he remained at the University of Freiburg as Staudinger's assistant for education and research.

Kern finished his habilitation on poly(acrylic acid) in 1937, after which he became *dozent* for organic and colloidal chemistry at the University of Freiburg.[36] He was then offered the opportunity to establish a plastics institute at the University of Frankfurt am Main in 1938 but was unsuccessful.

Instead, he became a chemist of the polymer department of IG Farbenindustrie at Farbwerke Hoechst. With the exception of a short interruption due to military duty, Kern stayed at IG Farben until the end of the Second World War. Following the war, he became professor at the newly founded Johannes Gutenberg University of Mainz in 1946. He then became the director of the Institute of Organic Chemistry at the University of Frankfurt am Main in 1952, after which he was made a full professor of organic chemistry in 1954.

Kern had a sustainable influence on the development of macromolecular science in Germany.[36] His contributions were recognized by the awarding of the Hermann Staudinger Prize of Gesellschaft Deutscher Chemiker in 1971, and the Carl Dietrich Harries Medal of Deutsche Kautschuk-Gesellschaft in 1977. He passed away on January 15, 1985.

It was while still at the University of Mainz that Kern turned to the investigation of polyphenylenes. His 1951 paper begins with a thorough review of previous research on the topic, most of which consisted of various *para*-linked oligomeric species up through the hexamer.[35] The previous 1949 paper of Goldfinger[29] was also discussed, although Kern doubted the assignment of *para*-linked species for the soluble, high–molecular weight products, considering that the tetramer and hexamer were known to exhibit low solubility.[35] Rather, Kern believed that Goldfinger's products were most likely branched to account for their solubility.

In order to produce larger soluble, linear polyphenylenes, Kern proposed the introduction of pendant methyl groups into *para*-polyphenylenes.[35] To accomplish this, he condensed 4,4'-diiodo-3,3'-dimethyldiphenyl with copper powder via Ullmann coupling (Figure 4.9). The reaction was carried out in α-methylnaphthalene at 240°C, which produced yellow to gray products, of which about 88% was soluble in benzene, toluene, THF, xylene, or chloroform. The use of benzene produced somewhat cloudy, brown solutions at room temperature, but clear solutions could be obtained after briefly warming to circa 40°C. The reaction products were then precipitated from benzene through the addition of polar solvents such as methanol.

Various fractions were then analyzed by various methods in order to evaluate the molecular weights of the products.[35] This included viscosity, cryoscopic, diffusion, sedimentation, ultracentrifugal, and osmotic measurements. Based on viscosity, molecular weights as high as 32,000 were determined, which Kern again attributed to either branched products or aggregates of smaller species. Ultracentrifugal and osmotic measurements suggested even higher molecular weights (up to 270,000), but this seems unlikely based on our modern understanding of these materials and likely corresponds to aggregates as proposed by Kern.[35]

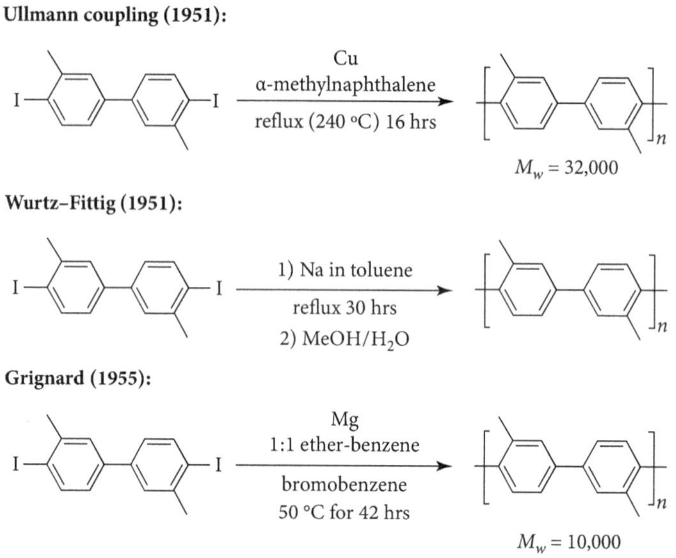

Ullmann coupling (1951):

$M_w = 32,000$

Wurtz–Fittig (1951):

Grignard (1955):

$M_w = 10,000$

Figure 4.9 Kern's preparation of polymethylphenylene via various condensation reactions.

In order to allow a more direct comparison with previous work, the condensation was then repeated using sodium in toluene via the Wurtz–Fittig reaction.[35] This resulted in products that melted at lower temperatures in comparison with those produced via Ullmann coupling, leading to the conclusion that the two coupling methods resulted in the production of significantly different polymer products. Kern proposed that this difference was most likely increased branching under Wurtz–Fittig conditions. He ultimately concluded that neither product represented linear, rigid-rod structures and that both condensation methods did not allow for suitable selectivity for the desired *para*-coupling.

This initial paper was then followed with a second in 1955 that focused on the preparation of polymethylphenylenes via the condensation of Grignard species.[37] Here, the hope was that condensation at lower temperatures could result in better selection for the linear, *para*-substituted polymer products. While no reaction was observed between 4,4′-diiodo-3,3′-dimethyldiphenyl and magnesium under standard conditions, the addition of bromobenzene resulted in the successful production of polymethylphenylene products (Figure 4.9). Analysis of these products suggested molecular weights up to 10,000. The rest of the paper focused on study of the Grignard reaction between magnesium and 4,4′-diiodo-3,3′-dimethyldiphenyl, as well as the reaction of the

Grignard products with various metal salts. In the end, however, Kern concluded that the Grignard reaction was still unsuitable for the production of linear polyphenylenes. Unfortunately, beyond melting points and efforts to determine their molecular weights, Kern never characterized any of the products produced via these various coupling methods.

4.4 Kovacic and the oxidative polymerization of benzene

Although the early investigation of polyphenylene focused on the polycondensation of dihalobenzenes, it was only natural that efforts to apply the oxidative polymerization methods previously applied to polyaniline and polypyrrole soon followed. Such efforts were first reported by Peter Kovacic at the Case Institute of Technology in 1960.[38] Kovacic was born on August 1, 1921,[39,40] in Wylandville, Pennsylvania, and grew up outside of Pittsburgh. After completing high school, he attended Hanover College in Indiana, where he majored in chemistry and mathematics, with a minor in physics. He graduated with a BA in 1943, after which he attended the University of Illinois at Urbana-Champaign, completing a Ph.D. in organic chemistry in 1946.[39] After a year of postdoctoral work at the Massachusetts Institute of Technology, he became an instructor at Columbia University in 1947. He only remained there for one year, however, in order to become a research chemist with DuPont, where he carried out polymer research until 1955.[39] He then returned to academics and accepted a position as assistant professor at Case Institute of Technology (now Case Western Reserve University), in Cleveland, Ohio. He then moved to the University of Wisconsin-Milwaukee in 1968, where he remained until his retirement in 1987.[39,40] However, he continued to teach and carry out research as a visiting professor at various universities and colleges in the United States, Australia, New Zealand, and Canada. Beginning in 1997, he then served as adjunct professor of chemistry at San Diego State University, working three days a week until he retired a second time in 2019. He died peacefully in his sleep on March 11, 2022.[39,40]

In his initial efforts, Kovacic treated benzene with anhydrous $FeCl_3$, which when heated at reflux for two hours yielded small amounts of an insoluble black material.[38] The amount of the material could be enhanced by either prolonged heating (up to nine hours) or the addition of various Bronsted acids, with small amounts of water or nitroethane giving the best results. The isolated product was characterized as a black, finely divided solid that was highly insoluble in organic solvents and was found to be difficult to ignite. The black solid was found to contain carbon, hydrogen, and chlorine, and X-ray diffraction results indicated that the material was not highly crystalline, but also not completely amorphous, as a few broad bands were observed.

Interestingly, Kovacic did not view the reaction as an oxidative process but, rather, a cationic polymerization initiated by protonation of benzene. Furthermore, his proposed structure was not a linear polymer but, rather, a polynuclear hydrocarbon resembling carbon black. However, he did admit that the formation of polyphenyl was also a possibility:[38] "According to the general reaction scheme advanced, it is reasonable to consider the possibility of polyphenyl formation."

This was then followed by a second paper in 1962 that focused on polyphenylene as the primary product.[41] Here, a mixture of benzene, $AlCl_3$, $CuCl_2$, and a minimal amount of water was heated at 36°C–37°C to give a brown solid after 30 minutes. The assignment of a *para*-polyphenylene structure was supported by the C/H atomic ratio obtained from elemental analyses, its thermal stability and insolubility, as well as characterization by IR spectroscopy and X-ray diffraction. In addition, pyrolysis under vacuum yielded lower oligomers including biphenyl, terphenyl, quaterphenyl, and quinquephenyl.

The mechanism of polymerization was now described as an oxidative cationic polymerization that involved initial protonation, coupling, and oxidative rearomatization. The previous material generated via $FeCl_3$ was now believed to undergo a similar process, although the stronger oxidant was viewed to result in cross-linking between polyphenylene chains. Additional oxidative cross-linking thus resulted in darker materials characteristic of a fused polynuclear hydrocarbon.[41]

A more substantial report of the polymerization of benzene using the $CuCl_2/AlCl_3$/water mixtures then followed in 1963.[42] By this point, stoichiometric amounts of $CuCl_2$ were specified, combined with catalytic amounts of $AlCl_3$ and water, to yield polyphenylene in yields of circa 60%. Again, the polyphenylene structure was supported by elemental analyses that gave an average C/H atomic ratio of 1.49, which was in excellent agreement with the theoretical C/H atomic ratio of 1.5 for *para*-polyphenylene. Furthermore, IR analysis showed a principal absorption band at 805–807 cm^{-1}, consistent with *para* substitution. Characterization by X-ray suggested that the polymer was quite crystalline, with d-spacing values of 4.53, 4.00, and 3.20 Å, in decreasing order of intensity. The d-spacing of 4.53 Å was viewed to correspond to the length of a phenyl unit, and based on the high intensity of this d-spacing, it was postulated that the backbone consisted of nearly coplanar, *para*-substituted rings. Lastly, the ultraviolet spectrum revealed a band at 259 nm,[42] fairly consistent with the previous results of Goldfinger,[34] although of somewhat higher energy.

Two additional 1963 papers then reported the polymerization of benzene with either $MoCl_5$[42] or $FeCl_3$,[44] respectively. As with the previous reports, small amounts of water were added as a potential cocatalyst, and the proposed mechanism was identical to that given for the copper-based polymerizations. Also in line with the previous reports, Kovacic based his assignment of a linear polyphenylene structure on elemental analyses, IR spectroscopy, X-ray diffraction, and thermal decomposition studies. Elemental analyses gave average C/H atomic ratios of 1.45–1.57 (for $MoCl_5$)[43] and 1.51 (for $FeCl_3$),[44] both in good agreement with the theoretical ratio of 1.5 for *para*-polyphenylene, while IR analysis of both samples revealed a band at 807 cm^{-1} that was consistent with *para* substitution.

Characterization by X-ray diffraction revealed d-spacings very similar to that previously reported for the copper-based polymerization products, and which were also in good agreement with *para*-quaterphenyl.[43,44] In both cases, the most intense d-spacing of 4.48 Å agreed well with the 4.5 Å value reported as the length of a phenyl unit. Based on this and the relative simplicity of the diffraction pattern, it was again believed that the backbone consisted primarily of a *para*-substituted structure.

Finally, Kovacic published a review of polyphenylenes in 1969.[45] While this focused heavily on his work with oxidative polymerization, it also included the previous work of both Goldfinger and Kern. Interestingly, he does mention that polyphenylene had also been prepared from the electrochemical oxidation of benzene, but that the details of that process had yet to be reported.

4.5 Jozefowicz and Buvet and the oxidative polymerization of dilithium benzene species

A related synthetic approach to polyphenylenes was also reported by Marcel Jozefowicz (b. 1934) and Rene Buvet (1930–1992) at the École Supérieure de Physique et de Chimie Industrielles de la ville de Paris (ESPCI Paris). Jozefowicz and Buvet are probably better known for their work on polyaniline, as previously discussed in Chapter 2 (see Section 2.9). However, that work on polyaniline was preceded by a brief study of polyphenylene in 1961–1962.[46,47] Rather than the direct oxidation of benzene, their approach involved the generation of the dilithium derivative of benzene by reacting 1,4-dibromobenzene with butyllithium (Figure 4.10). The analogous dilithium derivative of biphenyl was also prepared in the same way from 4,4′-dibromobiphenyl. These dilithium species were then treated with an oxidizing agent to produce the desired polyphenylenes.[46]

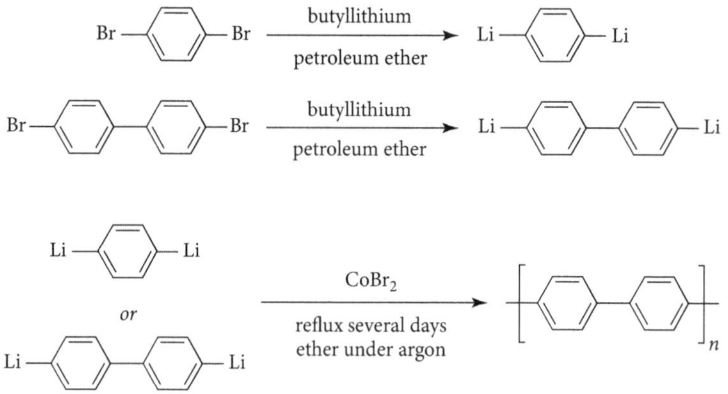

Figure 4.10 Jozefowicz and Buvet's preparation of polyphenylene via oxidation of dilithium species.

Due to the more facile oxidation of the dianions relative to the parent benzene, milder oxidizing agents could be applied here in comparison with the previous agents employed by Kovacic. Bubbling dry air into ether solutions of the dilithium benzene species generated predominately hydroquinone, while the biphenyl analog resulted in a mixture of hydroxylated derivatives of diphenyl and quaterphenyl.[46] More successful results were obtained by treating the dilithium species with an excess of $CoBr_2$ in ether. After heating at reflux for several days, water was added, followed by dilute HCl. The ether was then removed by distillation, and the insoluble polyphenylene products isolated by filtration. The solids were washed extensively with dilute HCl and water, after which they were dried and weighed to give yields of 80%.[46]

Attempts to fractionate the solid products were then made by sequential Soxhlet extractions, first with alcohol for 200 hours, followed by ether for 400 hours. Soluble fractions were then recrystallized, and all products characterized via IR and UV spectroscopy. The insoluble fraction made up 60% of the initially isolated solids, which was concluded to consist of phenylene chains comprising on average 7–8 benzene rings. The soluble products were determined to consist of biphenyl, *para*-terphenyl, and *para*-quaterphenyl. Although the insoluble fraction contained species higher than eight rings, Jozefowicz and Buvet never suggested a possible upper limit for chain lengths and referred to all products as oligomers.[46]

Of particular interest was the fact that Jozefowicz and Buvet were the first to exhibit an interest in polyphenylenes as potential semiconducting materials.[46] As such, pressed pellets were prepared by compression at 100 kg/cm^2, after which the pellets were placed between mercury electrodes at 25°C in order to measure their conductivity. The room temperature conductivities were thus determined to be between 10^{-10} and 10^{-12} mho cm^{-1} (i.e., S cm^{-1}). It was concluded that the low conductivities determined were due to the low degree of polymerization of the products. Of course, it should also be noted that these were neutral samples, and the importance of oxidation (i.e, doping) on the conductivity of such materials was not yet recognized.

This initial 1961 paper was then followed by a second the following year that focused on the spectroscopic characterization of the oligomers.[47] For characterization via IR, it was found that the number average molecular mass (M_n) of a polymolecular mixture could be estimated by examining the variation in the relative intensity of the bands of 5.13 and 5.32 µm (1949 and 1880 cm^{-1}), relative to that at 5.23 µm (1912 cm^{-1}). From such an analysis, a minimum M_n of 550 was determined, which would correspond to a minimum length of 7 benzene units.

UV spectroscopic studies in the solid state were carried out on circa 2-mm-thick pellets formed by compression of polyphenylene diluted in magnesia (MgO) at 100 kg/cm^2. For comparison, the solid-state spectra of diphenyl, *para*-terphenyl, and *para*-quaterphenyl were also recorded, which exhibited a single band with peaks located at 2500, 3000, and 3200 Å (250, 300, and 320 nm), respectively. Interestingly, the peaks for the ter- and quarter-phenyl samples in the solid state were redshifted by about 30 nm relative to their known solution spectra.[47] This may be the earliest known observation of the well-known redshift observed for conjugated polymers from solution to the solid state. In comparison, the solid-state spectra for the polymeric samples exhibited a single broad band with a lower-energy maximum at circa 3600 Å (360 nm) and a tailing out to 5000 Å (500 nm).[47] These results were in good agreement with the solution data previously reported by Goldfinger for soluble polyphenylenes thought to correspond to circa 34 repeat units.[34]

4.6 Pohl, Labes, Beck, and further electronic measurements of polyphenylene

Following the initial 1961 report by Jozefowicz and Buvet on the conductivity of polyphenylene, others soon followed with additional measurements of its electronic properties. The first of these reports came in 1962 by Herbert A. Pohl (1916–1986) and coworkers at Princeton University.[48] Born in Lisbon, Portugal, of American parents, Pohl completed undergraduate and graduate studies at Duke University, earning a Ph.D. in physical chemistry in 1939.[49,50] He then spent a year at John Hopkins Medical School as a National Defense Research Fellow, after which he served as a senior chemist at the U.S. Naval Research

Lab during World War II. Pohl then spent 12 years as a senior research associate at E. I. Du Pont de Nemours and Company, before moving to faculty positions at Princeton University and Brooklyn Polytechnic. He then moved to Oklahoma State University in 1964, where he remained until his retirement in 1981.[49,50] He was best known for his research in dielectrophoresis and polymer physics. Pohl fatally collapsed in MIT's Francis Bitter Magnet Laboratory,[49] where he was serving as a visiting professor.[49,50] He died on June 21, 1986.[49]

As part of a study of semiconducting properties across a series of organic polymers, Pohl and coworkers investigated samples of polyphenylene prepared via Goldfinger's methods.[48] In addition to separating the products into benzene-soluble and -insoluble fractions, the benzene solution portion was further fractionated by its solubility in various benzene:ethanol mixtures. Overall, however, there was very little difference in the measured resistivities at 25°C, with values ranging from 6.2×10^{10} to 7.0×10^{11} Ω cm. As these values would correspond to conductivities of 1.4×10^{-12} to 1.6×10^{-11} S cm^{-1}, the results agree well with the previous report by Jozefowicz and Buvet.[46] While Pohl stated that the determined conductivities were "somewhat unsatisfying," he gave no further commentary other than stating that further studies on the interrelation of extent of conjugation and electronic properties were actively underway.[48]

A second 1962 report then followed by Mortimer M. Labes (1930–2023) and coworkers at the Franklin Institute Laboratories in Philadelphia.[51] Labes is perhaps more well known for his contributions to the study of poly(sulfur nitride) in the early 1970s, as well as its place in the history of polyactylene, as outlined in Chapter 5 (see Section 5.9). However, his interest in the conductivity of organic polymers predates that work by over a decade, with the paper discussed here focusing on polyphenylene and polyvinylene (i.e., polyacetylene via the dehydrochlorination of polyvinylidene chloride).[51]

The polyphenylene samples analyzed by Labes were prepared by the previous methods of Goldfinger, with the benzene-soluble fraction isolated for electronic measurements.[51] Powdered samples were dried at low pressure (2 mm Hg) in a vacuum desiccator for 16 h, after which the sample was transferred under air to a simple annular stainless-steel die for compression. The walls of the die then functioned as one electrode, with a central wire acting as the other. The measured resistivity was found to be independent on the compression pressure between 600 and 10,000 atm. By this method, the resistivity of polyphenylene was determined to be 10^{11} Ω cm at 25°C. As this would correspond to a conductivity of 10^{-11} S cm^{-1}, this value was in excellent agreement with both of the previous studies.[46,48]

As there was concern with the effects of impurities and water on the material conductivity, the samples were studied again after more extensive drying methods. Here, the samples were dried at 1 μm Hg and 110°C for 16 hours in an Aberhalden drying apparatus. The polyphenylene powders were then transferred to the same annular die in a glove box, compressed immediately upon removal from the glove box, and measured to give a room temperature resistivity greater than 10^{15} Ω cm. The study was concluded with the comment that even in such an extremely dry state, these materials should still be regarded as quite impure.

Of particular interest is a brief study Labes reported the following year, in which polyphenylene was treated with iodine to produce what were described as polymer charge-transfer complexes.[52] This was accomplished by either treating the solid polymer with iodine vapor or the preparation of the charge-transfer complex in solution. As shown in Figure 4.11, the resistivity was found to decrease with increasing iodine content, dropping to a minimum of circa 2.5 $\times 10^4$ Ω cm, corresponding to a conductivity of circa 4×10^{-5} S cm^{-1}. It needs to be highlighted that this represents the earliest known example of the doping of a neutral conjugated polymer with halogen vapor. In fact, this is essentially the same process later used for the initial doping of polyacetylene films in 1977 (see Section 5.10).

A more in-depth study was then reported in 1964 by Fritz Beck (1931–2009) of Badische Anilin- & Soda-Fabrik A. G. (BASF).[53] Born in Stuttgart, Germany on March 5, 1931, Beck began the study of chemistry at the University of Stuttgart in 1951. After completing his undergraduate studies in 1956, he continued graduate studies at Stuttgart under Heinz Gerischer (1919–1994). Completing a thesis on electrochemistry, he received his doctorate in 1960, after which he began working at BASF. In 1978, he then moved from BASF to the University of Duisburg as the chair of electrochemistry.[54] He died in Essen on February 15, 2009.

The polyphenylene samples studied by Beck were prepared via the methods of Kovacic using $AlCl_3/CuCl_2$, although modified to some extent by his BASF colleague Herbert Naarmann, particularly in terms of the reaction temperature applied.[53] Unfortunately, while Kovacic had used temperatures of 36°C–37°C,[41,42] the specific temperatures used for the various samples reported in the BASF study are never given, only stating that the reaction temperature had been varied over a range of several hundred degrees Celsius. Of course, as the boiling point of benzene is only circa 80°C, temperatures above this would have required high-pressure conditions. It should be noted that some may recognize the name of Naarmann, who is generally recognized for his contributions to very-high-conductivity polyacetylenes as detailed in Chapter 5 (see Section 5.13).

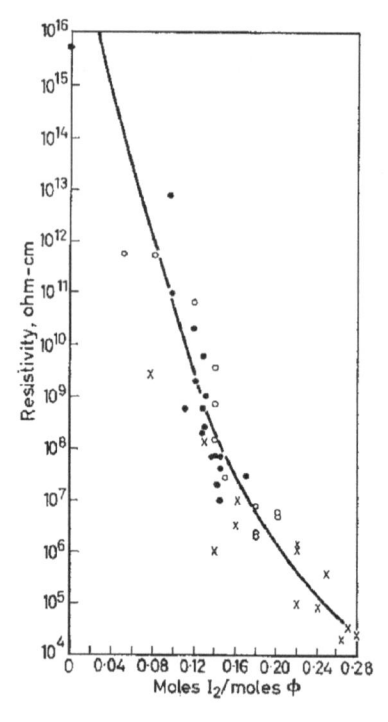

Figure 4.11 Resistivity of polyphenylene-iodine (o: solid-vapor preparation under room air; ●: solid vapor preparation under dry nitrogen; x: solution preparation under room air).

Used with permission of Walter de Gruyter and Company, from M. M. Labes. Conductivity in polymeric solids. *Pure Appl. Chem.* **1966**, *12*, 275–285; permission conveyed through Copyright Clearance Center, Inc.

For the various polyphenylenes studied, those produced at lower temperatures were characterized as yellow to dark brown in color, with room temperature conductivities of 10^{-15} to 1.7×10^{-12} S cm^{-1}.[53] The reported colors for the materials here were found to darken with reaction temperature, with the intermediate colors consistent with the brown solids previously reported by Kovacic.[41] In the same way, while the conductivities were slightly higher here, they generally agreed with the previous studies discussed earlier.[46,51] In contrast, polymer samples prepared at higher temperatures were found to be black, with corresponding conductivities of 1.1×10^{-11} to 1.7×10^{-4} S cm^{-1}.[53] Again, while specific temperatures were not given, the conductivities were found to increase with the reaction temperature applied. However, as Kovacic had believed that darker-colored samples were the result of cross-linking,[41] it is unclear the exact nature of these black samples. In addition to the effect of reaction temperature, it was also found that the oxygen content of the polymers also had a large influence on the conductivity, with lower oxygen content resulting in higher conductivities.

While all of the conductivities discussed so far were measured at room temperature, the BASF study also investigated the temperature dependence of the conductivity.[53] In all cases, it was found that the specific conductivity increased with temperature, consistent with the temperature dependence expected of a semiconducting material. Lastly, polymers of both naphthalene and anthracene were also studied, although the ring size of the monomer was found to have little influence on the conductivity.

4.7 Stille and new routes to polyphenylenes

A completely new approach to the synthesis of polyphenylenes via Diels-Alder reactions was then introduced by John K. Stille (1930–1989, Figure 4.12) in 1966.[55] Of course, Stille is familiar to most chemists due to the cross-coupling method that bears his name,[56] with Stille cross-coupling later becoming one of the most commonly applied methods for the production of modern conjugated polymers. However, in addition to his recognized contributions to organometallic chemistry, Stille also made important contributions to polymer chemistry, including work on the conjugated materials polyphenylene and polyacetylene (see Section 5.5).

Stille was born on May 8, 1930, in Tucson, Arizona.[57,58] Initially educated at the University of Arizona, he received BA and MA degrees in chemistry in 1952 and 1953,[58] respectively, before serving in the Navy during the Korean War.[57] He then attended the University of Illinois,[57,58] where he completed his Ph.D. under Carl Marvel (1894–1988) in 1957.[57] That same year, he joined the faculty of the University of Iowa,[57,58] becoming professor in 1965.[58] In 1977, he moved to Colorado State University, where he was appointed university distinguished professor in 1986.[58] During his career, his contributions in organometallic and polymer chemistry were recognized with numerous awards, including the American Chemical Society (ACS) Award in Polymer Chemistry (1982), the 1989 ACS Colorado Section Award, and the Arthur C. Cope Scholars Award (1989).[58] Unfortunately, Stille died prematurely on July 19, 1989, in the crash of United Airlines Flight 232 in Sioux City, Iowa.[57]

Stille first introduced the production of Diels–Alder polymers in 1966 via the reaction of bistetracyclones with diethynyl compounds (Figure 4.13).[55] While his initial report utilized various linking units, thus resulting in polymers containing only linked oligophenylene segments, this illustrated the viability of the approach for producing organic polymeric materials in this way. This was

Figure 4.12 John K. Stille (1930–1989).
Reproduced courtesy of University Historic Photograph Collection, Colorado State University.

then followed by a second paper in 1967 that extended the approach to the production of fully conjugated polyphenylenes.[59]

As outlined in Figure 4.13, the bistetracyclone and p-diethynylbenzene were placed in a 20 ml polymerization tube with 10 ml of toluene. After the sample was degassed by several freeze–thaw cycles, the tub was sealed and then heated in a Paar pressure reactor at 200°C for 24 hr. After the reactor had cooled to room temperature, the contents were poured into 200 ml of acetone and the precipitate filtered and dried at 60°C for 24 hr.[59] The resulting polymers were found to be appreciably soluble in common organic solvents, and clear films could be cast from chloroform. The polymer composition was confirmed with elemental analysis and molecular weights determined by osmometry to give M_n values of 19,000–36,000. The stoichiometry required to achieve the highest molecular weight required a 1% excess of bistetracyclone, which suggested possible impurities in the reactant or a potential competing side reaction. Finally, running the reaction at higher temperatures gave lower molecular weights, as well as a small amount of black, insoluble material.

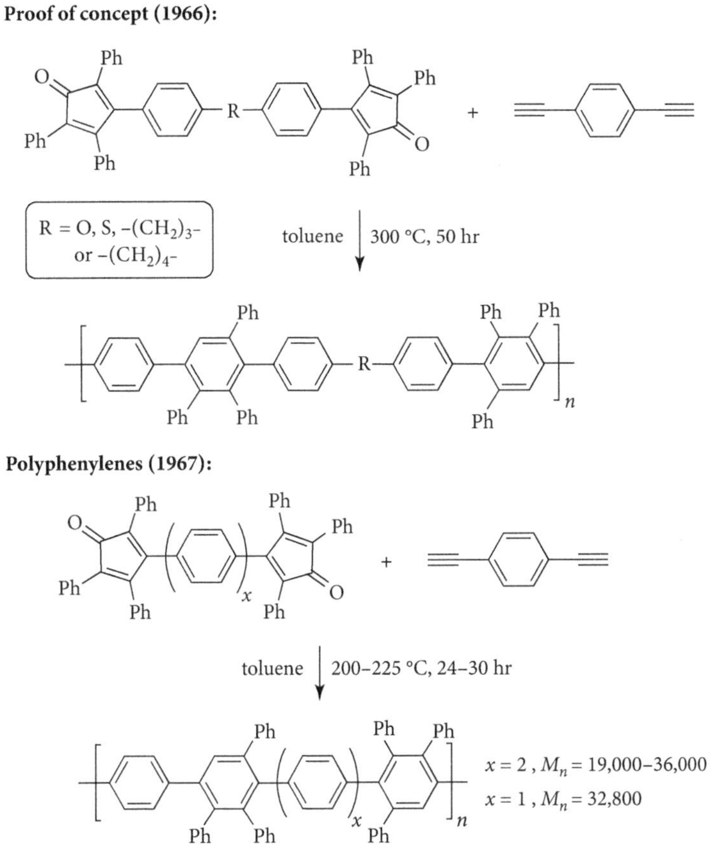

Proof of concept (1966):

R = O, S, –(CH$_2$)$_3$– or –(CH$_2$)$_4$–

toluene | 300 °C, 50 hr

Polyphenylenes (1967):

toluene | 200–225 °C, 24–30 hr

$x = 2$, $M_n = 19{,}000–36{,}000$

$x = 1$, $M_n = 32{,}800$

Figure 4.13 Stille's preparation of phenylated polyphenylenes via Diels–Alder reactions.

A third paper then followed in 1968, but it focused completely on polymers containing oligophenylene units connected via alkyl linkers.[60] Stille then returned to the production of conjugated polyphenylenes in 1969 with the report of a new approach utilizing the reaction of bispyrones and *para*-diethynylbenzene.[61] Unfortunately, this approach was not regiospecific and resulted in both *meta* and *para* linkages. In addition, the yields of polymer samples were considerably lower than the previous methods and resulted in lower-molecular-weight materials. A second 1969 paper then reported the preparation of polyphenylenes via the original bistetracyclone reactions, but with a bistetracyclone containing only a single phenylene unit, rather than the previously applied biphenylene analog.[62] As a result, this now produced polymers with a simple backbone comprised of alternating phenylene and triphenylphenylene units (Figure 4.13). The resulting molecular weights were found to be 32,800 and gave properties consistent with the previous polyphenylenes via bistetracyclone reactions.

A remaining question concerned the optical properties of Stille's polyphenylenes in comparison with those produced via other methods. The phenylated polyphenylenes were light yellow and showed a broad maximum near 340 nm, which was in good agreement with the limiting value predicted by data from oligophenylenes. Still, these polymers might be expected to exhibit reduced conjugation, as the pendant phenyl groups would introduce steric hindrance along the polymer backbone. In comparison, polyphenylenes via other synthetic methods had reported absorption maxima as high as 395 nm. In order to better address this question, Stille reported routes to unfunctionalized polyphenylenes in 1971.[63]

To accomplish this, Stille returned to the production of polyphenylene via bispyrones. While the previous phenylated bispyrones had resulted in a mixture of *para* and *meta* linkages, model reactions of 5-phenylpyrone with phenylactylene exclusively gave the *p*-terphenyl product (Figure 4.14). As such, *para*-phenylenebispyrone and *para*-diethynylbenzene were combined with trichlorobenzene in a sealed tube at 250°C, during which a quantitative yield of polyphenylene precipitated from the reaction solvent.

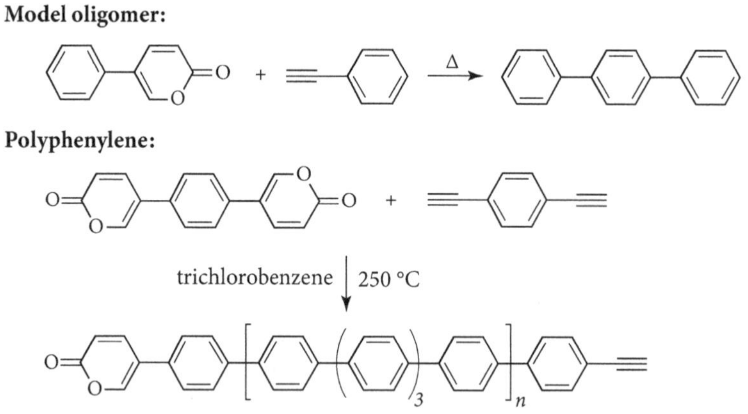

Model oligomer:

Polyphenylene:

Figure 4.14 Stille's preparation of unfunctionalized polyphenylene via Diels–Alder reactions.

The yellow product was insoluble in all solvents, but analysis by IR spectroscopy revealed only a weak carbonyl absorption from potential end groups and a strong C-H out-of-plane deformation characteristic of the *p*-phenylene moiety at 800 cm^{-1}. The product was also found to be crystalline, showing d spacings characteristic of *para*-quaterphenyl and polyphenylenes produced via other methods. Finally, an absorption maximum of 340 nm was determined, in good agreement with the theoretical value, thus suggesting that previous samples of darker color and lower-energy absorptions most likely contained impurities or additional cross-linking, and so on. This initial communication was then followed by a full paper in 1972, providing more experimental details for the production of the unfunctionalized parent polyphenylene.[64]

4.8 Yamamoto and polyphenylene via Kumada cross-coupling

A final advancement in the preparation of polyphenylene then came in 1977, with the introduction of catalytic cross-coupling of 1,4-dihalobenzenes by Takakazu Yamamoto (1944–2014) at the Tokyo Institute of Technology (Tokyo Tech).[65] Although Yamamoto is more well known for his contributions to polythiophenes (see Section 6.4), his independent research began here, with his first polyphenylene paper published essentially a year after he was made associate professor at Toyko Tech. A more detailed discussion of Yamamoto's biography can be found in Section 6.4.

The catalytic cross-coupling of Grignard reagents and aryl halides had been introduced in the early 1970s, which Yamamoto believed could provide a route to sterically regular polymers in high yield under mild conditions. This catalytic cross-coupling is now known as Kumada (or Kumada–Corriu) cross-coupling, as its application to the formation of C–C bonds was independently discovered by Makoto Kumada (1920–2007)[66] and Robert Corriu (1934–2016) in 1972. Yamamoto's 1977 report, however, was the first time this metal-catalyzed process had been applied to the polymerization of poly(arylene)s.[65]

As outlined in Figure 4.15, *para*-dibromo- or *para*-dichloro-benzene was first reacted with slightly more than one equivalent of magnesium in dry tetrahydrofuran (THF), thus resulting in the formation of an intermediate species containing both halo- and halomagnesium functionalities. The addition of a catalytic amount of a Ni species then allowed growth of the polymer via polycondensation with elimination of $MgBr_2$ or $MgCl_2$. Polymerization was nearly complete after an hour, although heating was continued for a minimum of three additional hours to ensure maximum yield of polymer products. The polymer was precipitated by addition of the reaction mixture to ethanol, after which it was collected by filtration and dried under vacuum to give a light-yellow material.

The polymer products exhibited poor solubility, although about 20% could be fractionated via Soxhlet extraction with hot toluene for 50 hours. Analysis via IR spectroscopy showed an out-of-plane vibration at 803 cm^{-1}, consistent with

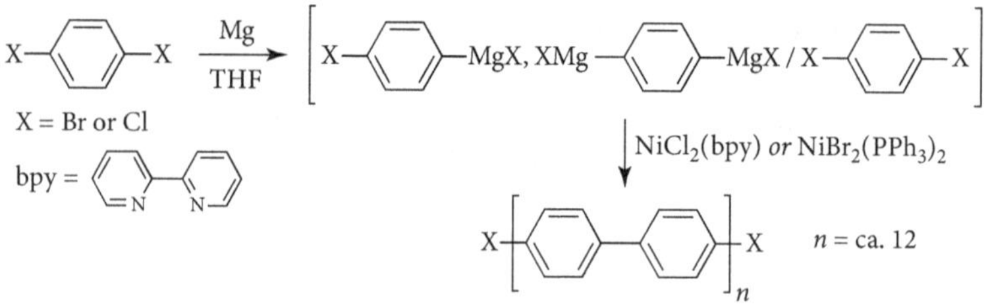

Figure 4.15 Polyphenylene via Kumada cross-coupling.

a linear polyphenylene. From elemental analysis data, the degree of polymerization (DP) was estimated to be 24, based on the assumption that both of the polymer end groups were halogens.[65]

A more detailed study then followed in 1978,[67] which expanded the dihalobenzene reactants to include *para*-iodobenzene in addition to the previous bromo and chloro analogs. In addition, the scope of catalytic species was also increased, although the complex $NiCl_2(bpy)$ (where bpy = 2,2'-bipyridine) still gave the best overall performance. More careful analysis of the products by IR spectroscopy revealed that in addition to the primary peak assignable to the *para*-phenylene units (ca. 805 cm^{-1}), weak C-X bands of the terminal *para*-halophenyl groups were also observed at 1085 (C-Cl), 1065 (C-Br), or 1055 (C-I) cm^{-1}. Furthermore, two out-of-plane vibrations due to terminal phenyl groups were also observed at 760 and 690 cm^{-1}, thus suggesting that the polyphenylene products contained both *para*-halophenyl and phenyl end groups. As such, it was concluded that the samples were of higher molecular weight than previously estimated from end group analysis. Finally, X-ray diffraction showed strong and sharp peaks, indicating the polymers to be of high crystallinity.

4.9 Campbell, Goldberg, and the introduction of poly(phenylene vinylene)

At least partially due to the limited conjugation exhibited by polyphenylenes, as well as the lack of solubility for unfunctionalized samples, interest developed in the analog poly(phenylene vinylene) (PPV, Figure 4.2), which can be viewed as an alternating copolymer of phenylene and acetylene.[68,69] In the same way, it can also be viewed as a polymeric isomer of poly(phenylacetylene). Although research on PPV became of significant interest in the 1970s, its origin can be traced back to 1960 and pioneering work on the application of the Wittig reaction to polymers by Tod W. Campbell (1919–1968) at DuPont.[70]

Although little is known about Campbell, he received his Ph.D. at the University of California, Los Angeles (UCLA) in 1946, under the supervision of James D. McCullough (1905–1985) and William G. Young (1902–1980).[71] He then went on to work with Max T. Rogers (1917–1985) at Michigan State University,[72] before returning again to UCLA. He moved again in 1950 to the Western Regional Research Laboratory of the U.S. Department of Agriculture,[73] before ultimately starting work at DuPont in 1957.[74]

Campbell and his coauthor Richard N. McDonald had previously demonstrated the production of distytylbenzenes via Wittig reactions in 1959[75] and felt that this could be extended to the production of polymeric analogs. As shown in Figure 4.16, this involved a bis-triphenylphosphonium intermediate that was prepared from *para*-xylylene dichloride.[75] The resulting intermediate was then combined with terephthaldehyde in ethanol, after which the mixture was treated with lithium ethylate to give a bright lemon-yellow-colored precipitate. The solid was collected by filtration, washed with ethanol, and dried to give the desired polymer in quantitative yield.[70]

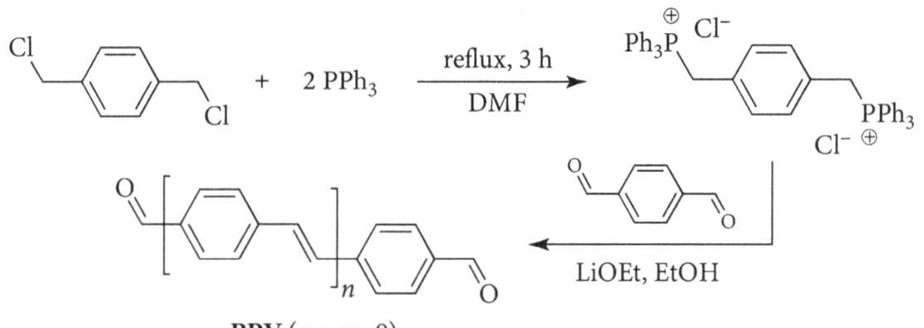

Figure 4.16 Poly(phenylene vinylene) via the Wittig reaction.

The polymer product, which the authors named poly-*para*-xylylidene, was found to be fairly insoluble, although some lower-molecular-weight material was removed by extraction with benzene.[70] The remaining insoluble material was then analyzed by IR spectroscopy and elemental analysis. The IR spectrum indicated the presence of aldehyde end groups, and elemental analysis was consistent with the structure given in Figure 4.16. Using the ratio of oxygen to C/H, the number average molecular weight was determined to be 1200, which would correspond to roughly nine repeat units ($n = 9$). Unfortunately, no further characterization was reported.

An alternate synthetic approach was then introduced four years later by Eugene P. Goldberg (1928–2019) in 1964.[76] Goldberg began his studies at the University of Miami, receiving a BS in chemistry in 1950.[77] Continuing his studies at Ohio University, he received an MS in organic chemistry in 1951, before moving to Brown University to work under Harold R. Nace (1921–1993).

After receiving his Ph.D. from Brown in 1953, he worked as a research chemist for General Electric. He then moved to the Borg-Warner Research Center in Des Plaines, Illinois, before being hired by Xerox in 1966 as director for the Chemistry Research Laboratory.

Goldberg left the industry in 1975 to become a professor of materials science and engineering and adjunct professor of chemistry at the University of Florida.[77] An additional faculty appointment in the Department of Pharmacology and Therapeutics was granted in 1985, followed by another appointment in Biomedical Engineering in 2001. In addition to being named a fellow of the American Institute for Medical and Biological Engineering and a Biomaterials Science and Engineering fellow, Goldberg was awarded the Florida Scientist of the Year Award in 1987, the Brown University Distinguished Alumnus Award in 1995, and the Society of Biomaterials' Clemson Award for Applied Research in 1998. He passed away on Friday, February 15, 2019, at his home in Florida.

It was while at the Borg-Warner Research Center that Goldberg and his coauthors reported on the synthesis and characterization of poly-*para*-xylylidene (i.e., modern PPV).[76] As with the previous work of Campbell, their synthesis started from *para*-xylylene dichloride but did not utilize the Wittig reaction. Rather, as outlined in Figure 4.17, the initial dichloride was deprotonated via treatment with sodium amide, after which the resulting anion polymerized to give an intermediate polymer containing saturated chloroethane units. Further reaction with sodium amide resulted in elimination of HCl to generate vinyl linkages between the phenyl units. The resulting product was produced in 90% yield as a bright yellow solid that fluoresced a green-yellow color under ultraviolet illumination.

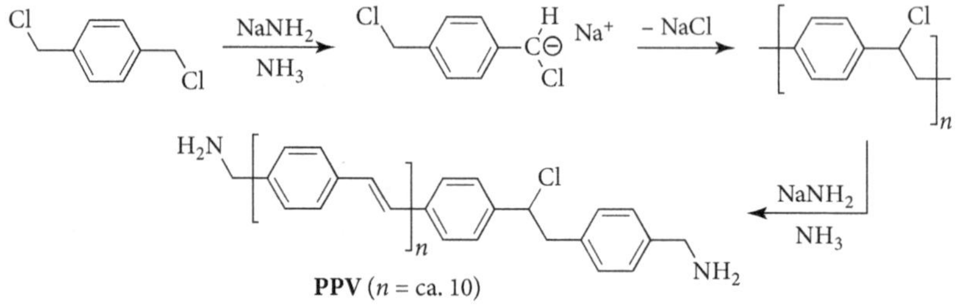

PPV (n = ca. 10)

Figure 4.17 Poly(phenylene vinylene) via dehydrochlorination.

This product was found to be infusible and insoluble in all the boiling solvents tried. Analysis revealed main distinguishing features at 825 cm^{-1}, consistent with *para*-disubstituted benzene, and at 960 cm^{-1}, consistent with *trans*-ethylene.[76] Preliminary electrical measurements also detected a low level of photoconduction. Elemental analysis found both nitrogen content, believed to be due to amine end groups, and chlorine content due to incomplete dehydrochlorination. Based on the total analysis results, the structure illustrated in Figure 4.18 was proposed, consisting of 10 conjugated repeat units.

In order to compare the product with that previously reported by Campbell,[70] Goldberg repeated the previously reported synthesis via the Wittig reaction. The two products were found to be very similar, although analysis of the Wittig material by IR spectroscopy revealed an additional absorption at 1690 cm^{-1}, presumably due to aldehyde end groups.[76] X-ray analysis also found the Wittig material to be more ordered, while the material produced via sodium amide was more amorphous.

4.10 Kossmehl and more detailed studies of PPV

Following these two initial reports, further studies on PPV did not appear until 1969 with a series of studies by Gerhard Kossmehl and coworkers at the Free University of Berlin. Kossmehl (or Koßmehl) was educated at the Fritz-Haber-Institut der Max-Planck-Gesellschaft, earning his Diplom in 1959 and a doctorate in 1964.[78] He then worked with Georg Manecke (1916–1990) at the Free University of Berlin,[79,80] completing his habilitation in 1970. The same year he then became professor in organic and macromolecular chemistry.

It was at the end of his habilitation that he began giving talks on PPV and its thiophene analog poly(thienylene vinylene) (PTV). This included the IUPAC (International Union of Pure and Applied Chemistry) Symposium on Macromolecular Chemistry, held in Budapest at the end of August 1969,[80] and the General Meeting of the Gesellschaft Deutscher Chemiker,[79] held in late September of 1969. Initial back-to-back papers were then published in 1970. The first paper focused on PTV,[81] followed by a second paper on PPV, PTV, and a mixed analog containing alternating phenylene vinylene and thienylene vinylene units, P(PV-TV).[80] The study of the thiophene analogs is particularly noteworthy, as these represent the earliest known thiophene-based conjugated polymers, predating even the parent polythiophene by a decade (see Chapter 6).

All three of these materials were prepared via the Wittig methods initially introduced by Campbell and utilized bis-triphenylphosphonium regents as outlined in Figures 4.16 and 4.18.[80,81] The resulting polymers were then extracted with ethanol, with the yield of the ethanol-soluble fraction amounting to circa 45%. The solubility of the materials in dimethylformamide (DMF) was so low that attempts to determine the molecular weights by vapor pressure osmometry were not possible. Although Campbell had not previously commented on the conformation of the vinylene units produced via the Wittig reaction, the

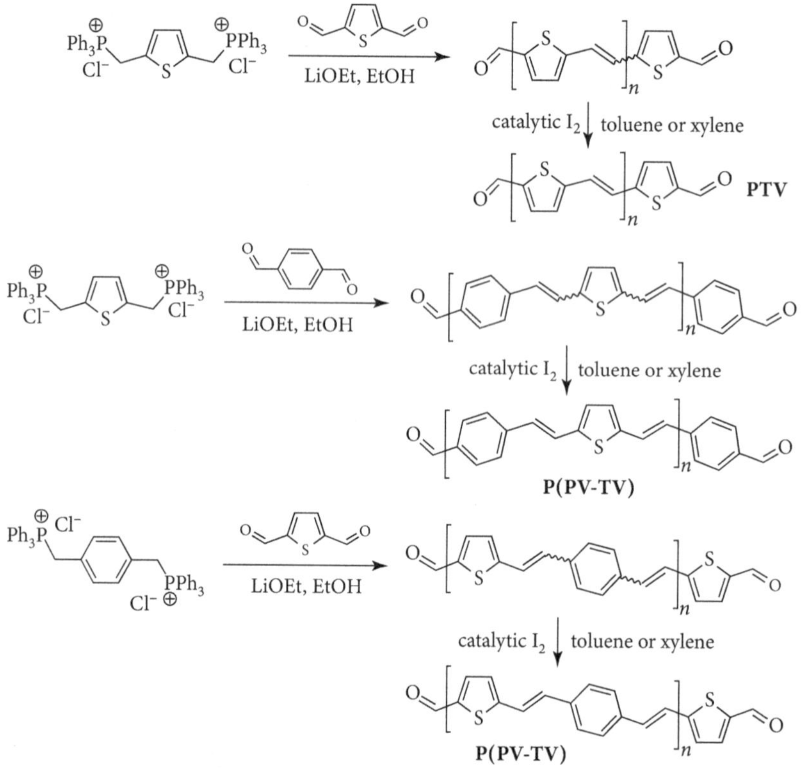

Figure 4.18 Poly(thienylene vinylene) (PTV) and its mixed phenylene-thienylene analog via the Wittig reaction.

materials here were determined to contain a statistical mixture of *cis*- and *trans*-vinylene linkages. Boiling the materials with a catalytic amount of iodine in either toluene or xylene, however, converted any *cis*-vinylene units into their *trans* isomers (Figure 4.18). This rearrangement was supported by the increase in the *trans*-vinylene IR band, a red-shifted visible absorption, and an increase in electrical conductivity. Still, it could not be conclusively determined whether the polymers consisted completely of *trans*-vinylene units and the presence of some *cis*-vinylene defects were possible.

The soluble fraction of the PPV sample exhibited an absorption maximum at 380 nm, similar to the highest values reported for polyphenylene. However, as these samples most likely consisted of smaller oligomers, the insoluble polymer fractions would be expected to exhibit lower energy values, as oligomer studies had shown the absorption energy decreased with oligomer chain length.[80] In comparison, the thiophene analog PTV exhibited a significantly redshifted

absorption maxima of 500–520 nm,[81] with the alternating copolymer of phenylene vinylene and thienylene vinylene units giving an absorption maxima
that fell between these two extremes at 432 nm.[80] As such, it was concluded
that the absorption shifted to longer wavelengths with increasing thiophene
content.

The DC dark conductivity of the various polymers was measured over a temperature range of 25°C–150°C, with a room temperature conductivity of 2.0 ×
10^{-14} Ω^{-1} cm^{-1} determined for the all-*trans* PPV. In comparison, the thiophene
analogs exhibited significantly higher conductivities, with values of 3.8 × 10^{-8}
Ω^{-1} cm^{-1} determined for PTV and 3.5 × 10^{-7} Ω^{-1} cm^{-1} determined for the alternating copolymer P(PV-TV).[80] The AC resistance was also measured for the
two thiophenes using a Wheatstone bridge to give values that agree very well
with the previous DC conductivities.

This initial work was then followed with another paper in 1971 that
reported the functionalized PPV analog poly(2,5-dimethoxyphenylene
vinylene) (DM-PPV).[82] This report was particularly noteworthy, as such
dialkoxy-functionalized PPVs later became the standard for high-performing
PPV materials.[83] The materials studied here were prepared using Wittig methods in the same way as their previous materials (Figure 4.19), which had been
previously reported by Manecke in 1969.[84] The following year, Hans-Heinrich
Hörhold and J. Offerman had also reported the material prepared via the
dehydrochlorination methods of Goldberg (as discussed in the next section).[85]

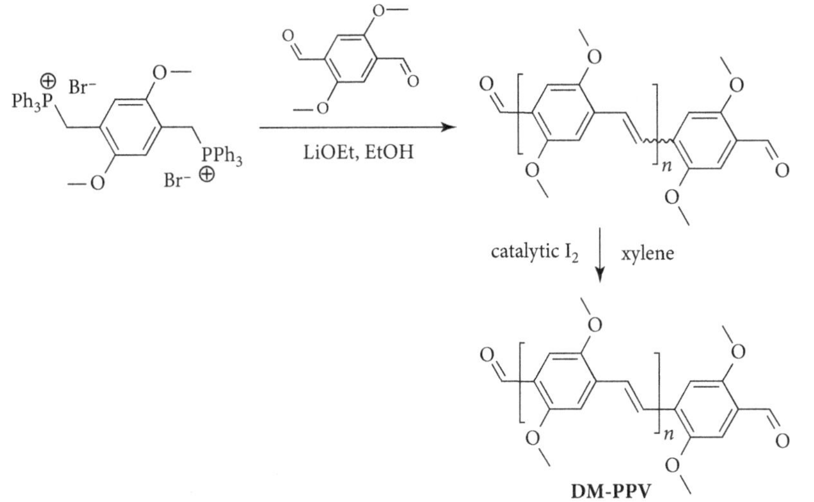

Figure 4.19 Poly(2,5-dimethoxyphenylene vinylene) (DM-PPV) via the Wittig
reaction.

The electrical conductivity of DM-PPV was measured at a pressure of 1500 kp cm^{-2} over a range from room temperature to about 120°C.[82] Under these conditions, the all-*trans* polymer resulting from the iodine treatment gave a conductivity of 3.0×10^{-10} Ω^{-1} cm^{-1}. In comparison, the initially prepared polymer prior to iodine treatment gave a lower conductivity of 10^{-15} Ω^{-1} cm^{-1}. The material prepared via the Wittig reaction also gave a better conductivity than the material reported by Hörhold, prepared by dehydrochlorination methods. While the value reported by Hörhold was essentially identical to that of the pretreated polymer here, it was reported at a temperature of 400 K, rather than at room temperature.[85] As the conductivity of semiconductors increases with temperature, the corresponding room temperature value would be even lower.

A more extensive study was then published in 1973 that significantly increased the family of poly(arylene vinylene)s, particularly in terms of various mixed copolymeric materials (Figure 4.20).[86] Thus, in addition to the previously reported PPV, poly(thienylene vinylene) (PTV), and poly(2,5-dimethoxyphenylene vinylene) (DM-PPV), this also included copolymeric species comprised of various combinations of phenyl, dimethoxyphenyl, nitrophenyl, thiophene, and 1,3,4-thiadiazole. These materials were all made via Wittig methods, followed by isomerization to the all-*trans* forms via treatment with catalytic iodine in xylenes. The exception to this was the thiadiazole-containing materials, which were produced via Knoevenagel condensation with an appropriate catalyst. Materials via Knoevenagel condensation were found to generate all-*trans* polymers.

The conductivity of all polymers was measured by DC methods, as summarized in Table 4.1.[86] In all cases, the materials studied were determined to be classic semiconductors in that the conductivity increased with the applied temperature. Furthermore, in all cases, this was determined to be electronic conductivity, with the existence of ionic conductivity excluded. The conductivities were also shown to increase when measured under light and were enhanced with increasing pressure.

In general, it was found that increasing the electron-donating strength resulted in higher conductivities.[86] For example, the addition of methoxy groups to PPV resulted in an increase in conductivity from 2.0×10^{-14} to 3.0×10^{-10} Ω^{-1} cm^{-1}, with the mixed copolymer P(PV-DMPV) exhibiting an intermediate conductivity of 2.5×10^{-11} Ω^{-1} cm^{-1}. In the same way, PTV, the thiophene analog of PPV, gives an even higher conductivity of 3.8×10^{-8} Ω^{-1} cm^{-1}. Interestingly, the copolymer P(PV-TV) with only half the thiophene content actually gives a higher value of 3.5×10^{-7} Ω^{-1} cm^{-1}. This break in the trend was attributed to the more crystalline nature of P(PV-TV) in comparison with PTV. The greatest overall conductivity was then found by further increasing the electron-donating nature of P(PV-TV). Thus, replacing the phenylene units

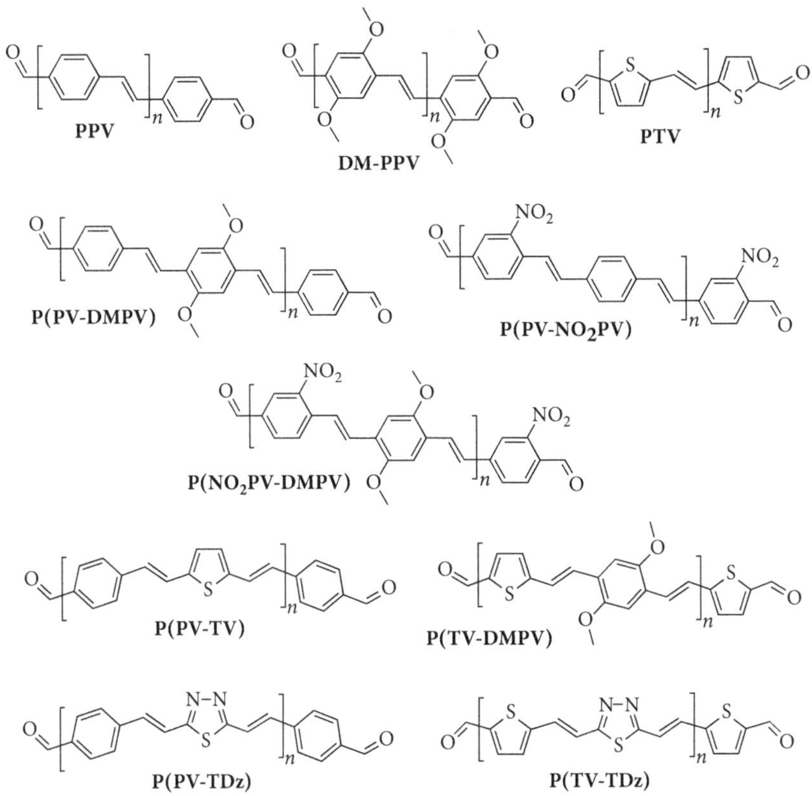

Figure 4.20 Kossmehl's growing family of PPV derivatives and analogs.

with dimethoxyphenyl, the copolymer P(TV-DMPV) exhibited a conductivity of 9.0×10^{-5} Ω^{-1} cm^{-1}.

The addition of electron-withdrawing units had the opposite effect and resulted in significantly diminished conductivity (Table 4.1).[86] Of such materials, the two thiadiazole-containing polymers gave the lowest conductivities of 10^{-19}–10^{-20} Ω^{-1} cm^{-1}. The effects of the electron-rich and electron-poor units in the copolymer P(NO$_2$PV-DMPV) seem to cancel one another to give a conductivity nearly identical to the parent PPV.

Kossmehl continued to publish on poly(arylene vinylene)s throughout the 1970s, although the bulk of this work tended to move further away from the parent PPV in order to focus on new heterocyclic analogs. His attention also shifted to other material classes, including polythiophenes in the 1980s (see Chapter 6). While Kossmehl and Manecke dominated the study of PPV and its derivatives throughout the 1970s, significant studies were also reported by Hans-Heinrich Hörhold at the Friedrich-Schiller-Universität Jena.

Table 4.1. Conductivities of various homopolymeric and copolymeric poly(arylene vinylene)s

Material[a]	$\sigma\,(\Omega^{-1}\,cm^{-1})^{b}$	Material[a]	$\sigma\,(\Omega^{-1}\,cm^{-1})^{b}$
PPV	2.0×10^{-14}	P(NO$_2$PV-DMPV)	5.6×10^{-14}
DM-PPV	3.0×10^{-10}	P(PV-TV)	3.5×10^{-7}
PTV	3.8×10^{-8}	P(TV-DMPV)	9.0×10^{-5}
P(PV-DMPV)	2.5×10^{-11}	P(PV-TDz)	10^{-20}
P(PV-NO$_2$PV)	8×10^{-16}	P(TV-TDz)	2×10^{-19}

[a]Data collected from Reference 86.
[b]At 298 K.

4.11 Hörhold and additional studies of PPV

About the same time that Kossmehl and Manecke began their initial reports on PPV, Hörhold and his colleagues at Jena also began publishing their own PPV studies, beginning with a 1970 report of a series of oligomers consisting of up to seven phenylene vinylene repeat units.[87] This was then followed with a more extensive report on PPV, as well as its dimethyl and dimethoxy derivatives, as back-to-back papers.[85] This more extensive study began with a comparison of various synthetic methods for the production of the parent PPV. This included the Wittig methods introduced by Campbell and used extensively by Kossmehl, as well as the dehydrochlorination methods introduced by Goldberg. In addition, Hörhold investigated several other methods including a variant of the Wittig methods using an asymmetric monomer capable of coupling with itself, the coupling of a bis(phosphonic acid diethyl ester) with terephthaldehyde in the presence of potassium *tert*-butoxide, and a variant of Goldberg's dehydrochlorination methods that replaces the NaNH$_2$ in liquid ammonia with NaH in DMF (Figure 4.21).[85] In all cases, the products were insoluble, and thus it was acknowledged that nothing could really be said about the corresponding molecular weights of the materials.

For both of the Wittig methods, the elemental analysis of the products agreed well with the idealized structure, with circa 0.1% phosphorous content. The alternate diethyl phosphonate method, however, results in significantly higher phosphorous and oxygen content. The PPV prepared by Goldberg's dehydrohalogenation methods exhibited both nitrogen and chlorine content in agreement with previous reports, with the chlorine indicative of incomplete dehydrochlorination. In contrast, the modified methods utilizing NaH removed

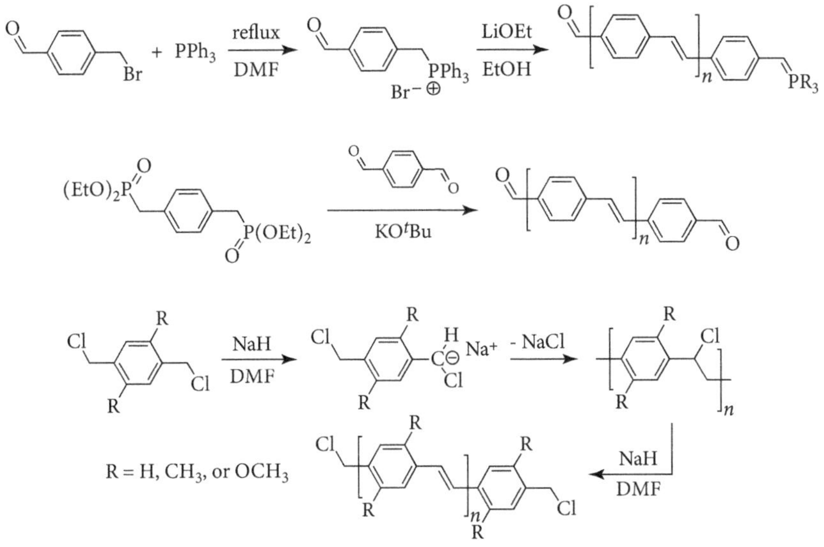

Figure 4.21 Variations on the synthesis of PPV as introduced by Hörhold.

the issue of nitrogen content and produced materials with considerably less chlorine content.

Analysis by IR spectroscopy was then used to probe the relative *trans*-vinylene content, with the traditional Goldberg dehydrohalogenation methods giving the lowest such content. The modified NaH methods were found to be improved, with no evidence of *cis*-vinylene content. The two polymers made via the Wittig methods, however, exhibited significant *cis*-vinylene content, in agreement with the reports of Kossmehl discussed earlier. It should be noted that Hörhold did not seem to be yet aware of the *cis*-to-*trans* isomerism methods introduced by Kossmehl to overcome this limitation.[80,81]

The various materials were also characterized in the solid state via diffuse reflection spectroscopy and electrical conductivity. In terms of the optical spectra, it was determined that the onset was more important than the corresponding maxima, with the bulk of the polymers all exhibiting an absorption onset of 2.45 eV. The two exceptions were the PPV produced via the alternate diethyl phosphonate method, which gave an onset of 2.35 eV, and Hörhold's modified NaH method, which gave an onset of 2.32 eV. In terms of conductivity, however, the material via the alternate diethyl phosphonate method gave the lowest conductivity (2×10^{-15} Ω^{-1} cm^{-1} at 400 K), while the modified NaH method and the asymmetric Wittig methods gave identical values (2×10^{-14} Ω^{-1} cm^{-1} at 400 K). The traditional Wittig methods gave a slightly higher conductivity (3×10^{-14} Ω^{-1}

cm^{-1} at 400 K). It should be pointed out that Kossmehl reported nearly identical room temperature values for his postisomerized PPV at 2×10^{-14} Ω^{-1} cm^{-1} (see Table 4.1).[80,86] As the conductivities for all of these materials increase with temperature, the extrapolated room temperature conductivity for Hörhold's PPV via NaH would be below 10^{-17} Ω^{-1} cm^{-1}.[85]

The modified NaH methods were then applied to the synthesis of the functionalized PPV derivatives poly(2,5-dimethylphenylene vinylene) and poly(2,5-dimethoxyphenylene vinylene) (DM-PPV),[85] with DM-PPV also produced by Kossmehl as described earlier. Characterization by diffuse reflection spectroscopy exhibited a noticeable bathochromic shift caused by the addition of the methoxy groups, resulting in an absorption onset of 2.10 eV. In comparison, the addition of simple methyl groups had no observable effect on the absorbance. In terms of the measured conductivity at 400 K, however, both derivatives exhibited reduced values of 1×10^{-15} and 2×10^{-17} Ω^{-1} cm^{-1} for the dimethoxy and dimethyl derivatives, respectively. As previously discussed, Kossmehl reported essentially the same value at room temperature for DM-PPV. Again, extrapolation of the data reported by Hörhold would correspond to a room temperature conductivity below 10^{-17} Ω^{-1} cm^{-1}.[85]

This was then followed by a 1972 review of PPV and its derivatives.[88] Although this primarily presented the various materials discussed earlier, a few new derivatives were also included that had been previously covered in German patents and less common journals. Unlike previous derivatives, these included either cyano or phenyl functionalities on the vinylene linkage (Figure 4.22). Although only limited data was provided for these less common derivatives, the addition of one cyano group to the vinylene linkage resulted in a slight redshift in absorption onset. In comparison, the addition of a single phenyl group caused the onset to shift to significantly higher energy. He later returned to the phenyl derivatives in 1977,[89,90] reporting conductivities of 10^{-14} and 4×10^{-17} Ω^{-1} cm^{-1} at 400 K for the phenyl and diphenyl species, respectively.

4.12 Conclusions

Although the modern conjugated polymer literature rarely includes much discussion of polyphenylene, it is hoped that the discussion in this chapter illustrates its important place in the history of these materials. Not only does polyphenylene represent the first conjugated polymer not to be synthesized primarily by oxidative polymerization, but it was also the first to introduce the metal-catalyzed polycondensation methods that dominate the synthesis of modern conjugated materials. Furthermore, the report by Labes in 1963 represents the very first example of a neutral conjugated polymer undergoing doping to

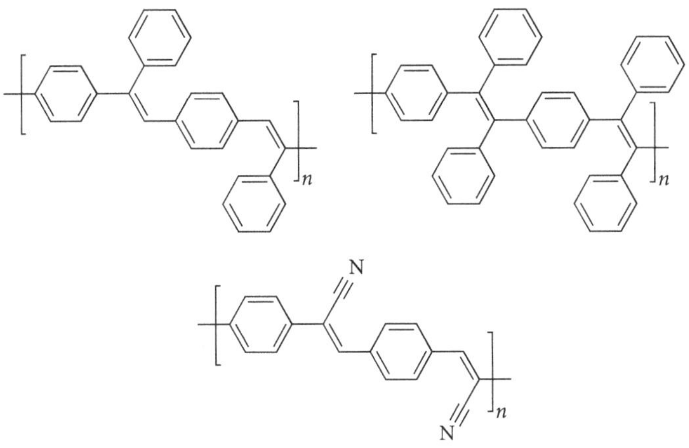

Figure 4.22 PPV derivatives with functionalized vinylene linkages.

enhance its conductivity.[52] While the conductivities achieved were considerably lower than that later obtained for polyacetylene, the relative enhancement in conductivity was nearly identical (i.e., ca. six orders of magnitude), with the primary difference being that the neutral polyphenylene was much less conductive than neutral polyacetylene.

Of course, polyphenylene also paved the way for the later materials PPV and polyfluorene, both of which became more significant materials in the field. While polyfluorene was not introduced until the mid-1980s, PPV also has a fairly long history dating back to 1960, only five years after the initial report of polyacetylene. As with polyphenylene, the synthesis of PPV did not rely on oxidative polymerization and introduced a number of new synthetic techniques for the production of conjugated polymers. Furthermore, PPV was the first conjugated polymer to introduce side chain functionalization as early as 1969. Such side chain functionalization later allowed the production of soluble materials and became a mainstay of conjugated polymer design. Although work in PPV slowed down in the later 1970s, PPV and its analogs made a resurgence in the late 1980s due to the development of new synthetic methods utilizing soluble nonconjugated polymeric intermediates.[68] Such new synthetic methods and the introduction of solubilizing side chains then led to the application of PPV derivates in some of the earliest electronic device applications, including organic solar cells and organic light-emitting diodes.

References and Notes

1. Noren, G. K.; Stille, J. K. Polyphenylenes. *J. Polym. Sci. D Macromol. Rev.* **1971**, *5*, 385–430.

2. Schlüter, A.-D. Synthesis of poly(*para*-phenylenes). In *Handbook of Conducting Polymers*, 2nd ed. Skotheim, T. A.; Elsenbaumer, R. L.; Reynolds, J. R.; Eds.; Marcel Dekker: New York, 1988, pp. 209–224.

3. Leising, G.; Tasch, S.; Graupner, W. Fundamentals of electroluminescence in paraphenylene-type conjugated polymers and oligomers. In *Handbook of Conducting Polymers*, 2nd ed. Skotheim, T. A.; Elsenbaumer, R. L.; Reynolds, J. R.; Eds.; Marcel Dekker: New York, 1988, pp. 847–880.

4. Moratti, S. C. The chemistry and uses of polyphenylenevinylenes. In *Handbook of Conducting Polymers*, 2nd ed. Skotheim, T. A.; Elsenbaumer, R. L.; Reynolds, J. R.; Eds.; Marcel Dekker: New York, 1988, pp. 343–361.

5. Toshima, N.; Hara, S. Direct synthesis of conducting polymers from simple monomers. *Prog. Polym. Sci.* **1995**, *20*, 155–183.

6. Wnek, G. W.; Capistran, J.; Chien, J. C. W.; Dickinson, L. C.; Gable, R.; Gooding, R.; Gourley, K.; Karasz. F. E.; Lillya, C. P.; Yao, K.-D. Studies in conducting polymers. In *Conductive Polymers*; Seymour, R. B., Ed.; Polymer Science and Technology, Vol. 15; Plenum Press: New York, 1981, pp. 183–208.

7. Leclerc, M. Polyfluorenes: Twenty years of progress. *J. Polym. Sci. Part A. Polym. Chem.* **2001**, 39, 2867–2873.

8. Faraday, M. On new compounds of carbon and hydrogen, and on certain other products obtained during the decomposition of oil by heat. *Philos. Trans. Royal Soc. London* **1825**, 115, 440–446.

9. Inde, A. J. *The Development of Modern Chemistry*. Harper & Row: New York, 1964, p. 171.

10. Modern characterization of benzene gives a specific gravity of 0.884 at 60°F, a melting point of 5.53°C (41.95°F), and a boiling point of 80.1°C (176.2°F).

11. Mitscherlich, E. Über das Benzol und die Säuren der Oel- und Talgarten. *Ann. Pharm.* **1834**, *9*, 39–48.

12. Inde, A. J. *The Development of Modern Chemistry*. Harper & Row: New York, 1964, p. 455.

13. Roscoe, H. E.; Schorlemmer, C. *A Treatise on Chemistry*, Vol. 3, Part 3. D. Appleton and Company: New York, 1897, pp. 65–66.

14. Leigh, J. On a new product obtained from coal naphtha. *Brit. Assoc. Report* 1842, 39.

15. Hofmann, A. W. Ueber eine sichere Reaction auf Benzol. *Ann. Chem. Pharm.* **1845**, 55, 200–205.

16. Cramb, J. A. Mansfield; Charles Blachford. In *Dictionary of National Biography*, Vol. 36. Lee, S., Ed.; Smith, Elder & Co.: London, 1893, p. 90.

17. Mansfield, C. B. An improvement in the manufacture and purification of spiritous substances, and oils applicable to the purposes of artificial light and various useful arts, and in the application thereof to such purposes; and in the construction of lamps and burners applicable to the combustion of such substances. British Patent 11,960, November 11, 1847.

18. Mansfield, C. B. Untersuchung des Steinkohlentheers. *Ann. Chem. Pharm.* **1849**, 69, 162–180.

19. Kekulé, F. A. Sur la constitution des substances aromatiques. *Bull. Soc. Chim. Paris* **1865**, *3*, 98–110.

20. Kekulé, F. A. Ueber die Constitution der aromatischen Verbindungen. *Z. Chem.* **1865**, *8*, 176–184.

21. Rocke, A. J. Hypothesis and experiment in the early development of Kekulé's benzene theory. *Ann. Sci.* **1985**, *42*, 355–381

22. Rocke, A. J. *Image and Reality: Kekule, Kopp, and the Scientific Imagination*. University of Chicago Press: Chicago, 2010, pp. 186–227.

23. Lonsdale, K. The structure of the benzene ring in $C_6(CH_3)_6$. *Proc. R. Soc. A* **1929**, *123*, 494–515.

24. Rittman, W. F.; Dutton, C. B.; Dean, K. W. *The manufacture of gasoline and benzene-toluene from petroleum and other hydrocarbons*, Bulletin 114. United States Bureau of Mines: Washington, DC, 1916.

25. Gembicki, S. Vladimir Haensel 1914–2002. *Biogr. Mem. Natl. Acad. Sci.* **2006**, *88*, 1–14.

26. Haensel, V. Process of Reforming a Gasoline with an Alumina-Platinum-Halogen Catalyst. U.S. Patent 2479110, August 16, 1949.

27. R. Fittig, Ueber das Monobrombenzol. *Ann. Chem. Pharm.* **1862**, *121*, 361–365.

28. Busch, M.; Weber, W. Über Kohlenstoffverkettungen bei der katalytischen Hydrierung von Alkylhalogeniden. *J. Prakt. Chem.* **1936**, *146*, 1–55.

29. Goldfinger, G. Polyphenyl. *J. Polym. Sci.* **1949**, *4*, 93–96.

30. Othmeralia. Othmer Library of Chemical History, Science History Institute. https://othmeralia.tumblr.com/post/181321028395/george-goldfinger-was-born-in-budapest-hungary/amp (accessed December 2, 2022).

31. *American Men of Science. A Biographical Directory*, 10th ed. The Jaques Cattell Press, Inc.: Tempe, AZ, 1961, p. 916.

32. Goldfinger, G.; Shulman, S. Electrophoresis of polyphenylene. *J. Polym. Sci.* **1957**, *25*, 479.

33. Edwards, G. A.; Goldfinger, G. Influence of a magnetic field on the viscosity of polyphenyl. *J. Polym. Sci.* **1951**, *6*, 125–126.

34. Edwards, G. A.; Goldfinger, G. Polyphenyl. *J. Polym. Sci.* **1955**, *16*, 589–597.

35. Claesson, S.; Gehm, R.; Kern, W. Uber methylsubstituierte Polyphenylene. *Makromol. Chem.* **1951**, *7*(1), 46–61.

36. Höcker, H. Professor Werner Kern 1906–1985. *Macromol. Chem. Phys.* **2006**, *207*, 145–147.

37. Kern, W.; Gehm, R.; Seibel, M. Uber methylsubstituierte Polyphenylene. IV. Mitt. *Makromol. Chem.* **1955**, *15*, 170–176.

38. Kovacic, P.; Wu, C. Reaction of ferric chloride with benzene. *J. Polym. Sci.* **1960**, *47*, 45–54.

39. Peter Kovacic. University of Illinois Urbana-Champaign. https://chemistry.illinois.edu/system/files/2022-04/Peter%20Kovacic%20obituary.pdf (accessed November 26, 2023).

40. Kovacic, Peter. Chemistry & Biochemistry, University of Wisconsin - Milwaukee. https://uwm.edu/chemistry/our-people/kovacic-peter (accessed November 26, 2023).

41. Kovacic, P.; Kyriakis, A. Polymerization of benzene to *p*-polyphenyl. *Tetrahedron Lett.* **1962**, 467–469.

42. Kovacic, P.; Kyriakis, A. Polymerization of benzene to *p*-polyphenyl by aluminum chloride-cupric chloride. *J. Am. Chem. Soc.* **1963**, *85*, 454–458.

43. Kovacic, P.; Lange, R. M. Polymerization of benzene to *p*-polyphenyl by molybdenum pentachloride. *J. Org. Chem.* **1963**, *28*, 968–972.

44. Kovacic, P.; Koch, F. W. Polymerization of benzene to *p*-polyphenyl by ferric chloride. *J. Org. Chem.* **1963**, *28*, 1864–1867.

45. Kovacic, P.; Koch, F. W. Poly (phenylenes). In *Encyclopedia of Polymer Science and Technology: Plastics, Resins, Rubbers, Fibers.* Mark, H. F.; Bikales, N. M., Eds.; Interscience Publishers: New York, 1969, pp. 380–389.

46. Jozefowicz, M.; Buvet, R. Préparation de poly-p-phénylènes oligomères. *C. R. Acad. Sc.* **1961**, *253*, 1801–1803.

47. Jozefowicz, M.; Buvet, R. Caractérisation spectrophotométrique des mélanges de p-oligophénylènes. *C. R. Acad. Sc.* **1962**, *254*, 284–286.

48. Pohl, H. A.; Bornmann, J. A.; Itoh, W. Studies of some semiconducting polymers. In *Organic Semiconductors*, Brophy, J. J.; Buttrey, J. W.; Eds.; The MacMillan Company: New York, 1962, pp. 142–158.

49. Westhaus, P. A.; Swamy, N. V. V. J.; Herbert A. Pohl. *Physics Today* **1987**, *40*, 90–91.

50. Westhaus, P. A.; Swamy, N. V. V. J. Tribute biography. Professor Herbert A. Pohl. *IEEE Trans. on Electrical Insulation* **1986**, *EI-21*, 682.

51. Mainthia, S. B.; Kronick, P. L.; Labes, M. M. Electrical measurements on polyvinylene and polyphenylene. *J. Chem. Phys.* **1962**, *37*, 2509–1510.

52. Originally published as Mainthia, S. B.; Kronick, P. L.; Ur, H.; Chapman, E. F.; Labes, M. M. Electronic conductivity of complexes of poly-p-phenylene. *Polymer Preprints* **1963**, *4*(1), 208–212. Data was later republished in Labes, M. M. Conductivity in polymeric Solids. *Pure Appl. Chem.* **1966**, *12*, 275–285.

53. Beck, F. Elektrische Messungen zur Charakterisierung von polymeren organischen Halbleitern. *Ber. Bunsenges.* **1964**, *68*, 558–567.

54. Kreysa, G.; Wragg, A. A. Special issue in honor of the 60th birthday of Professor Dr. Fritz Beck. *J. Appl. Electrochem.* **1991**, *21*, 847.

55. Stille, J. K.; Harris, F. W.; Rakutis, R. O.; Mukamal, H. Diels-Alder polymerizations. Polymers containing controlled aromatic segments. *J. Polym. Sci. C, Polym. Lett.* **1966**, *4*, 791–793.

56. Espinet, P.; Echavarren, A. M. The mechanisms of the Stille reaction. *Angew. Chem. Int. Ed.* **2004**, *43*, 4704–4734.

57. Lenz, R. W. In memory of John Kenneth Stille. *Macromolecules* **1990**, *23*, 2417–2418.

58. Hegedus, L. S. John K. Stille. *Organometallics* **1990**, *9*, 3007–3008.

59. Mukamal, H.; Harris, F. W.; Stille, J. K. Diels-Alder polymers. III. Polymers containing phenylated phenylene units. *J. Polym. Sci. A-1, Polym. Chem.* **1967**, *5*, 2721–2729.

60. Stille, J. K.; Rakutis, R. O.; Mukamal, H.; Harris, F. W. Diels-Alder polymerizations IV. Polymers containing short phenylene blocks connected by alkylene units. *Macromolecules* **1968**, *1*, 431–436.

61. Schilling, C. L., Jr.; Reed, J. A.; Stille, J. K. Diels-Alder polymerizations. VI. Phenylated polyphenylenes from bis-2-pyrones and p-diethynylbenzene. *Macromolecules* **1969**, *2*, 85–88.

62. Stille, John K.; Noren, G. K. Diels-Alder polymers. Polyphenylenes containing alternating phenylene and triphenylphenylene units. *J. Polym. Sci. B, Polym. Lett.* **1969**, *7*, 525–527.

63. Stille, J. K.; Gilliams, Y. Poly(*p*-phenylene). *Macromolecules* **1971**, *4*, 515–517.

64. VanKerckhoven, H. F.; Gilliams, Y. K.; Stille, J. K. Poly(p-phenylene). The reaction of 5,5'-p-phenylenebis-2-pyrone with p-diethynylbenzene. *Macromolecules* **1972**, *5*, 541–546.

65. Yamamoto, T.; Yamamoto, A. A novel type of polycondensation of polyhalogenated organic aromatic compounds producing thermostable polyphenylene type polymers promoted by nickel complexes. *Chem. Lett.* **1977**, *6*, 353–356.

66. Tamao, K.; Sumitani, K.; Kumada, M. Selective carbon-carbon bond formation by cross-coupling of Grignard reagents with organic halides. Catalysis by nickel-phosphine complexes. *J. Am. Chem. Soc.* **1972**, *94*, 4374–4376.

67. Yamamoto, T.; Hayashi, Y.; Yamamoto, A. A novel type of polycondensation utilizing transition metal-catalyzed C-C coupling. I. Preparation of thermostable polyphenylene type polymers. *Bull. Chem. Soc. Japan* **1978**, *51*, 2091–2097.

68. Wnek, G. W.; Capistran, J.; Chien, J. C. W.; Dickinson, L. C.; Gable, R.; Gooding, R.; Gourley, K.; Karasz, F. E.; Lillya, C. P.; Yao, K.-D. Studies in conducting polymers. In *Conductive Polymers*. Seymour R. B., Ed.; Polymer Science and Technology, Vol. 15, Plenum Press: New York, 1981, pp. 183–208.

69. Feast, W. J.; Tsibouklis, J.; Pouwer, K. L.; Groenendaal, L.; Meijer, E. W. Synthesis, processing and material properties of conjugated polymers. *Polymer* **1996**, *37*, 5017–5047.

70. McDonald, R. N.; Campbell, T. W. The Wittig reaction as a polymerization method. *J. Am. Chem. Soc.* **1960**, *82*, 4669–4671.

71. Campbell, T. W. Organometallic compounds: I. The synthesis and properties of some organic selenium compounds. II. The reactions of sodium allylbenzene. Ph.D. dissertation, University of California, Los Angeles, 1946.

72. Rogers, M. T.; Campbell, T. W. The electric moments of some aromatic selenium compounds. *J. Am. Chem. Soc.* **1947**, *69*, 2039–2041.

73. Campbell, T. W.; Burney, W.; Jacobs, T. L. The reaction of Grignard reagents with di-*t*-butyl peroxide. *J. Am. Chem. Soc.* **1950**, *72*, 2735–2736.

74. Campbell, T. W. 1-Hydroxy-2,4-di-t-butylphenazine. *J. Org. Chem.* **1957**, *22*, 1731–1732.

75. Campbell, T. W.; McDonald, R. N. Synthesis of hydrocarbon derivatives by the Wittig synthesis. I. Distyrylbenzenes. *J. Org. Chem.* **1959**, *24*, 1246–1251.

76. Hoeg, D. F.; Lusk, D. I.; Goldberg, E. P. Poly-p-xylylidene. *J. Polym. Sci. B, Polym. Lett.* **1964**, *2*, 697–701.

77. Department of Materials Science & Engineering, University of Florida. In Memory of Dr. Eugene P. Goldberg, 1928–2019. https://mse.ufl.edu/in-memory-of-dr-eugene-p-goldberg/ (accessed February 9, 2024).

78. Freie Universität Berlin. Gerhard Koßmehl. https://www.bcp.fu-berlin.de/en/chemie/chemie/forschung/Emeriti/OC/Kossmehl.html (accessed February 9, 2024).

79. Koßmehl, G.; Hartel, M.; Manecke, G. Preparation and electrical properties of polymers with conjugated polyene structures containing aromatic, substituted aromatic, or thiophene rings. *Angew. Chem. Inter. Ed.* **1969**, *8*, 906.

80. Koßmehl, G.; Hartel, M.; Manecke, G. Uber Polyenarylene und Polyenheteroarylene 2. Mitt.: Poly-[thienylen-(2.5)-athenamer-alt-phenylen-(p.m,o)-athenamer] und einige Oligomere. Synthese, spektrales Verhalten und elektrische Leitfahigkeit. *Makromol. Chem.* **1970**, *131*, 37–54.

81. Kossmehl, G.; Hartel, M.; Manecke, G. Über polyenarylene und polyenheteroarylene. 1. Mitt.: Poly-[thienylen-(2.5)-äthenamer] und einige seiner oligomeren. Synthesen, spektrales verhalten und elektrische leitfähigkeit. *Makromol. Chem.* **1970**, *131*, 15–36.

82. Manecke, G.; Zerpner, D.; Kossmehl, G. Über die elektrische Leitfahigkeit oligomerer und polymerer 2.5-Dimethoxyphenylen-vinylene. *Makromol. Chem.* **1971**, *147*, 35–39.

83. Heeger, A. J. Semiconducting and metallic polymers: The fourth generation of polymeric materials (Nobel Lecture). *Angew. Chem. Int. Ed.* **2001**, *40*, 2591–2611.

84. Manecke, G.; Zerpner, D. Über oligomere und polymere 2.5-dimethoxyphenylenvinylene. *Makromol. Chern.* **1969**, *129*, 183–202.

85. Hörhold, H.-H.; Opfermann, J. Poly-*p*-xylyliden. Synthesen und Beziehungen zwischen Struktur und elektrophysikalischen Eigenschaften. *Makromol. Chem.* **1970**, *131*, 105–132.

86. Hartel, M.; Kossmehl, G.; Manecke, G.; Wille, W.; Wohrle, D.; Zerpner, D. Struktur und elektrische Leitfahigkeit von polymeren organischen Halbleitern. *Angew. Makromol. Chem.* **1973**, *29*, 307–347.

87. Drefahl, G.; Kühmstedt, R.; Oswald, H.; Hörhold, H.-H. Oligomere als Modelle für Poly-*p*-xylyliden. Untersuchungen über Stilbene. XLIX. *Makromol. Chem.* **1970**, *131*, 89–103.

88. Hörhold, H.-H. Neue Polykondensationsreaktionen zur Synthese von Polymeren mit Halbleitereigenschaften. *Z. Chem.* **1972**, *12*, 41–52.

89. Hörhold, H.-H.; Gottschaldt, J.; Opfermann, J. Untersuchungen uber Poly(arylenvinylen)e. XVII. Poly(1,4-phenylen-1,2-diphenylvinylen). *J. Prakt. Chem.* **1977**, *319*, 611–621.

90. Hörhold, H.-H.; Ozegowski, J.-H.; Bergmann, R. Untersuchungen uber Poly(arylenvinylen)e. XVIII. Synthese von Poly(1,4-phenylen-1-phenylvinylen) durch dehydrochlorierung. *J. Prakt. Chem.* **1977**, *319*, 622–626.

Chapter 5
Polyacetylene

5.1 Introduction

When it comes to the history of conjugated polymers, no other material receives the focus given to polyacetylene, and it is usually portrayed as the origin of the field. However, as the previous three chapters have shown, conjugated polymers had been studied for over 100 years before the first true report of linear polyacetylene. This is not to say, however, that polyacetylene does not hold an important place in the overall history of conjugated and conducting materials.[1-7] Not only was doped polyacetylene the first example of an organic polymer with a conductivity solidly in the metallic regime, but it still holds the place as the most conductive species in the field, with reported values as high as 170,000 S cm^{-1}. In addition, polyacetylene was the first conjugated polymer to be produced via polymerization methods more commonly used for traditional nonconjugated polymers. Thus, rather than oxidative polymerization or aryl-aryl cross-coupling, polyacetylene was generated via Zigler–Natta polymerization (Figure 5.1).[8,9]

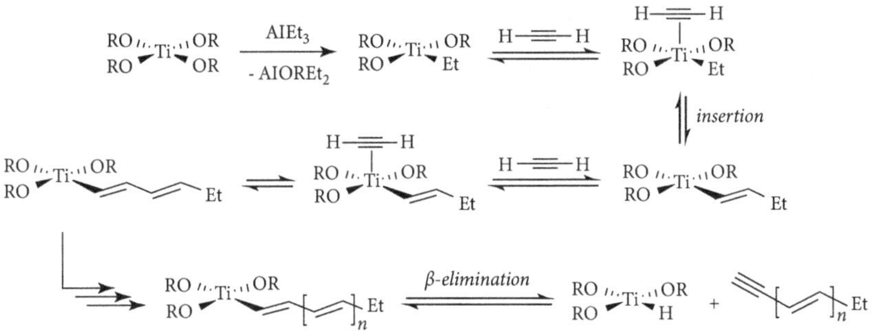

Figure 5.1 Simple mechanism for the Ziegler–Natta polymerization of acetylene.

As a consequence of its method of synthesis, polyacetylene was one of the earliest conjugated polymers that could be produced directly in its neutral undoped form, rather than an oxidized form that required dedoping to isolate the neutral material. This means that while the initial doping of previous materials such as polyaniline and polypyrrole was primarily studied via reducing the extent of doping, neutral polymers such as polyphenylene and polyacetylene focused on

The Origins and Early History of Conjugated Organic Polymers. Seth C. Rasmussen, Oxford University Press.
© Oxford University Press (2025). DOI: 10.1093/9780197638194.003.0005

the effect of increased doping, a process that was, for many, conceptually more straightforward. In addition, polyacetylene was the first conjugated polymer that could be doped via both oxidation and reduction of the neutral material. It was then the focused study of such positive doping processes in polyacetylene that led to much of our current understanding of doping of conjugated polymers in general. Of course, all of this has also contributed to the emphasis on polyacetylene as the origins of doped conducting polymers. Still, to place all of this in context, we first need to review the discovery of acetylene and the initial attempts at its polymerization.

5.2 A brief history of acetylene

The production of acetylene can be accomplished via the hydrolysis of alkali and alkaline-earth metal carbides, by the elimination of halides and haloacids from various organohalides, and through the direct synthesis from elemental carbon and hydrogen. The earliest known production of acetylene dates back to 1836 when the English chemist Edmund Davy (1785–1857), cousin of the more famous Humphry Davy (1778–1829), first presented results of a new gaseous compound of carbon and hydrogen on August 26, 1836, at the Sixth Meeting of the British Association for the Advancement of Science in Bristol.[10,11] The details of this presentation were then published in late 1836,[12] followed quickly by French and German versions published in 1837.[13,14]

Davy's discovery had resulted from efforts to produce potassium metal on a large scale in January of 1836, which involved the high-temperature heating of a mixture of tartar (potassium hydrogen tartrate, $KC_4H_5O_6$) and charcoal powder in an iron bottle.[11–14] In the process, Davy obtained a dark-gray, granular material that was described as being soft enough that it could be easily cut with a knife. When added to water, this solid decomposed vigorously to evolve large amounts of gas, along with occasional inflammations on the water surface. Analysis of the gas produced led to the conclusion that it consisted of a mixture of hydrogen and "a new bi-carburet of hydrogen,"[12] the latter consisting of carbon and hydrogen in nearly equal volumes.[11] As such, Davy believed the original gray substance to be a mixture of potassium and carburet of potassium.[11–14]

In one attempt, no potassium was produced, generating only a small quantity of a dark black substance that Davy believed to be the carburet of potassium previously produced.[11–14] When added to water, the new bicarburet of hydrogen was again produced, although this time as the only gaseous product. Davy then concluded that pure carburet of potassium was a binary compound consisting of one proportion of carbon and one of potassium (what is now known as potassium carbide or potassium acetylide, K_2C_2).[11–14]

Davy then went on to investigate his new gaseous bicarburet of hydrogen obtained via the addition of carburet of potassium to water. The gas was highly flammable and burned in the presence of air with a bright flame,[12–14] described to be "denser and of greater splendour than even olefiant gas,"[12,15] with *olefiant gas* an early name for ethylene. It could be stored over mercury for an indefinite time without reaction, but if stored over water, the gas was slowly absorbed. Upon heating, the gas evolved from the solution without apparent reaction.[12–14]

The gas detonated in the presence of oxygen to give water and carbonic acid, with its complete combustion requiring 2.5 volumes of oxygen, two volumes of which were converted into carbonic acid, and the remaining volume into water.[12–14] Davy thus concluded that the new gas was composed of one volume of hydrogen and two volumes of the of carbon vapor,[12–14] stating,[12]

> It is, in fact, a bi-carburet of hydrogen composed of two proportions of carbon and one of hydrogen, and may be represented by the formula C2 + H, or 2 C + H; and its constitution seems to differ from that of any other known gas.

Davy then gave a second presentation at a Scientific Meeting of the Royal Dublin Society, entitled "On a New Gaseous Compound of Carbon and Hydrogen," on June 26, 1837.[15] Later that same year, in September, he also gave a very brief update of his investigations on the new gas at the Seventh Meeting of the British Association for the Advancement of Science in Liverpool.[16] These were then followed by his final and most detailed paper on the topic in 1839.[17]

The 1837 publication from the Proceedings of the Royal Irish Academy did not offer much new data and largely repeated material from the previous publications. However, it did report a value of 0.917 for the specific gravity of the gas, in comparison with 1.000 for air as a standard, and formalized Davy's proposed name *bicarburet of hydrogen* for the gas.[15] The following British Association report was very brief and only gave results from passing electrical sparks through the gas, which resulted in the deposition of carbon with no change in gas volume.[16] Although he later stated that he thought the gas volume after the reaction to be comprised of hydrogen,[17] he ultimately decided that it was another new hydrocarbon gas. Unlike the original bicarburet of hydrogen, this new gas did not ignite in contact with chlorine and required only 1.5 volumes of oxygen for complete combustion, again producing only carbonic acid and water vapor. Davy believed this new gas to be a binary compound described by the formula C + H.[16]

In the final paper, Davy provided detailed procedures for generating and isolating the gas from the previously described carburet of potassium, for which Davy found that six grains of the solid produced about two cubic inches of the gas.[17] He then continued in describing the properties of the gas, stating that the new gas burned with a bright white flame, in comparison with the bluish flame of olefiant gas. Additional new properties reported also included the fact that aqueous solutions of the gas had no smell or taste, nor did it cause any effect to litmus paper, although the remaining characterization details had been previously reported in his earlier papers.[17] Lastly, Davy concluded with detailed experiments to determine the composition of his bicarburet of hydrogen, leading to the same formula as previously reported (i.e., 2C + H). He did correctly determine, however, that his gas had one less volume of hydrogen than olefiant gas, which he viewed as 2C + 2H (rather than the correct C_2H_4).[17] As previously mentioned in Chapter 2, such doubling of carbon was typical of the time period, usually resulting from the use of an atomic weight of 6 for carbon.

Although Davy had provided detailed characterization of this new gas, he was not able to establish its structure, nor did his name for the gas endure. Furthermore, although he had suggested its use for lighting, it was not applied as such until the 1890s. As a result, Davy's gas was essentially forgotten until it was rediscovered by the French chemist Marcelin Berthelot (1827–1907) (Figure 5.2), who reported studies on what he believed to be a new hydrocarbon gas in 1860.[18] Berthelot gave his new gas the name *acétylène*.[10,18,19]

Figure 5.2 Pierre Eugène Marcelin Berthelot (1827–1907).

It should be pointed out that at the time of his 1860 publication, Berthelot was unaware of Davy's previous discovery. By 1863, however, he knew of Davy's work and did recognize the previous discovery in a review of this own work on acetylene, stating,[19]

> Edm. Davy obtained this gas in 1836 by treating with water the black mass which occurs in the preparation of potassium by means of calcium tartrate and charcoal. But his observation, which had remained isolated, had disappeared from science: I was not aware of it when I found the same gas by very different methods.

Although Davy's had reported extensively on the gas over the period of 1836–1839, this had not been followed up with any further studies and had thus faded from memory as pointed out by Berthelot. Berthelot's rediscovery of the gas in 1860 then marked the beginning of the significant study and application of acetylene.

Berthelot produced the gas by passing various organic gases or vapors (ethylene, alcohol, ether, aldehyde, etc.) through a red-hot tube, which produced a complex mixture of gaseous species. Bertholet, however, was able to trap the acetylene by passing the gaseous mixture into an ammonia solution of cuprous chloride (Figure 5.3), thus resulting in the precipitation of a red copper acetylide that could then be collected and purified. The acetylene gas was then liberated by treating the solid with hydrochloric acid, which was then purified by washing with a little potash.[18,19]

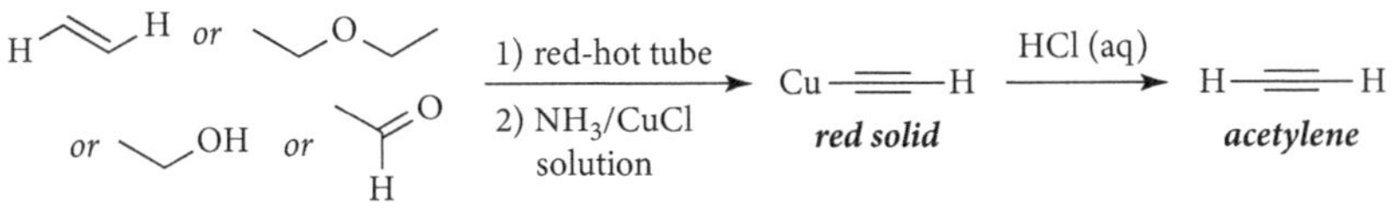

Figure 5.3 Berthelot's initial synthesis of acetylene.

Berthelot described the gas as colorless, endowed with a characteristic and unpleasant odor, and which was sparingly soluble in water.[18,19] Although he was not able to liquefy the gas by either cold or pressure, he did determine its relative density (specific gravity) to be 0.92.[18] He then continued his characterization with study of the combustion of the gas, which burned with a very bright and smoky flame. As Davy had previously determined,[12–14] he found that the complete combustion of one volume of acetylene required 2.5 volumes of oxygen, the product of which was two volumes of carbonic acid. Combined with the previously determined density, this data led Berthelot to conclude that acetylene was the species hydrogen tetracarbide, represented by the formula C_4H_2.[18]

Lastly, he investigated its chemical reactivity, as well as the production of various derivatives. As Davy had previously described,[12–14] Berthelot found that the gas detonated almost immediately when mixed with chlorine, even in the absence of direct light.[18] Overall, however, he found that acetylene possessed most of the essential properties of ethylene, furnishing paralleled derivatives via reaction with bromine and sulfuric acid. Furthermore, he showed that acetylene could be converted to ethylene via its treatment with hydrogen at low temperature.[18,19] Berthelot did not study this reactivity or the resulting products in great detail, however, asserting that difficulties in preparing large quantities of the gas prevented more extensive investigations.

Berthelot then published a second paper in early March of 1862 that attempted to place acetylene in the context of other known carbon-hydrogen species of the time (methane, ethylene, propylene, etc.).[20] This focused on various synthetic conversions, during which he showed that methane could be used to generate acetylene via either heat or spark. Furthermore, the isolated acetylene could then be converted to ethylene by treating the copper acetylide intermediate with zinc in aqueous ammonia (Figure 5.4).[19,20]

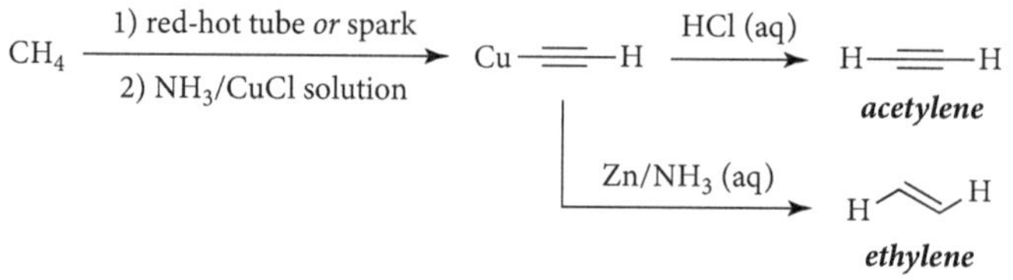

Figure 5.4 Interconversion of various hydrocarbons.

A few weeks later, Berthelot then reported the production of acetylene via an electric discharge between two carbon rods in the presence of hydrogen in late March of 1862.[19,21–24] This utilized what has been referred to as Marcelin's *d'œuf électrique* (electric egg, Figure 5.5),[23,25] which was an ellipsoid-shaped device equipped with openings at the ends of the long axis. These ends were both sealed with fat plugs (labeled M in Figure 5.5) that were each fitted with two glass tubes, one of which (t and t′) allowed H_2 gas to be fed into the device. The other tube (T and T′) was fitted with a metal rod that could slide within the glass tube with only gentle friction, while still maintaining a seal between the vessel interior and the outer atmosphere. The inner end of the metal rod was equipped with a connector (P and P′) fitted with a carbon rod (Q and Q′), while the exterior end was then connected to a battery.[23] Stressing the importance of the carbon purity, Berthelot indicated that the carbon used to form the rod should first be heated in a porcelain tube under a stream of dry chlorine until it becomes red-white.[19,21–23]

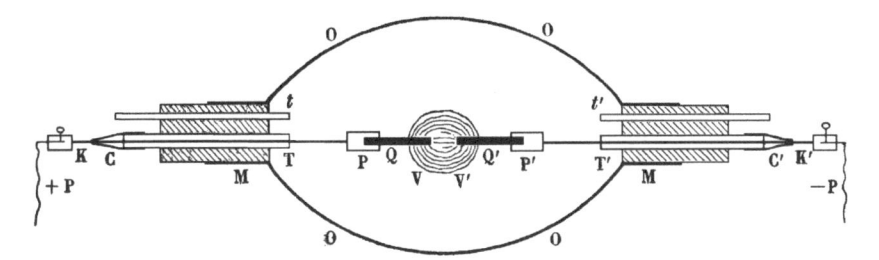

Figure 5.5 Marcelin's *d'œuf électrique* (electric egg).

Reproduced from Berthelot, M. *Leçons sur les méthodes générales de synthèse en chimie organique.* Gauthier-Villars: Paris, 1864.

In order to generate the desired acetylene, the device is first filled with H_2 gas, and the metal rods connected to the appropriate poles of the battery. The carbon rods are then brought together until they touch, after which their position is adjusted to provide a small gap capable of allowing an electric arc between them. This process thus converts approximately half of the carbon rod into acetylene, while the other half is dispersed into dust, which adheres to the interior device walls. As the ends of the carbon rods are consumed during the acetylene production, the distance between the carbon rods must be periodically readjusted to maintain the necessary electric arc.[23]

As before, the generated acetylene gas was trapped by passing it through an ammonia solution of cuprous chloride to generate the red copper acetylide intermediate.[19,21–23] The copper acetylide was then isolated and purified, before treatment with HCl to liberate the pure acetylene gas. This new method now allowed Berthelot to generate significant quantities of acetylene, with a reported production rate of 10–12 mL per min.[19,21–23]

Bertholet then followed this with three additional papers in May of the same year.[26–28] The first of these focused on the effectiveness of various carbon sources for production of the carbon electrodes in the electrolytic acetylene synthesis,[26] while the second was a review of the known methods for the production of acetylene, as well as those methods that had failed to generate acetylene.[27] The final paper was then a study of what was commonly known as *lighting gas.*[28] Lighting gas was obtained from the pyrolysis of carbon species such as coal or wood to give a mixture of gases that had then been applied to early gas lighting since the late 18th century. Bertholet was able to confirm that lighting gas contained acetylene, although its relative proportion in the mixture was fairly low. Still, it was believed that it was acetylene that provided much of the flame's considerable illumination, as well as the distinctive smell of lighting gas.

Although Berthelot could now use the electric discharge process to produce greater amounts of acetylene than previously possible, it was still not suitable for the mass production of the gas. Such large-scale manufacture of the gas would have to wait until the 1890s, when the Canadian inventor Thomas L. Willson (1860–1915) (Figure 5.6) developed a process for the large-scale production of calcium carbide from lime and coke. The calcium carbide from this process could then be reacted with water to generate acetylene in a process analogous to that reported by Davy for potassium carbide.

Figure 5.6 Thomas Leopold Willson (1860–1915).
Reproduced courtesy of the Library and Archives Canada/C-53499.

It was on May 2, 1892, that Willson accidentally discovered the processes for making calcium carbide and acetylene in commercial quantities in Spray, North Carolina.[29–32] The work leading up to these discoveries, however, began in 1888 with efforts to develop an economical way to make aluminum. Willson's approach involved the reduction of aluminum ore with carbon in a high-temperature, electric-arc furnace,[29,33,34] a process also being explored in the laboratory of the French chemist Henri Moissan (1852–1907).[31] To his disappointment, these efforts were not that successful, and he was only able to produce a few globules of aluminum in this way. He then reasoned that if he could first produce a more chemically active metal, such as calcium, he could then use that to reduce the aluminum ore.[29,31] Through the application of an improved electric furnace (Figure 5.7) with a current of 2000 amperes and 36 volts, he thus began attempts to produce metallic calcium through the electrothermal reduction of lime (CaO) with various carbon sources.[29,30,33,35] As a result, Willson was able to produce a dark molten mass that became a heavy, brittle, dark-colored solid upon cooling. As this was clearly not the desired metallic calcium, it is said that Willson discarded the material by throwing it into a neighboring stream, which unexpectedly liberated a great quantity of gas.[35,36]

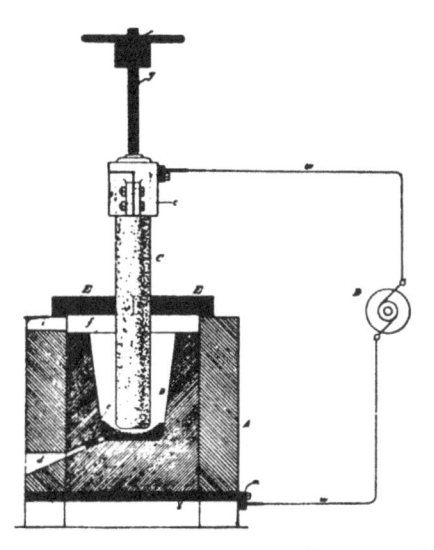

Figure 5.7 A schematic of the original electric-arc furnace used by Willson Aluminum Company at Spray, North Carolina.

Reproduced from T. L. Willson, J. J. Suckert. The carbides and acetylene commercially considered. *J. Franklin Inst.* **1895**, *139*, 321–341.

Willson then retained Francis P. Venable (1856–1954) of the University of North Carolina as a consultant, who repeated and refined the process to find the solid substance to be calcium carbide and the gas to be acetylene.[29–33,35,36] On August 9, 1892, Willson filed a U.S. patent for the reduction of refractory ores or compounds (including lime) via electric smelting,[29,31,37] which was then granted on February 21, 1893.[37,38] This was then followed with additional U.S., British, and Canadian patents on the production of calcium carbide and acetylene.[30]

Of course, calcium carbide was not a new species and is thought to have been first prepared by Robert Hare (1781–1858) in 1839,[39] just a few years after Davy's initial work on acetylene.[12] Hare had exposed calcium cyanide to a strong current, resulting in particles of metallic character that effervesced in water.[39,40] The first to recognize the nature of the compound and to identify the resulting gas as acetylene, however, was the German chemist Friedrich Wöhler (1800–1882),[38] who had prepared calcium carbide by heating carbon with an alloy of zinc and calcium in 1862, after which he treated it with water to produce calcium hydroxide and acetylene.[33,35,41,42]

Still, the ability to make the production of calcium carbide practical did not come about until after development of electric-arc furnaces capable of providing the necessary high temperatures.[36,42] The electric-arc furnace was first demonstrated by William Siemens (1823–1883) in 1879, with further advances by the 1890s allowing even higher temperatures, and it was a modified electric-arc furnace of this type that was utilized by Willson in 1892. Of course, others were also working along similar lines, with Henri Moissan presenting a new furnace capable of 3000°C in December of 1892, along with its use to generate calcium carbide from quicklime and carbon.[43,44] Calcium carbide generated by similar methods had also been reported by L. K. Böhm and L. M. Bullier.[38,45] Moissan in particular had conveyed his results to the world,[43,44] resulting in legitimate controversy with regard to the priority of the discovery.[34-36] Luckily, Willson had shared his discovery with Lord Kelvin via a letter and specimen of calcium carbide sent to Glasgow on September 16, 1892,[34,35] with Kelvin acknowledging receipt of the sample in a return letter to Willson dated October 3, 1892, along with confirmation of its reaction with water.[30,33] As a result, the issue of priority was ultimately decided in Willson's favor,[31,35,36] as his correspondence with Kelvin predated Moissan's December publication. Furthermore, the Imperial High Court of Germany acknowledged this conclusion by annulling a German patent previously granted to Bullier.[35,45]

Recognizing the discovery's commercial potential, Willson continued developing the technology for the large-scale production of calcium carbide using common materials, as well as finding suitable applications.[32] After initially failing to find anyone willing to buy their calcium carbide and acetylene patents, however, Willson and his commercial partner James Turner Morehead (1840–1908) turned their focus to finding and promoting uses for the manufactured products. These efforts began with applications in lighting, and after illustrating that acetylene could produce a flame 10–12 times brighter than that of coal gas, it was concluded that acetylene outshone either coal gas or incandescent electric lights (prior to the development of the tungsten filament), resulting in its rapid development as an illuminant.[31,32]

In August 1894, they sold their patents to a new firm, the Electro-Gas Company of New York City, but retained the rights for chemical manufacturing (Electro-Gas was ultimately absorbed into the Union Carbide Company formed in 1898).[29,31] The Electro-Gas Company then made arrangements with the Niagara Falls Power Company to apply 1000 electrical horsepower to the manufacture of calcium carbide, with the plan to quickly increase this to 5000

horsepower.[36] The Electro-Gas Company then proceeded to sell carbide manufacturing rights worldwide, although Willson reserved all Canadian rights as
part of this agreement.[29,31] In the process, Morehead also purchased a manufacturing franchise.[31] Moving back to New York, Willson set up a laboratory
at Eimer and Amend to explore chemical uses for acetylene in the fall of 1893.
After making small quantities of chloroform and various aldehydes, he filed for
a patent for the use of acetylene in the manufacture of "hydrocarbon products"
in February 1894.[29,31] In August of 1894, Morehead then completed the first
commercial calcium carbide plant in Spray.[31]

As publicity regarding the significant potential of acetylene applications
increased, so did the demand for the carbide precursor. The plant in Spray began
to operate around the clock on May 1, 1895, with the months following providing continued success.[31,34] On March 29, 1896, however, disaster struck when
the plant was destroyed by fire. Undaunted, Morehead persevered and built
a much larger plant on the James River near Lynchburg, Virginia, which was
eventually sold to the Union Carbide Company.[31]

By this point, significant amounts of the acetylene were now available for a
variety of applications.[46] While much of the focus was on fuel for lighting, its
application for oxyacetylene welding and cutting was also developed in 1903.[32]
At the same time, acetylene had become a feedstock chemical for the further
production of more complex chemical species,[46] as well as a polymerizable
monomer for polymeric products.

5.3 Early studies of acetylene polymerization

As access to acetylene became more available, it was only a matter of time until
the polymerization of this unsaturated species was investigated. This did not
require the mass production facilitated by Willson, however, with the history
of acetylene polymerizations dating to just a few years after Berthelot's development of acetylene production via electric discharge.[47,48] It is thus not surprising
that the earliest reported examples of acetylene polymerization also came from
Berthelot, with a series of studies detailing the action of heat upon acetylene
beginning in 1866.[49,50]

Berthelot found that upon placing acetylene in a bell over mercury and then
heating the gas at extreme temperatures (described as the softening temperature
of glass), a decrease in the gas volume could be observed, with the production
of a mixture of liquid and tarry products.[49,50] As described by Berthelot,[49]

> Almost all of the elements of acetylene are found in the liquid and fixed prod-
> ucts of the reaction. The latter consist mainly of two carbides: one volatile and

> which exhibits the properties and reactions of styrene ... the other almost fixed, resinous, and which appears to be melastyrol.

Introduced by John Blyth and August Wilhelm Hofmann (1818–1892) in 1845, the term *metastyrol* refers to the solid resulting from the polymerization of styrene (i.e., modern polystyrene).[51,52] It could then be inferred that the resinous product described by Berthelot is a polymeric material of some form. Further study of the volatile product described it as a yellowish liquid, which was revealed to be a mixture consisting primarily of benzene, along with styrene and other hydrocarbons.[50] The benzene content was determined to make up circa 50% of the total liquid products formed. It is important to note that at the time, the definition of the term *polymer* did not yet have conditions of molecular weight or macromolecular nature, but only specified species of higher multiples of a common empirical formula.[52,53] As such, Berthelot considered the benzene, styrene, and resinous solid produced here to all be polymers of acetylene.

Berthelot continued his investigations by heating gaseous acetylene in the presence of various solids such as elemental carbon or iron. It was found that the temperature required to cause reaction was decreased significantly in the presence of these solids, while also increasing the overall reaction rate. In addition, these added solids influenced the specific nature of the products generated. Finally, he repeated the process using mixtures of acetylene with equal volumes of either N_2, CO_2, methane, or ethane. In all cases, he observed the same results, although with slower reaction times in comparison with samples of pure acetylene. Ultimately, he concluded,[49]

> In summary, the transformation of acetylene by heat is not comparable to the phenomena of dissociation: it is not the result of a destruction of the affinity that holds together carbon and hydrogen; but it shall be by a very different mechanism, which is not incompatible with the stability of acetylene. What the heat determines here, it's not a decomposition, it is rather a combination of a higher order, developed by the mutual union of several acetylene molecules.

Further studies of acetylene polymerization then had to wait until 1874, when the French chemist Paul Thenard (1819–1884) and his son Arnould Thenard (1843–1905) published a brief communication in January of 1874 on the study of the influence of electric discharge on acetylene.[54] Using an undescribed discharge device designed by Arnould, the electric discharge was found to cause rapid condensation of acetylene to (4–5 cm^3 min^{-1}) to give a solid deposit on the inner walls of the device.[54] The solid was described as very hard, glassy material, with a color that was compared to the dregs of wine and an elementary formula identical to that of acetylene. The material was found to be insoluble in all solvents investigated, and treatment with fuming nitric acid had no effect. Furthermore, attempts to dry-distill the solid failed, leading to the conclusion that the material was analogous or similar to bitumen (a highly viscous liquid or semisolid form of petroleum, now also known as asphalt).

Changing the conditions of the experiment, they also produced a liquid that gave a composition equivalent to acetylene, although it could only be produced in very small amounts, so its identity was never determined. Although this brief report was never followed up with additional studies, this transformation of acetylene has been cited as the earliest known example of the generation of an organic polymer by electric discharge.[55] This report was then quickly followed up with a closely related study from the Belgian P. De Wilde later that same year,[56] along with another study from Berthelot in 1877.[57]

In the first case, De Wilde observed the condensation of an oily yellow liquid on the discharge tube walls, which solidified after a few hours to produce a hard, brittle material.[56] The brown, amorphous material was found to be completely insoluble, but De Wilde did find that it burned to leave behind a coal-like residue. Berthelot then followed up the work of both the Thenards[54] and De Wilde[56] with his own study in which he confirmed that he was able to verify the accuracy of the Thenards' report but sought to provide some additional insight.[57] Berthelot went on to characterize the solid brown material as a polymer with the formula $(C_4H_2)_n$ (the carbon content again doubled as for his previous formula of acetylene). In addition, he found that heating the material under N_2 resulted in exothermic decomposition to give styrene, a carbonaceous residue, and other gaseous byproducts. This led Berthelot to state that this reactivity "distinguishes it from all other known acetylene polymers,"[57] although he did not directly compare these results with his previous report on the thermal polymerization of acetylene.[49]

After this spate of studies between 1866 and 1877, no further detailed studies on acetylene polymerizations appeared for the next 20 years. However, Willson did briefly mention in 1895 that when exposing acetylene to electric spark, he observed the formation of a "solid poly-acetylene" that resembled horn and was insoluble in the ordinary solvents.[58] More detailed efforts began again in 1898 with a new report by Hugo Erdmann (1862–1910) and Paul Köthner (1870–1932).[59]

It was in June of 1898 that Erdmann and Köthner reported the results of previous studies carried out in 1895, which had involved heating acetylene gas over copper metal.[59] They began with the statement that pure acetylene underwent thermal reaction at a temperature of 780°C. While this value appears to be in agreement with Berthelot's vague description of the temperature used for his acetylene polymerization,[49] Erdmann and Köthner imply that this primarily resulted in the generation of graphite. More critically, however, this temperature could be significantly reduced when carried out in the presence of copper, with small crystals of graphite formed from acetylene over copper powder at 400°C–500°C. In addition, they found that if the temperature was maintained below 250°C, no graphite formation was observed, with instead the production of a light brown solid.[59] Lastly, it was found that the brown solid could be produced at a much faster rate, and in larger scale, by replacing copper powder with copper oxide.

The light brown material was characterized as very light and bulky, with a density of circa 0.023 g/mL. To test for the presence of copper, the material was boiled in dilute HCl and filtered, after which yellow copper hydroxide was precipitated from the colorless filtrate via treatment with NaOH. This detection of copper then led to the view that the brown solid was a copper compound of some type, with a formula of $C_{44}H_{64}Cu_3$ suggested by combustion analysis. This ultimately led to the conclusion that[59] "There have been analyzes of different preparations, which give such well-matched values, that we should not hesitate to address this light brown copper acetylene compound as a uniform, albeit very complex composite compound." This work was then followed by a closely related conference report in May of 1899 by Paul Sabatier (1854–1941) and Jean P. Senderens (1856–1937), which presented initial results obtained by heating acetylene over copper at circa 180°C to produce a yellow-brown material.[60] Described as a complex hydrocarbon consisting mainly of ethylenic carbides, this material was characterized as very light and voluminous with small traces of dispersed copper. In a more detailed study reported the following year, it was stated that at ambient temperature, a stream of acetylene passed through a tube containing copper generated no appreciable reaction.[61] If the temperature was raised to 180°C, however, the copper turned brown with observable condensation of acetylene. Over time, the copper took on a darker hue, and the growing mass swelled to completely fill the tube, closing off the flow of gas. Curiously, it was found that if a small amount of the yellow-brown substance was smeared into a fresh tube, a stream of acetylene heated to 180°C–250°C would cause resumed expansion to once again fill the entire tube. This resumed process could be repeated several more times before failing to give further reaction.[61]

The material prepared in this way was described as a dark yellow solid, which under a microscope appeared to be composed of thin twisted filaments.[61] The material was reported to be lightweight and soft, yet a minor compression could give it the consistency and look of wood. Although insoluble, the material could be burned to give off a smoky, aromatic flame, resulting in a black residue. Ultimately, it was concluded to be a hydrocarbon with an empirical formula of

C_7H_6, in which small quantities of copper (1.7%–3%) were dispersed. Due to the origin of the material, they proposed to name it *cuprene*.[61]

By the early 1920s, it had become clear that cuprene was not a copper-based species and contained no copper when fully purified. Nevertheless, the name cuprene stuck, and over time it was found that in addition to thermal production over copper or electric discharge, cuprene could also be generated via UV polymerization, treatment with α particles, or γ-ray irradiation, with all of these methods producing the same or nearly the same material, with an empirical formula very close to that of acetylene. Finally, it was concluded in the 1970s that cuprene was formed by the initial generation of polyenes consisting of 4–20 carbons, after which secondary polymerization and cross-linking led to the final yellow solid.[48] A simplified representation of this process is outlined in Figure 5.8. Of course, it must be stressed that the likely structural complexity of cuprene, coupled with its insolubility, made detailed structural determination impossible.

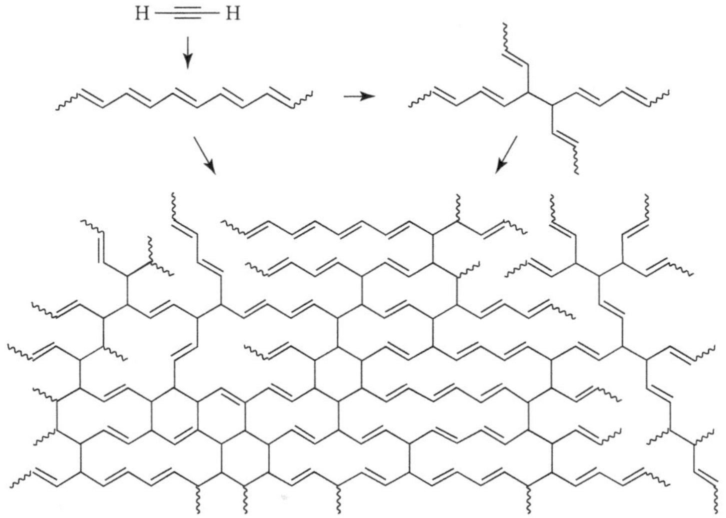

Figure 5.8 A simplified representation of the proposed formation of cuprene.

Adapted with permission from Springer Nature from S. C. Rasmussen. *Acetylene and Its Polymers. 150+ Years of History*. Springer Briefs in Molecular Science: History of Chemistry. Springer: Heidelberg, 2018.

It was during the later period of the study of cuprene that the production and study of true, linear polyacetylene was finally reported. This advancement was made possible by the development of metal-based polymerization catalysts by Karl Ziegler (1898–1973) in the early 1950s. The application of these catalysts to the polymerization of first olefins and then acetylenes was then carried out by Giulio Natta (1903–1979) of Milan, Italy.[1–4]

5.4 Natta and linear polyacetylene

Giulio Natta (Figure 5.9) was born on February 26, 1903, in the small Italian city of Porto Maurizio (now Imperia), on the Italian Riviera near the French border.[62-69] His father was a judge in the larger city of Genoa (ca. 118 km to the east), where the family spent the winters.[63,68] Like other members of his family, it was expected that Giulio would study law.[63,65,67] However, it has been said that he decided to study science as the result of reading a chemistry book at the age of 12.[65,67] After first attending Genoa's Christopher Columbus School,[63,67] Natta entered the University of Genoa to study mathematics.[63,67,68] Two years later, in 1921, he then moved to the Polytechnic University of Milan to study chemical engineering.[63,68,69] At Milan, Natta began research in 1922 under Giuseppe Bruni (1873–1946), director of the Institute of General Chemistry, and Giorgio Renato Levi (1895–1965).[68] Natta graduated *dottore*[70] in chemical engineering in 1924,[62-63,65,66,69] after which he continued on as Bruni's assistant.[66,68] In 1927, he attained the *libero docente* qualification,[62-63,65,67] which then allowed him to teach. Natta visited Freiburg in 1932 to study the work of Hugo Seeman on the new technique of electron diffraction, during which he met Hermann

Figure 5.9 Giulio Natta (1903–1979).

Reproduced courtesy of the Giulio Natta Archive.

Staudinger (1881–1965), who suggested that he study polymer structures via X-ray crystallography.[66]

After time as an assistant lecturer in chemistry at Milan,[63] he was appointed full professor and director of the Institute of General Chemistry at the University of Pavia in 1933.[62–63,65–67,69] Two years later, however, he again moved to the University of Rome to occupy the chair of physical chemistry in 1935.[62–63,65–67,69] The same year he also married Rosita Bead, a graduate of the arts.[65,68] Again, his time at Rome was brief, moving to the Polytechnic University of Turin to take the chair of industrial chemistry in either 1936[62] or 1937.[63,66,67,69] Finally, he returned to his *alma mater* to become the chair of industrial chemistry of the Polytechnic University of Milan in 1938.[62–67,69] It was at Milan that Natta turned his focus to polymers and where he remained for the remainder of his career. Upon his return to Milan, he also became a consultant for the Montecatini chemicals company, and his continued financial support from this firm has been cited as a significant factor in his impending success.[64,67–69]

Unfortunately, Natta began to suffer from Parkinson's disease in 1956,[63–68] which severely restricted his mobility. Still, he insisted on continuing his research to the extent that he was able.[66,68] Natta was ultimately awarded the Nobel Prize in Chemistry in 1963, which he shared with Karl Ziegler, for their work on the catalytic polymerization of macromolecules.[63–68] Due to his diminishing health, he had to be supported by his son Giuseppe as he received the Nobel, and his Nobel Lecture was read by a colleague.[63,67] In addition to the Nobel, Natta received the gold medal of the Union of Italian Chemists in 1964, the Lomonosov Gold Medal of the USSR Academy of Science in 1969, and the Carl Dietrich Harries Plakette of the Deutsche Kautschuk Gesellschaft in 1971, as well as honorary degrees from numerous universities.[62,67] His wife died in 1968, after which he lived with his daughter Frances in nearby Bergamo.[65,67,68] Natta then finally retired from Milan in 1973.[63,67–69] At the age of 76, only six years after his retirement, Natta died on May 2, 1979[62,64,65,67] (although some sources give it as May 1).[63,66]

Natta is of course best known for his work in high-molecular-weight polymers, which began with work on butadiene and synthetic rubber upon his return to Milan in 1938.[62,63,65,66] It was also in 1938 that he began work on the polymerization of olefins,[62,63] which eventually led to the extension of Karl Ziegler's work on metal-based polymerization catalysts in the early 1950s. It was these efforts that led to the discovery of new classes of polymers with a sterically ordered structure—*isotactic*, *syndiotactic*, and *di-isotactic* polymers (Figure 5.10), as well as linear nonbranched olefinic polymers and copolymers with an atactic structure.[62,63,65–68] It was for these accomplishments that Natta shared the 1963 Nobel Prize in Chemistry with Ziegler.[63–68]

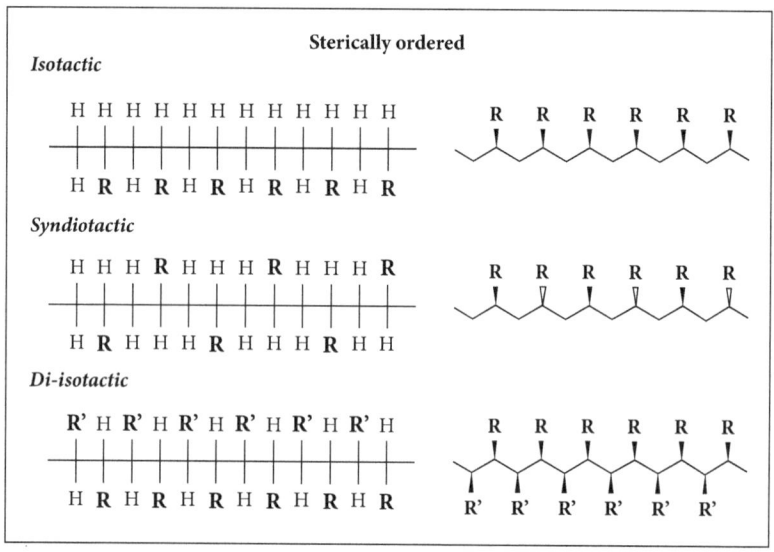

Figure 5.10 Fischer projections and line drawings of various sterically ordered structural types of polyolefins.

After his successful application of catalytic polymerization to α-olefins and diolefins in the early 1950s, Natta expanded his scope to the analogous polymerization of acetylenes.[71–73] While these efforts acted as an extension of his previous work with olefins, they were also motivated by the fact that metal-catalyzed polymerizations of acetylene were already known, as previously reported for the generation of cuprene.[47,48,71,72] This work resulted in an initial Italian patent granted in July of 1955,[71] which comprised the polymerization of acetylene, along with its alkyl or aryl derivatives, using various organometallic catalysts generated from transition metals of Groups 4–8. Specific examples of catalytic systems included various titanium species such as titanium trichloride, titanium tetrachloride, or titanium alkoxides, coupled with either triethylaluminum or diethylaluminum chloride.[71]

In July of 1957, Natta and his coworkers then presented the results of some of the initial acetylene polymerizations at the XVI International Congress for Pure and Applied Chemistry in Paris, a summary of which was included as part of a report of the meeting published in *Angewandte Chemie* later that same year.[72] In the process, they reported the successful polymerization of various acetylenes using catalyst combinations of either titanium halides or titanium alkoxides with alkylaluminum species. While the parent acetylene gave a solid product that was insoluble in organic solvents, the methyl, ethyl, and butyl derivatives gave rubber-like polymers soluble in organic solvents such as diethyl ether. Of the alkyl derivatives, the polymerization product of 1-hexyne was then investigated in more detail in order to draw conclusions about the structure of the polymer.[72]

The crude material obtained from the polymerization of 1-hexyne was first fractionated via various solvent washes. Thus, the polymer fraction that was found to be insoluble in acetone, but soluble in ether, was isolated and then used in the sequential investigations. Colorimetric wet chemical tests were then used to probe the sample for double bond content, the results of which were consistent with the presence of double bonds as expected. The hydrogenation of the sample was then investigated in the presence of Raney nickel. Although it was found to be difficult to hydrogenate the polymer in this way, the resulting hydrogenated product gave an infrared (IR) spectrum very similar to that of polyhexane synthesized under similar conditions. In contrast, the sample was easily oxidized with perbenzoic acid, consuming circa 0.85 equivalents of perbenzoic acid per repeat unit. This produced what was believed to be the corresponding peroxide derivative, which was then hydrolyzed and further oxidized with lead(IV) acetate. Analysis of the oxidative degradation products revealed valeric acid along with smaller amounts of butyric and propionic acid. Lastly, the IR spectrum of the polymer product revealed peaks consistent with butyl groups bonded to unsaturated carbon atoms, which are no longer present after hydrogenation. This then led to the conclusion that[72] "from these preliminary studies, it appears that the polyhexyne is an essentially linear, highly unsaturated polymer containing pendant C_4 groups, mainly butyl groups."

Natta then followed this initial report with a full publication in 1958 that detailed the successful catalytic polymerization of acetylene via combinations of titanium alkoxides and triethylaluminum (Et_3Al).[73] As outlined in Figure 5.11, the optimal conditions consisted of the bubbling of acetylene gas into a heptane solution of Et_3Al (0.2 M) and titanium(IV) propoxide (0.08 M) over a period of 15 hours at 75°C. The application of these conditions resulted in a 98.5% conversion of acetylene to give a black polymer that was described as a crystalline material with a metallic luster, and that was completely insoluble in organic solvents. In some cases, the formation of a shiny black mirror of the polymer adhering to the walls of the glass reaction vessel was also observed.[73]

Characterization of the isolated black polymer samples by X-ray diffraction (CuKα) revealed a material found to be ~90%–95% crystalline, with low amorphous content. Furthermore, lattice spacings of $d_1 = 3.65$ Å and $d_2 = 2.11$ Å were determined, which were noted to be related by the expression $d_1 = \sqrt{3}d_2$. It was thus concluded that this relationship suggested that these reflections came from lattice planes parallel to the axes of linear polymer chains of a small cylindrical

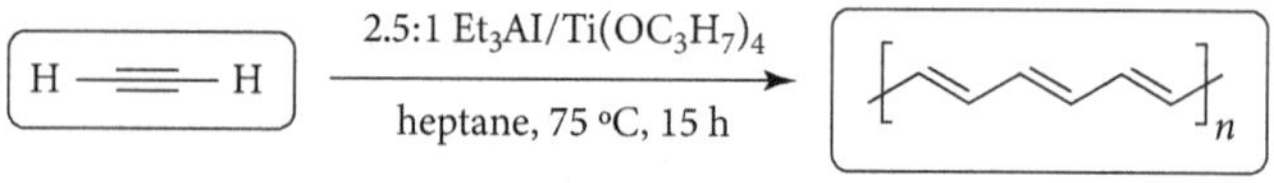

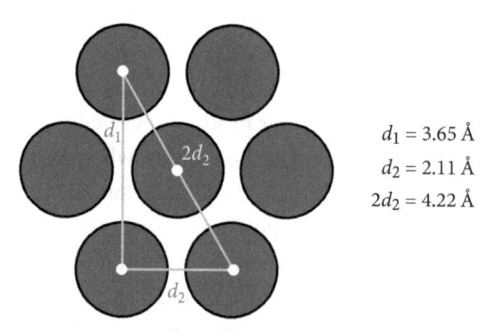

Figure 5.11 Natta's optimized conditions for the polymerization of acetylene.

enclosure, densely packed in a pseudohexagonal lattice at relative distances of $4.22\,AA = (2/\sqrt{3})d_1 = 2d_2$, as illustrated in Figure 5.12.[73] This conclusion led from the known fact that for linear polymers in the solid state, the chains tend to achieve the best space occupation in which they are arranged parallel to each other. In addition, the fact that the chains appear to pack in a pseudohexagon pattern indicates that they have a cylindrical footprint, which makes the *trans* configuration more likely. Taken all of these points as a whole, it was determined that the collected X-ray data were consistent with an assembly of linear chains of polyacetylene, in which the double bonds were concluded to be predominantly *trans* in configuration.[73]

$d_1 = 3.65\ \text{Å}$

$d_2 = 2.11\ \text{Å}$

$2d_2 = 4.22\ \text{Å}$

Figure 5.12 Proposed polyacetylene packing looking down the long axis of the polymer chains.

Adapted with permission from Springer Nature from S. C. Rasmussen, *Acetylene and Its Polymers. 150+ Years of History.* Springer Briefs in Molecular Science: History of Chemistry. Springer: Heidelberg, 2018.

The relatively low electrical resistivity of the samples (ca. $10^{10}\ \Omega$ cm, in comparison with 10^{15}–$10^{18}\ \Omega$ cm for typical polyhydrocarbons), coupled with the material's black color and metallic luster, led Natta to conclude that these polyacetylene products consisted of long sequences of conjugated double bonds. As such, these polyacetylene materials were viewed to be structurally identical to a very long *polyene*, a term dating back to 1928 to describe oligomeric species of an idealized formula $(-CH=CH-)_n$ ($n = 2$–10). However, Hideki Shirakawa later stated during his 2000 Nobel Lecture that this conclusion was not accepted widely at the time.[74-77]

The complete insolubility of the polymer precluded any possibility of determining its molecular weight. Even attempts to use solvents to fractionate the material were unsuccessful, with solvent treatments at high temperature only resulting in decreased crystallinity. For example, treating samples with boiling tetralin (1,2,3,4-tetrahydronaphthalene; bp 206°C–208°C) under N_2 caused a reduction in crystallinity to only circa 25%. The loss of crystallinity under such conditions was attributed to cross-linking.[73]

Finally, Natta also attempted some limited study of the chemical reactivity of the polymer samples, during which it was noted that polyacetylene exhibited higher reactivity than comparable polyolefins, particularly with oxidants such as O_2 and Cl_2. For example, at high temperatures, the samples rapidly absorbed atmospheric oxygen to generate more lightly colored products. In the case of chlorine, reaction occurred even at room temperature to give a white solid that was amorphous by X-ray characterization. Heating this white solid at 70°C–80°C induced a rapid darkening of the material and a loss of HCl. Alternately, nearly all of the chlorine could be removed by treatment of the white solid with potassium metal in hot ethanol, resulting in the production of a black amorphous powder. Natta highlighted the fact that this high reactivity with chlorine clearly differentiated polyacetylene from cuprene.[73]

This was then followed up with a second report in 1959 that focused on the polymerization of 1-hexyne.[78] When applied to 1-hexyne, the conditions previously used for the parent acetylene largely generated only low-molecular-weight materials. It was found, however, that a combination of $TiCl_3$ and Et_3Al at temperatures between –10°C and 20°C produced polymer yields as high as 86%, with only small quantities of low-molecular-weight material. Such conditions produced a viscous mass that was quite swollen by the heptane solvent used in the polymerization, which was then fractionated by washing with acetone. This removed oily, low-molecular-weight material to give polyhexyne samples described as a rubbery polymer that was insoluble in acetone, but completely soluble in ether, benzene, CCl_4, and cyclohexane. The resulting polyhexyne was found to be quite reactive to oxygen, even at room temperature, thus requiring samples to be kept under a nitrogen atmosphere.

The solubility of the polyhexyne samples allowed characterization not possible for the completely insoluble parent polyacetylene. This included attempts to determine the molecular weight of these samples by viscosity, which provided intrinsic viscosity values of circa 0.7100 cm^3 g^{-1} in toluene solutions and led to the conclusion that these samples were comprised of linear macromolecules of considerable length.[78] Of particular interest was also the ability to characterize the UV-visible absorption properties of the polymers in solution. Such measurements in either heptane or chloroform revealed a broad absorption without any definable maxima, with significant absorbance ranging from the UV to an onset of circa 420 nm. Both the limited absorbance in the visible region and lack of any clear maxima were attributed to a lack of coplanarity of the polymer backbone due to side chain–induced steric hindrance.

Perhaps the most important results, however, were revealed from the IR spectra of heat-treated samples.[78] The analysis of samples prepared at low temperatures exhibited largely just a single wide C=C band at circa 1630 cm^{-1}, along with a weaker band at circa 1660 cm^{-1}. Samples treated at temperatures above 70°C, however, exhibited the growth of an additional band at circa 1600 cm^{-1}, with an associated decrease in the original signal at 1630 cm^{-1}. This change was concluded to be due to a temperature-induced isomerization of the polymer, with the most likely isomerization proposed to be a conversion of the conjugated backbone from a *cisoid* conformation to a corresponding *transoid* confirmation, as illustrated in Figure 5.13. Alternative explanations also included the formation of new conjugated double bond systems onto the side chains and formation of cyclic units within the chain. This final possibility, however, was considered to be unlikely.

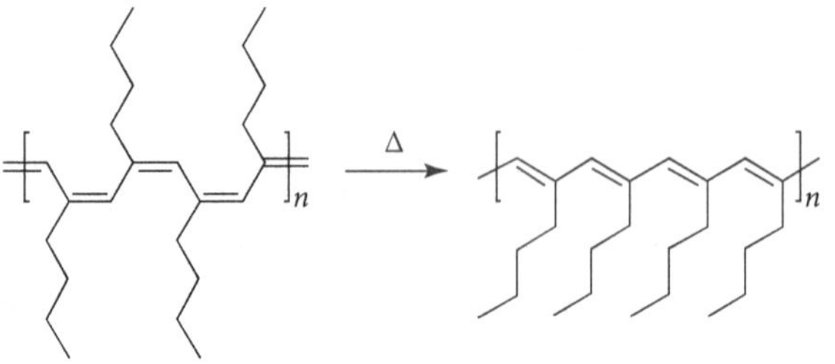

Figure 5.13 Proposed thermal isomerization of polyhexyne.

Characterization via X-ray diffraction found the initially prepared polymer samples, the acetone-insoluble fractions, and the thermally treated samples discussed earlier, to all be amorphous.[78] It was stated that it was not possible to establish whether the noncrystalline nature was due to structural irregularities of the main chains or simply due to steric hindrance of the butyl side chains. However, it was stated that the acetone-insoluble fractions were remarkably similar to that of polyhexene, with nearly identical dimensions and intermolecular distances.

While the thermally treated samples were characterized via both IR spectroscopy and X-ray diffraction, none of the initial solution measurements were repeated in order to further determine other possible effects of the proposed isomerization. The results of the UV-visible spectroscopy would have been particularly interesting. Finally, the samples were treated with various chemical reagents and/or catalytic hydrogenation, all of which gave results consistent with a conjugated material.[78]

Although Natta states that his 1958 paper represented only an initial communication, with additional papers planned,[73] no further reports on polyacetylenes were published beyond the single 1959 paper on polyhexyne. Still, others did not hesitate to continue what Natta had started. As more studies began to utilize Natta's polymerization methods to generate polyacetylenes, the formal name *polyacetylene* gradually replaced the term *polyene*, regulating the older term strictly to oligomeric species.

5.5 Additional studies of acetylene polymerizations

The initial extension of Natta's work on polyacetylene was focused not only on the further study of the polymerization reaction itself or the new polymeric materials produced, but also on new catalytic systems applied to acetylene polymerizations. Such efforts were initially led by L. B. Luttinger of the American Cyanamid Company beginning in 1960.[79-81] The general system reported by Luttinger consisted of a combination of a hydride source with either a metal salt or complex from Groups 8–10. Typical combinations consisted of sodium borohydride ($NaBH_4$) with either nickel halide (or analogous complexes such as $(Bu_3P)_2NiCl_2$) or cobalt nitrate.[79,80] These catalytic mixtures presumably produced a catalytic metal hydride species that then facilitated the smooth and rapid polymerization of either the parent acetylene or its monosubstituted derivatives, even at room temperature.

Of the catalytic mixtures studied, the $NaBH_4/Co(NO_3)_2$ mixture gave the best results, with its application of the polymerization of acetylene producing a black, high-molecular-weight polymer of acetylene.[79,80] As with the samples produced by Natta, this polymer was completely insoluble. Furthermore, attempts to determine its melting point were unsuccessful, with the material exhibiting a slow decomposition at temperatures as low as 130°C, although not being complete even at temperatures as high as 300°C.[80] The IR spectrum of the polymer was virtually featureless, except for the presence of a strong band at circa 1010 cm^{-1}, which was attributed to *trans* double bonds. X-ray diffraction of the materials revealed degrees of ordering ranging from completely amorphous to results consistent with those previously reported by Natta.[73]

Application of Luttinger's catalysts to monosubstituted acetylenes gave primarily only oligomeric species, while disubstituted acetylenes gave little to no product at all.[79,80] In addition to nickel and cobalt, other metals under study included ruthenium, osmium, platinum, and palladium, while alternate sources included lithium aluminium hydride ($LiAlH_4$) and diborane (B_2H_6).[79,81] The primary strengths of these catalysts as advocated by Luttinger were their ease of preparation and their insensitivity to moisture and oxygen in comparison with previously applied Ziegler-Natta-type catalysts.[79] It was later shown by W. E. Daniels that nickel phosphine complexes alone could catalyze the polymerization of acetylene, although the products lacked the characteristic luster of polyacetylene samples produced from other catalyst systems.[82]

One of the more recognizable names to follow up on Natta's discovery was John K. Stille (1930–1989), best known for the cross-coupling method that bears his name. However, as previously detailed in Section 4.7, he also contributed to research on conjugated polymers, including both polyphenylene and polyacetylene. It was shortly after Luttinger introduced his new catalyst system that Stille reported the extension of Natta's methods to the polymerization of nonconjugated diynes such as 1,6-heptadiyne. These efforts were first presented at the Spring National Meeting of the American Chemical Society (ACS) in late March of 1961, the details of which were published as a preprint of the ACS Division of Petroleum Chemistry.[83] This was followed up with a more formal report later that same year in the *Journal of the American Chemical Society*.[84]

The polymerization of 1,6-heptadiyne via titanium halides or alkoxides coupled with alkylaluminum species resulted in the production of dark red to black products.[83,84] The use of titanium(IV) ethoxide and either Et_3Al or tri(isobutyl)aluminium ($^iBu_3Al$) produced black, insoluble products. In contrast, the use of either $TiCl_3$ or $TiCl_4$ resulted in red, soluble products, with the combination of $TiCl_4$ and $^iBu_3Al$ giving the highest conversions (80%–90%). The color of the materials indicated a significant amount of conjugation, with the analysis of the soluble materials from $TiCl_4/^iBu_3Al$ giving a number average molecular weight (M_n) of 13,500. IR spectroscopy revealed a C=C stretch at 1605 cm^{-1} and, when combined with analysis of the polymer via hydrogenation and oxidative decomposition studies, led to the proposed structure given in Figure 5.14.[83,84]

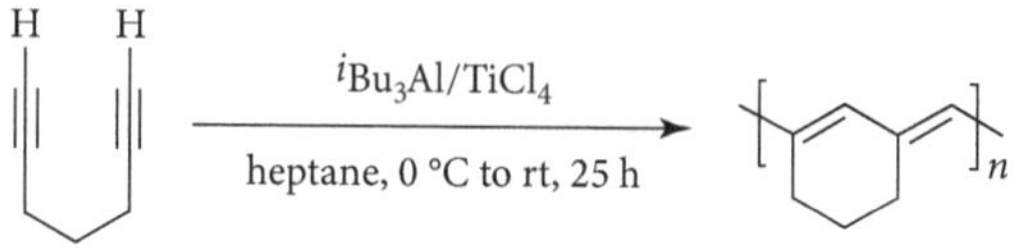

Figure 5.14 Polymerization conditions and proposed structure for poly(1,6-hexadiyne).

Analysis of the polymer by UV-visible spectroscopy in CCl_4 revealed a maxima at 270 nm, followed by a broad, featureless absorption in the visible region out to circa 640 nm.[84] The extended absorption in the visible here in comparison to that previously reported by Natta for polyhexyne[78] could be due to reduced steric interactions from the alkyl segments as a result of the cyclic structure proposed. It was found that either heating or prolonged exposure to air caused a loss of the visible absorption, with the polymer solution becoming pale yellow.

Polymer samples exhibited a resistivity of 10^{10}–10^{13} Ω cm,[83,84] which is in good agreement with that previous reported by Natta for polyacetylene.[78] It was stated that this high resistance was probably due to the instability of the polymer to oxygen, which could result in breaks in the conjugated system. Further attempts to polymerize the additional nonconjugated diynes 1,7-octadiyne and 1,8-nonadiyne gave only lightly colored products in low yields.[83,84]

Later that same year, William H. Watson, Jr. (1931–2022) of Texan Christian University published a study that revisited the polymerization of acetylene via Ziegler–Natta catalysts in order to provide a more detailed study of the process.[85] Born in Houston, Texas, on September 2, 1931, Watson attended Rice University where he received a BA in 1953, followed by a Ph.D. in 1958. He then took a faculty position with Texas Christian University, where he remained for 50 years. Throughout his career, he also spent time as a visiting professor at Cambridge University, the University of Southampton, the University of Heidelberg, and the University of Bonn, as well as a visiting scientist at the Argonne National Laboratory. He died on March 6, 2022.[86]

It was just a few years after joining Texas Christian University that he turned his attention to the polymerization of acetylene. In this 1961 report, he cited as his motivation the growing interest in organic semiconductors and the emphasis being placed upon the electronic properties of long-chain polyenes. For his study, Watson utilized $TiCl_4$ coupled with various metal alkyls, including $^iBu_3Al$, butyl lithium, and diethylzinc.[85] The most effective catalyst combinations and conditions (Figure 5.15) were both in close agreement with Natta's previous results[78] and nearly identical to those utilized by Stille.[84] However, Watson did provide more detailed accounts of the experimental conditions of the polymerization methods. Such new details included the need to use cooling baths to maintain a constant reaction temperature, the specific flow rate of the added acetylene gas (350 mL min^{-1}), the fact that the addition of $TiCl_4$ to the metal alkyl led to the formation of new species, and that extensive washing of the product was required in order to remove large amounts of inorganic salts trapped in the polymeric material. Surprisingly, though, the reaction time employed never seems to be specified.

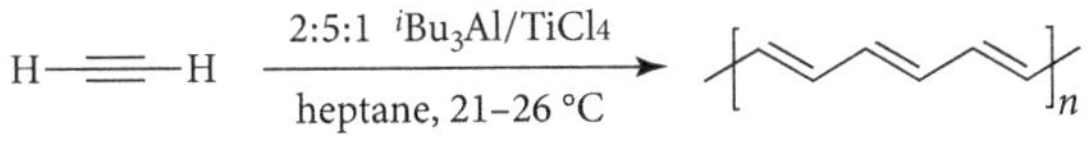

$$H \!-\!\!\equiv\!\!-\! H \quad \xrightarrow[\text{heptane, 21–26 °C}]{2:5:1\ ^iBu_3Al/TiCl_4} \quad$$

Figure 5.15 Watson's optimized conditions for the polymerization of acetylene.

The bulk of the polymers generated were insoluble black materials that did not melt below 400°C, with elemental analysis indicating an empirical formula of $(CH)_n$. The report of some limited electronic characterization was also detailed, including room temperature electron paramagnetic resonance (EPR) measurements to give values of 0.94–14.6 × 10^{18} spin g^{-1} and electrical measurements to give a resistance of circa 10^4 Ω cm at 25°C (considerably lower than the circa 10^{10} Ω cm reported by both Natta[78] and Stille[84]). Furthermore, the electrical measurements exhibited a temperature dependence typical of semiconductors, with resistance decreasing with increased temperature.[85]

Although Watson had stated that more detailed reports of the polymer electronic properties would be forthcoming, he never did publish any further reports on the polymerization of acetylene or the corresponding polymeric materials. This may have been because of a paper reporting exactly the electronic characterization alluded to by Watson that was published shortly thereafter by Masahiro Hatano (b. 1930) at the Tokyo Institute of Technology.

5.6 Hatano, Ikeda, and the Tokyo Institute of Technology

Masahiro Hatano was born in 1930 and received his doctorate in 1959 from the Tokyo Institute of Technology (Tokyo Tech), located in the Ookayama neighborhood of Ota Ward, Tokyo. After completing his doctorate, he joined the Chemical Resources Laboratory at Tokyo Tech as a research associate under Shu Kanbara (1906–1999, sometimes given as Kambara).[87] In 1967, he then moved to the Research Institute for Non-Aqueous Solution at Tohoku University as an associate professor. He remained at Tohoku University, becoming professor in 1969.[3] The Institute for Non-Aqueous Solution then became the Institute for Chemical Reaction Science in 1991,[88] where Hatano remained until he became professor emeritus in 1994.[3]

It was just a couple years after receiving his doctorate that Hatano, working with Kanbara, reported the first detailed study of the semiconducting properties of polyacetylene.[89] The 1961 study began with an exploration of the Ziegler–Natta catalyzed polymerization parameters. Unlike previous studies by other researchers, however, the focus here was not on increasing the polymer yield but, rather, on the effects of various reaction conditions on the crystallinity and electronic properties of the polyacetylene products. This included the effects of $TiCl_4$ versus the analogous propoxide as the titanium species, the effect of reaction temperature, and to a lesser extent, how changes in the ratio of Et_3Al to the titanium salt impacted the polymerization process. The most important conclusions were that the application of titanium(IV) propoxide gave more crystalline polymers than $TiCl_4$, and that higher reaction temperatures generally resulted in increased crystallinity.[89]

The polymer products were then characterized via electron spin resonance (ESR) spectroscopy and pressed-pellet DC conductivity measurements, with a focused emphasis on the effect of crystallinity.[89] As demonstrated by the results summarized in Table 5.1, a correlation was found between sample crystallinity and either the concentration of unpaired electrons or the measured resistance, with an increase in crystalline nature giving lower resistance and a greater number of unpaired electrons. The value for the room temperature resistance of the highly crystalline sample here is in relatively good agreement with the previous measurements of Watson,[85] while the resistance determined for the amorphous sample is in good agreement with the value previously reported by Natta.[73] The resistance of the amorphous sample was additionally determined over a range of temperatures (ca. 20°C–125°C), while under an argon atmosphere. The exhibited temperature dependence (Figure 5.16) was found to be typical of intrinsic semiconductors, in good agreement with similar measurements previously reported by Watson.[85]

Table 5.1 Effects of crystallinity on select electronic properties of polyacetylenes

Polymer crystallinity	Concentration of unpaired electrons (spin g^{-1})	Resistance (Ω cm)
Amorphous	4.4×10^{18}	3.7×10^{9}
Low crystallinity	11×10^{18}	1.6×10^{8}
Medium crystallinity	36×10^{18}	4.2×10^{5}
High crystallinity	47×10^{18}	1.4×10^{4}

Note: Values collected from Reference 89.

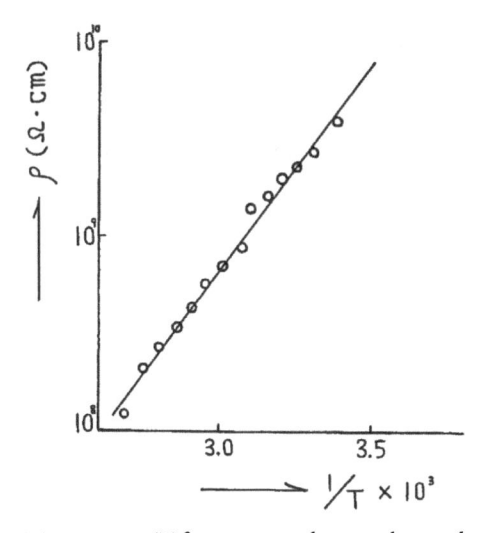

Figure 5.16 Resistivity versus 1/T for an amorphous polyacetylene sample.

Finally, the effects of air oxidation on the polymer samples were also investigated.[89] A sample of amorphous polyacetylene underwent oxidization at room temperature over 15 days to give a material that was characterized as "oxidized almost completely." In the process, the sample underwent a color change from greenish black to pale orange, with a corresponding increase in oxygen content (from 14.2% to 22.9%) as determined by chemical analysis. Characterization of the oxidized polymer sample indicated a decrease in the concentration of unpaired electrons (from 3.5×10^{18} to 3×10^{17} spin g^{-1}), with an increase in room temperature resistivity (from ca. 3.7×10^9 to 1.4×10^{12} cm).

Hatano, along with Kanbara and Tatsuo Hosoe, continued these efforts with a report in early 1962 that detailed an investigation into new types of polymerization catalysts with the overall goal of producing highly crystalline forms of polyacetylene.[89] Based on their previous study,[89] they concluded that the use of transition metal halides ($TiCl_4$, $VOCl_3$, VCl_3, or VCl_4) generated amorphous polyacetylene products. In contrast, the application of species of low electrophilicity (such as the previously applied alkoxides) gave crystalline materials. In an effort to produce even more crystalline forms of polyacetylene, this follow-up study then focused on the application of various transition metal acetylacetonates as active polymerization catalysts.[90] The study investigated metal acetylacetonate (acac) complexes, in which the metals used included Ti, V, Cr, Fe, Co, and Cu, that were then combined with Et_3Al to form the potential catalyst systems. Of the systems studied, however, only two species, $TiO(acac)_2$ and $VO(acac)_2$, were deemed to be effective catalysts and generated crystalline products when combined with Et_3Al. All of other metal species investigated failed to polymerize acetylene.

This was then followed by a 1963 collaboration with his colleague Sakuji Ikeda (1920–1984), who we will return to again here shortly. In this team effort, the two researchers published a review of known organometallic complexes as polymerization catalysts.[91] Although this review did include the catalytic polymerization of acetylene, the focus was primarily on the polymerizations of olefins.

After a bit of a break in publications, Hatano returned to publish two additional reports in 1967, at the very end of his time at Tokyo Tech.[92,93] The first of these reports again included Kanbara, along with Kaoru Shimamura from the University of Kyoto and Ichiroh Nakada from the University of Tokyo, and focused on the study of pressure effects on the conductivity of polyacetylene samples. Highly crystalline samples of polyacetylene were first produced via $Ti(OC_4H_9)_4/Et_3N$ and then pressed into solid pellets with a thickness of 0.07 mm that were fitted into a high-pressure cell for the conductivity measurements. The cell was held at one of several selected constant pressures (0, 9, 33, 71, or 108 kbar) while the conductivity was measured over range of temperatures from room temperature to that of liquid nitrogen (ca. −196°C).[92] The current-voltage linearity was examined for an applied voltage of 0–50 V, which revealed an Ohmic relationship. Furthermore, the temperature dependence was found to be typical of an intrinsic semiconductor at all applied pressures (Figure 5.17), consistent with that previously observed at atmospheric pressure. However, it was found that the overall conductivity increased with increasing pressure. These results ultimately led to the conclusion that the charge carrier density changed under pressure, but that any pressure dependence on the charge carrier mobility itself could not be determined.[92]

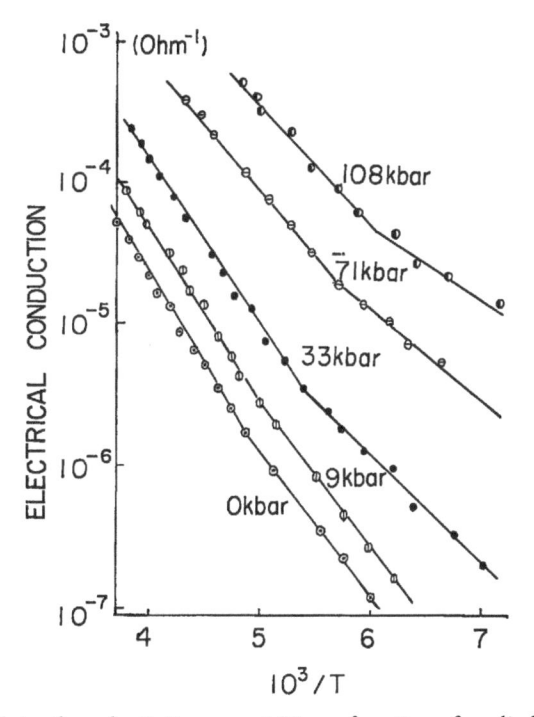

Figure 5.17 Plots of conductivity versus 1/T as a function of applied pressure.

Reproduced with permission from K. Shimamura, M. Hatano, S. Kanbara, I. Nakada. Electrical conduction of polyacetylene under high pressure. *J. Phys. Soc. Jpn.* **1967**, *23*, 578–581. © 1967 The Physical Society of Japan.

Hatano's final polyacetylene paper was a sole-author contribution that attempted to use the bulk of his research to date to present a more refined model of the polymer structure, a proposed mechanism for the polymerization that could account for these structural aspects, and a summary of the known structure–function relationships for polyacetylene.[93] Much of these efforts involved detailed IR spectroscopic studies of both polyacetylene and its deuterated analog. This led to the conclusion that the refractive index of the material was quite high for an organic substance (close to 2), and that the specific gravity was also large. The most significant vibration was a very strong absorption found at 1010 cm^{-1}, which was attributed to an out-of-plane bending vibration of a C–H unit contained within a *trans* –CH=CH– moiety. At the same time, a second transition at 740 cm^{-1} was concluded to correspond to *cis* –CH=CH– units, and it was found that the relative *cis* and *trans* content in the polymer was dictated by the polymerization temperature.[93]

These structural details and the corresponding temperature dependence then led to the proposed polymerization mechanism given in Figure 5.18. As proposed by Hatano, the addition of acetylene to the catalyst resulted in the cleavage of one of its π-bonds to give a *cis*-alkene unit, and additional acetylene additions then resulted in the formation of a trimeric intermediate (A) consisting of a *cis-cisoid* conformation. At this point, the oligo-ethylene chain could either cycle back to eliminate a molecule of benzene (Process 1), or growth could continue (Process 2) to generate an open chain of double bonds with a *cis-transoid* conformation (B). The *cis* conformation could also undergo isomerization (Process 3) to a corresponding all-*trans* form (C). This final process, however, was believed to have a sufficiently large activation energy compared with the other processes and was proposed to be mediated by the polymerization catalyst.[93]

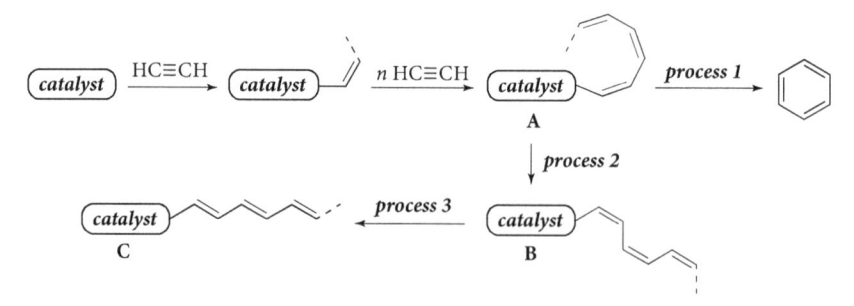

Figure 5.18 Hatano's proposed acetylene polymerization mechanism.

Efforts to probe the polyacetylene structure was then continued via X-ray diffraction of samples polymerized at 80°C, which was believed to consist of a near pure all-*trans* conformation based on the IR spectroscopic studies discussed earlier. The X-ray diffraction observed for these polymer samples suggested a hexagonal packing and exhibited features similar to that previously observed for carbon black. Hatano thus concluded that polyacetylene had a planar molecular structure and a layered crystal structure.[93] Surprisingly, no effort was made to compare these results with the previous X-ray studies of Natta.[73]

Hatano did, however, utilize the X-ray diffraction of various polyacetylene samples consisting of different degrees of crystallinity to develop a more detailed relationship between polymer crystallinity and resistivity.

While Hatano continued the study of polymers and polymerization processes after his move to Tohoku University in 1967, he did not pursue any further studies of polyacetylene. The established tradition of polyacetylene studies at Tokyo Tech, however, did not end with Hatano's departure. Rather, these efforts were continued by his colleague Sakuji Ikeda, who was first introduced earlier in this chapter.

Born in 1920, Sakuji Ikeda graduated from the Department of Applied Chemistry of Tokyo Tech in 1942.[3] He went on to receive his doctorate from the same institution in 1961, after which he joined the university's Chemical Resources Laboratory, becoming professor in 1967. He served as the director of the Chemical Resources Laboratory from 1973 to 1976,[94] after which the laboratory was relocated in 1977 to the Nagatsuta Campus of Tokyo Tech (now known as the Suzukakedai Campus).[95] In 1980, Ikeda moved to the National Institute of Technology, Nagaoka College, to become university president and remained so until his death in 1984, at the relatively young age of 64.

After the completion of his doctorate, Ikeda turned to the study of Ziegler–Natta polymerizations, beginning with a 1962 review of recent advances in polymer synthesis.[96] Although the bulk of this review focused on olefin polymerizations, Section V of the review covered the polymerization of triple bonds, including alkynes such as acetylene and its derivatives, as well as nitrile species. The polyacetylene work reviewed included that of Natta, Stille, and Hatano, as discussed earlier. This was then followed with the related 1963 review coauthored with Hatano on organometallic complexes as polymerization catalysts.[91] Again, as stated earlier, this focused primarily on olefin polymerizations but did include some acetylene polymerizations.

As discussed by previous authors such as Hatano,[93] the production of polyacetylene via Ziegler–Natta methods produces benzene as a byproduct. Ikeda took advantage of this fact in 1963 to investigate the synthesis of the isotopically labeled species benzene-$^{14}C_6$ and benzene-$^2H_6$ from either acetylene-$^{14}C_2$ or acetylene-$^2H_2$.[97] As it was known that the Al:Ti ratio of $TiCl_4$/Et_3Al catalyst mixtures could affect the extent of benzene production relative to polyacetylene, Ikeda began with finding the optimal ratio to maximize benzene production. This was accomplished by holding the amount of Et_3Al constant and varying the $TiCl_4$ content, from which he determined that a benzene yield of circa 70% could be reached by using Al:Ti ratios of 1.0–1.5. Using such conditions, he then showed that acetylene-$^{14}C_2$ and acetylene-$^2H_2$ could be polymerized at room temperature and low pressure to obtain the desired benzene-$^{14}C_6$ and benzene-$^2H_6$ in yields of about 70%. Analysis of the isotopic products then verified that all carbon and hydrogen content in the benzene products originated from acetylene and no exchange processes were involved in the polymerization process.[97]

This then followed with a 1964 report that applied acetylene-$^{14}C_2$ to probe the acetylene polymerization mechanism by following its polymerization via radio gas chromatography.[98] This began by polymerizing acetylene-$^{14}C_2$ using a $TiCl_4/Et_3Al$ catalyst mixture (using an Al:Ti ratio of 2.0), during which the residual gas was examined. It was found that this gas was primarily ethane, with a small amount of acetylene and just a trace of ethylene. While the ^{14}C level of the residual acetylene agreed with that of the acetylene-$^{14}C_2$ starting material, no detectable ^{14}C was found in the ethylene, and only a minor amount was detected in the ethane. The active catalyst was then quenched by adding propanol to the polymerization mixture, after which the evolved gas was again examined. Here, the gas was found to be primarily ethane, with a trace amount of ethylene. The ^{14}C level of the ethylene was nearly that of the initial acetylene-$^{14}C_2$, with the ethane again giving only a minor detectable amount. A series of reactions were then proposed as potential processes occurring during the polymerization of acetylene that could account for the observed results (Figure 5.19).[98] As the focus here was on the various small-molecule gases, no attempts were made to discuss any potential isomeric forms of the associated polymer species.

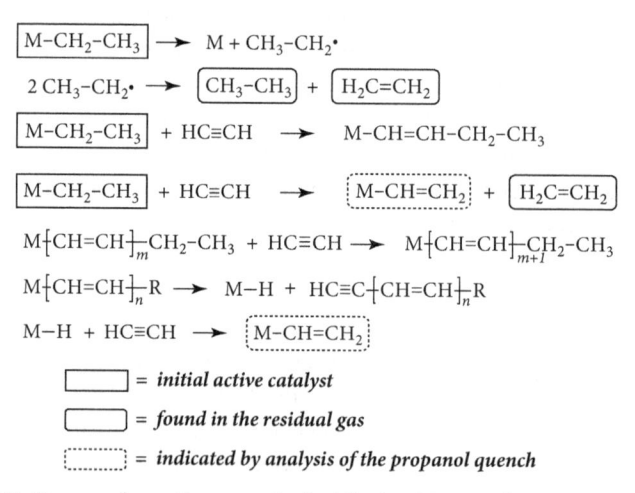

Figure 5.19 Proposed reaction steps in the Ziegler–Natta polymerization of acetylene to account for the results of radio gas chromatography.

Ikeda then followed this in 1966 with a more detailed study of the acetylene polymerization mechanism, with a focus on the formation of the benzene byproducts.[99] The study was initiated due the detection of alkylbenzene as a minor byproduct, which was thought to potentially be the key to elucidating the polymerization mechanism. For example, it was found that when TiCl$_4$/Et$_3$Al catalyst systems were used with an Al:Ti ratio of 2:1, benzene and ethylbenzene were formed in respective yields of 72 and circa 0.2%. In contrast, if trimethylaluminum (Me$_3$Al) was used in place of Et$_3$Al, toluene was produced rather than ethylbenzene. It was thus ultimately concluded that a bond must be formed between acetylene and the initial alkyl group of the aluminum cocatalyst before cyclization to produce the benzene ring. This conclusion was further confirmed by experiments utilizing ^{14}C and ^{2}H tracers, which revealed that both carbon and hydrogen from the ethyl groups of Et$_3$Al were introduced into benzene. As a result, the reaction steps given in Figure 5.20 were thus proposed to account for the observed benzene products.[99]

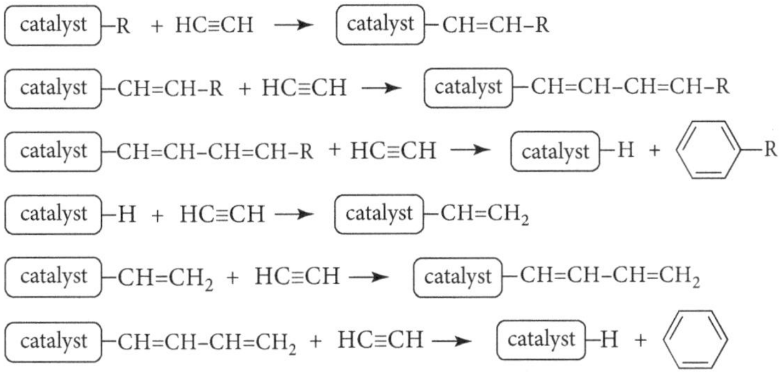

Figure 5.20 Proposed reaction steps to account for the formation of benzene byproducts during acetylene polymerization.

Ikeda continued with the study of Ziegler–Natta catalyzed polymerizations the following year, with a focus on accounting for the stereochemistry of the resulting polyolefin and polyacetylene products.[100] Extending the previous conclusions explaining the formation of alkylbenzenes during acetylene polymerization, it was proposed that the polyacetylene end groups could consist of either alkyl moieties incorporated from the AlR$_3$ (R = Me or Et) cocatalyst or phenyl groups formed via a cyclization step that removed the growing polymer from the active catalyst. In order to probe the relative *cis* versus *trans* configuration of the polyacetylene backbone, an investigation using IR spectroscopy was undertaken in a similar manner to the previous efforts of Hatano.[93] Polyacetylene samples polymerized at different temperatures varying from –78°C to 80°C were analyzed to reveal a clear trend of increasing *trans* configuration with increasing temperature. Unlike the previous study, however, Ikeda also quantified the relative amount of benzene formed at each temperature to find no real temperature effect on the ratio of benzene to polyacetylene.[100] Similar to

Hatano, Ikeda concluded that the *cis* configuration was initially formed as the acetylene was added to the growing polymer backbone. Such *cis* units, however, could be thermally converted to the *trans* configuration, although with a relatively large associated activation energy. Benzene formation was then thought to be dependent on whether the growing chain takes a *cisoid* structure or a *transoid* structure near the active catalyst.[100] These results, along with the resulting proposed processes, were then assembled to lead to a refined mechanism for the polymerization of acetylene (Figure 5.21).

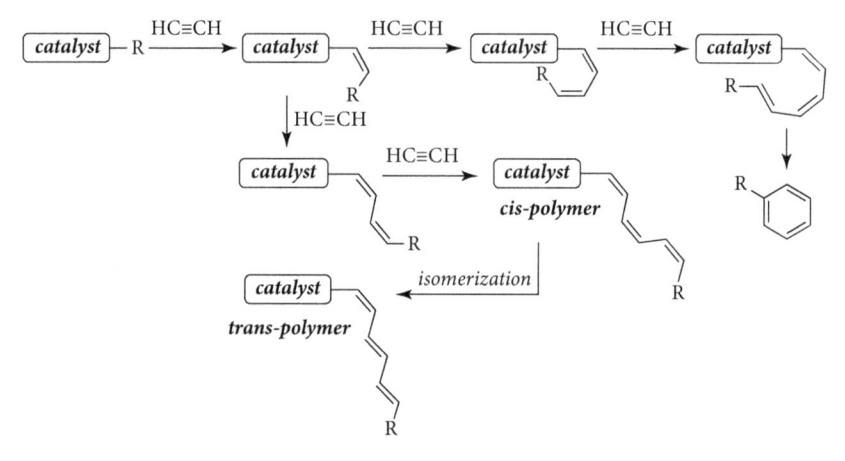

Figure 5.21 Proposed polymerization mechanism to account for both *cis*- and *trans*-polyacetylene, as well as the formation of benzene.

While the bulk of Ikeda's efforts focused on advancing mechanistic details of olefin and acetylene polymerizations, he also worked to develop new catalytic species.[101–104] In terms of applications to acetylene polymerizations, the most successful of these was a new iron complex, Fe(bpy)$_2$Et$_2$ (where bpy = 2,2'-bipyridine), first reported in 1965 as a collaboration with Akira Misono and coworkers from the University of Tokyo.[101] This complex was reported to catalyze the cyclization of butadiene (Figure 5.22) but was also capable of catalyzing the polymerization of acetylene to give both benzene and polyacetylene. It was stated that a detailed study of this polymerization would be reported at a later date, but this does not appear to have happened. A more detailed study was reported later in 1968, but this focused only on the reactions of butadiene.[104]

Of course, the most well-known work to have come from Ikeda's group was the discovery of methods for the generation of polyacetylene thin films. These synthetic methods and the corresponding characterization of the new form of polyacetylene were then reported throughout the 1970s.[105–110] This work was

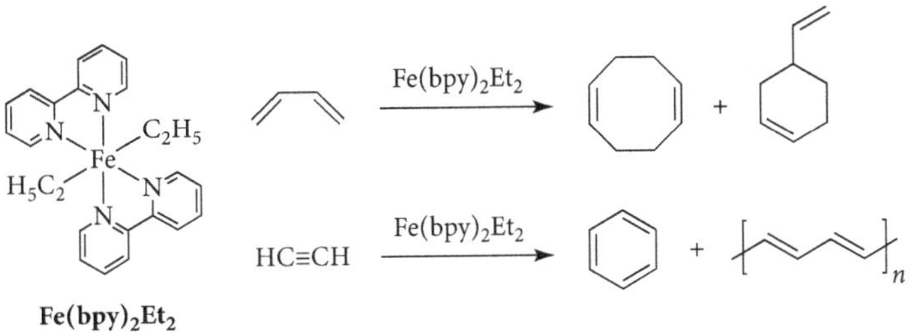

Figure 5.22 Reactions catalyzed by Fe(bpy)$_2$Et$_2$.

primarily carried out by Hideki Shirakawa (b. 1936), who joined Ikeda's group at Tokyo Tech as a new research associate in April of 1966.[3,5,6,111,112]

5.7 Shirakawa, Pyun, and polyacetylene films

Hideki Shirakawa (Figure 5.23) was born in Tokyo on August 20, 1936, to parents Fuyuno and Hatsutarou Shirakawa, the third of five children.[3,5,74,111,112] His father was a physician, and the family moved several times after Hideki's birth as his father transitioned from one position to another. They finally settled

Figure 5.23 Hideki Shirakawa (b. 1936).

Adapted from N. Hall. Twenty-five years of conducting polymers. *Chem. Commun.* **2003**, 1–4, with permission of the Royal Society of Chemistry.

in the small city of Takayama, his mother's hometown, in 1944. With a population of less than 60,000 and located in the middle of Honshu, Japan, it is there that Hideki spent his childhood.[3,5,111,112]

In April of 1957, he entered Tokyo Tech,[111,112] receiving a BS (1961), MS, and a doctorate (1966) during his time there.[6,75] For his undergraduate studies, he focused primarily on applied chemistry and had applied to work in a laboratory carrying out polymer synthesis for his undergraduate thesis during his final year. Unfortunately, the lab had too many applicants at the time, and he ended up joining a polymer physics laboratory instead. While Shirakawa was initially reluctant to work on polymer physics, he realized in hindsight what an important impact this experience had on his later career.[3,5,111,112] However, he was finally able to focus on his primary interest of polymer synthesis during his five years of graduate studies,[113] receiving a Doctor of Engineering (Dr. Eng.) degree in March of 1966,[5,111,112,114] with a dissertation on the study of block copolymers.[114] That same year, he also married his wife Chiyoko Shibuya.[5,112]

Following the completion of his graduate studies, he stayed at Tokyo Tech to join the group of Sakuji Ikeda as a research associate.[5,6,74,111,113,115] During 1976–1977, he also spent one year working with Alan G. MacDiarmid (1927–2007) and Alan J. Heeger (b. 1936) at the University of Pennsylvania.[5,6,75,111,112] He then moved to the University of Tsukuba in November of 1979, where he was appointed associate professor of the Institute of Materials Science.[5,6,75,111,112] In October of 1982, Shirakawa was then promoted to full professor, and he formally retired from Tsukuba at the end of March 2000 after nearly 21 years.[5,75,111-113] Although retaining the rank of professor emeritus, he then withdrew from further research or educational activities.[112]

In addition to being awarded the 2000 Nobel Prize in Chemistry along with Heeger and MacDiarmid,[5,75,111,112] Shirakawa also received a number of other significant awards. This included the 1982 Award of the Japan Society of Polymer Science in May of 1983 and the 1999 Award for Distinguished Service in Advancement of Polymer Science from the Japan Society of Polymer Science in May of 2000. Finally, he was honored by the Japanese government in November of 2000 with the Order of Culture.[5,75,112]

The story behind the critical discovery of generating polyacetylene as lustrous, silvery films is well known within the field of conducting polymers. In the process, it has achieved legendary or mythical status,[3,116] having been told and retold by many different authors over the years, with the story rarely told the same way twice.[2-6,74,77,111,112,115,126] What is generally agreed upon is that this discovery was made in October of 1967, approximately a year and a half after Shirakawa had joined Ikeda's group at Tokyo Tech.[74-77,112,117] However, even this fact has been distorted in some accounts, with the discovery incorrectly dated to the "beginning of the 1970s,"[6] "the early 1970s,"[119,126] 1974,[118] and

1975.[120] According to Shirakawa,[74-77,112,117] a visiting scientist was preparing a sample of polyacetylene, but rather than the insoluble black precipitate normally generated, "ragged pieces of a film"[112] had been produced on the surface of the catalyst solution instead. It was eventually determined that the researcher had mistakenly used a circa thousand-fold excess of catalyst, and it was the usually high catalyst concentration that had caused the surprising result, what Shirakawa has referred to as a "fortuitous error."[74-77]

Of course, the various accounts given by Shirakawa prior to 2022 were not very detailed, which has led others to fill in specifics based on their own preconceptions. It is such embellishments that have resulted in the multiple, and often erroneous, versions of this event. One of the most common variable aspects given in the many retellings of this event involves the identity and nature of the researcher who made the critical error. While the researcher's gender is generally viewed as male, and Shirakawa refers to him as a "visiting scientist" in his Nobel autobiography,[112] he has been described in various accounts as a "student,"[6,119,122,123] "Shirakawa's student,"[119] "foreign student,"[122] "graduate student,"[119,121,123] "Korean graduate student,"[123] "Korean visitor,"[124,126] and "visiting Korean researcher."[6] It is only in the acknowledgment of his Nobel Lecture that Shirakawa specifically reveals the name of the researcher to be Dr. Hyung Chick Pyun.[74-77]

Hyung Chick Pyun (more commonly given in Korea as Byun Hyung Jik or Byeonhyeongjik) was born on December 23, 1926, in Bongsan county of Hwanghae province, located in the northwest of Korea.[3,116,127] In 1953, at the end of the Korean War, this region became part of North Korea. His family moved to Seoul in 1936, and it was there that he received the bulk of his education.[128] He attended Kyungdong High School[3,121] until April 1945, when he continued his education in Japan at the Sixth High School (which later become part of Okayama University in 1949). Pyun then returned to Seoul in October 1945 to enter the preparatory school of Kyungsung University.[128] Kyungsung University became part of Seoul National University the following year, with Pyun finishing his preparatory studies in 1947.[127] He then entered Seoul National University's Department of Chemical Engineering, where he completed a BS in chemical engineering in 1951. With the onset of the Korean War (1950–1953), Pyun began work at the Science Research Institute of the Ministry of National Defense in December of 1950, while also working to complete his university studies.[3,115,127,128]

Pyun continued to work at the Science Research Institute until 1960, at which point he moved to the newly established Korea Atomic Energy Research Institute[129] (KAERI).[3,127] He then visited the University of Kansas in February 1961,[127,128] where he spent a year working with William E. McEwen (1922–2002).[3,116] His first published papers then followed in 1964,[130,131] the first of which was based on his work with McEwen.[130] In 1967, he received funding from the International Atomic Energy Agency (IAEA)[3,121,132] for a nine-month visit to Tokyo in order to work on a joint project between Sakuji Ikeda at Tokyo Tech and Yoneho Tabata (b. 1928) of the Nuclear Engineering Department at the University of Tokyo from May 1967 to March 1968.[116,128,132,133] Although it has been reported that Pyun had already acquired his doctorate before his work with Ikeda,[115] this is incorrect, and it was after his return to Korea, in 1970, that he received his Ph.D. in nuclear engineering from Seoul National University, based on work he had previously published in 1964.[3,116,127,134] Although he expressed a desire to pursue polyacetylene research after his return from Japan, he stated that his superiors discouraged him from doing so.[132] As such, the only publication that appears related to his time in Tokyo was a 1969 report comparing the copolymerization of phenylacetylene and styrene via gamma irradiation to Ziegler–Natta catalyzed methods.[135] By 1970, however, his work turned primarily to polymeric materials and composites, which seemed to be the focus of this research for the rest of his career. Pyun was made chief of the Radiation Chemistry Laboratory in 1973, and director of the Radiation Division in 1979, and he ultimately retired from KAERI in 1991.[127] On March 8, 2018, Pyun passed away after an extended illness.[116,128]

Based on the biographical material mentioned earlier, Pyun was already 17 years into his career at the time of his visit to Ikeda's laboratory and thus clearly not a student as often stated. Although he did later receive a Ph.D., the topic of his dissertation was not at all related to his work in Ikeda's lab but, rather, appears to be connected to his previous work with McEwen at the University of Kansas.[134] Thus, at the time of his visit to Japan, he would more correctly be considered a visiting researcher or scientist, although very few accounts describe him this way.

At the time of the fortuitous error, Pyun was attempting to polymerize acetylene using methods nearly identical to that previously used by both Natta[73] and Hatano[89] (Figure 5.24). In an attempt to make sense of the unusual result, Shirakawa analyzed the polymerization conditions again and again to ultimately determine that the amount of catalyst was nearly 1000 times greater than intended.[6,74–77,112,115,113,122–124] The excess catalyst resulted in a highly concentrated catalyst solution, which accelerated the rate of the polymerization to the point that, rather than polymerizing in solution to give the typical black precipitate, polymerization occurred at the air–solvent interface or along the walls of the vessel that had been wetted by the catalyst solution to give the observed polymer films.[6,102,117]

Of course, another variable point in the many retellings of this event is the specific reason for the error itself. While most accounts point to miscommunication as the cause, the nature of the miscommunication differs even within

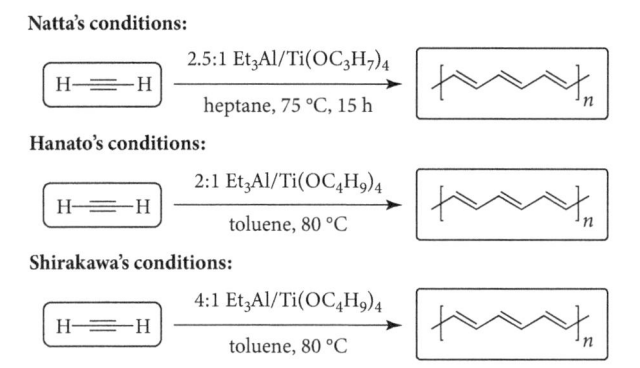

Natta's conditions:

$$H\text{—}\!\!\!\equiv\!\!\!\text{—}H \xrightarrow[\text{heptane, 75 °C, 15 h}]{\text{2.5:1 Et}_3\text{Al/Ti(OC}_3\text{H}_7)_4}$$

Hanato's conditions:

$$H\text{—}\!\!\!\equiv\!\!\!\text{—}H \xrightarrow[\text{toluene, 80 °C}]{\text{2:1 Et}_3\text{Al/Ti(OC}_4\text{H}_9)_4}$$

Shirakawa's conditions:

$$H\text{—}\!\!\!\equiv\!\!\!\text{—}H \xrightarrow[\text{toluene, 80 °C}]{\text{4:1 Et}_3\text{Al/Ti(OC}_4\text{H}_9)_4}$$

Figure 5.24 Comparative experimental conditions for the polymerization of acetylene.

the accounts of the three Nobel laureates. For example, Shirakawa has said, "I might have missed the 'm' for 'mmol' in my experimental instructions, or the visitor might have misread it. For whatever reason, he had added the catalyst of some molar quantities in the reaction vessel."[112] In contrast, MacDiarmid has explained, "I asked him how he [Shirakawa] had made this silvery film of polyacetylene and he replied that this occurred because of a misunderstanding between the Japanese language and that of a foreign student."[122] Heeger too has pointed to issues with the Japanese language, stating "he [Shirakawa] had a Korean visitor who misunderstood what he said in Japanese."[124] The claims that Pyun might have limitations with the Japanese language, however, are at odds with the fact that he had grown up during the years that Korea was under Japanese occupation (1910–1945), and thus spoke fluent Japanese, a fact confirmed by Shirakawa for the chemist and historian István Hargittai.[115]

Pyun's comments on the event were finally reported in 2021,[116] after focused efforts to find Pyun in Korea had proved successful in 2017. Although it was found that Pyun was still alive, he was quite ill at the time and had been in the hospital since the fall of 2016, which made it impossible to speak with him directly. However, Pyun had prepared a written account[132] prior to his failing health, which his son Dr. Joongmoo Byun provided so that his father's memories of the event could be documented and reported.[116] A previous version of this account had also been published in a KAERI publication in 2002.[133]

Of particular interest is the fact that Pyun's version of the events differs drastically from the many other accounts discussed earlier. According to Pyun,[132,133] he arrived at Tokyo Tech in 1967 to begin work on a collaborative project between Ikeda and Tabata, with the goal of studying the copolymerization of ethylene and tetrafluoroethylene (TFE) via the IR analysis of isotopically labeled species. Deuterated ethylene was to be prepared in Ikeda's laboratory, which would then be copolymerized with TFE at the University of Tokyo. The work in Ikeda's lab had been successfully completed, but work at the University of Tokyo was postponed, as Tabata was visiting the United States and his return had been delayed. While waiting for Tabata's return, Pyun said that he became interested in the polyacetylene studies being carried out in Ikeda's lab and felt that its properties could be improved if produced as larger polymer particles. He proposed that this could be accomplished by decreasing the stir speed during polymerization and claims that he began investigating this on his own, stating,[132]

> One day . . . the stirring motor stopped during the experiment because the stirring speed had been excessively reduced. I was very embarrassed at first, but after a closer look, I found it surprisingly to see a silver film on the surface of the reaction solution. It was nothing more than a polyacetylene film . . . I realized that the scientists who were studying this field had not synthesized the acetylene in the film state because the polymerization reaction had proceeded with stirring. That is, if it is not stirred, it is allowed to polymerize in the film state. Stirring thereby was hindering film formation.

Pyun went on to state that he repeated the film production multiple times and, thinking that Ikeda would be pleased by the results, went to his office to share his progress. According to Pyun,[132,133] however, Ikeda was not pleased and reminded him of the assigned project with Professor Tabata, who had now returned from the United States. Pyun was then sent to Tokyo University to finish his project with Tabata there. Pyun stated that Shirakawa later asked him to demonstrate the film formation during a return to Ikeda's lab to collect a sample of deuterated ethylene. Pyun claimed that he showed Shirakawa his method in detail, after which Shirakawa was then able to reproduce his results. Meanwhile, the research in Tabata's lab was proceeding smoothly, but he was running out of time due to the initial delay, so Pyun decided to apply to the IAEA for an extension to his visit. However, this required Ikeda to provide a recommendation letter, which he declined to give, thus denying any extension.[132] Pyun thus returned to Korea in March of 1968.

While Pyun's account can offer new insight into the aforementioned "fortuitous error," it also contains unfortunate inconsistencies not supported by our modern understanding of the polyacetylene film formation.[116] Most notably, Pyun insists that he did not make any errors in the experimental conditions, that he did not use excess catalyst, and and that he had rather purposely reduced the stirring rate, thus resulting in production of polyacetylene films.[132,133] His claim that this innovation was merely the result of reduced stir rate, however, contradicts a wealth of available evidence. To begin with, the original production of linear polyacetylene by Natta in 1955 had used very similar conditions to those of Ikeda's group, as illustrated in Figure 5.25.[71-73] More critically, Natta had reported polymerizations performed both with and without stirring, all of which gave only crystalline powders, never anything that could be considered a freestanding film. In addition to this, the critical requirement of catalyst concentration for generation of the polyacetylene films was independently confirmed by multiple groups after Shirakawa and Ikeda finally reported the detailed experimental procedure in 1974.[3,107,136] Shirakawa and Ikeda then reported the macroscopic morphology of polyacetylene prepared at different catalyst concentrations in 1980.[137] In the process, it was confirmed that only powders were formed at low catalyst concentration, with the production of films requiring roughly 100–1000 times that normally used for the synthesis of powders. Even so, the films produced at the lower end of these concentrations were better classified as gels, with no real clear border between gel and film formation.

Finally, Herbert Naarmann at BASF, along with coworkers from the University of Montpellier, published a detailed study in 1987 that probed the effect of polymerization conditions on the resulting properties of the polyacetylene formed.[138] The first of the conditions analyzed was the catalyst concentration, with the morphology of the polyacetylene product directly related to the concentration of the $Ti(OC_4H_9)_4$ applied. Consistent with both the 1980 Shirakawa–Ikeda study discussed earlier and the results obtained prior to the discovery of polyacetylene films, only powder products were found to form from unstirred solutions at low catalyst concentrations. In contrast, true films were only formed via the application of high catalyst concentrations, with intermediate concentrations affording what was described as a gel. As such, there is no experimental or literature support for Pyun's belief that he produced films without an excess of catalyst in the original experiments. Of course, it must be pointed out that the lack of stirring does play a role, as the film is formed at the gas–solvent interface. As such, an unstirred solution provides a calm, undisturbed surface optimal for the production of smooth, uniform films, and it is perhaps not coincidence that Pyun did not observe film formation until the stirrer failed.

In response to the publication of Pyun's account in 2021, Shirakawa then elected to provide his own further recollections of the event in 2022, largely contesting various points claimed by Pyun.[87] According to Shirakawa, Pyun came to his office and asked to try an acetylene polymerization, as he was interested in its polymerization and the polyacetylene product. Shirakawa thus provided an experimental protocol for the polymerization as a trial experiment in his research laboratory at Tokyo Tech. Unfortunately, as this trial was not Shirakawa's own work, any relevant data, including the date and specific conditions for the polymerization, were not recorded in his laboratory notebook and the fate of Pyun's notebook is unknown, but long lost. Shirakawa, however, states that he gave Pyun normal guidelines for the polymerization, including the solvent, the catalyst components, and their corresponding concentrations. As such, he expected a powder product as usual. After briefly instructing Pyun on the handling of lab apparatus and equipment, Shirakawa recalls returning to his office, a preparatory room next to the lab.

According to Shirakawa,[87] Pyun returned to his office later that same day to say that the polymerization had stopped. As the experiment was considered relatively easy to perform, Shirakawa was puzzled by what could have gone wrong. After carefully checking the apparatus, he found that the magnetic stirrer had stopped, and the manometer did not indicate any decrease in acetylene pressure. After confirming that the acetylene was not actively undergoing polymerization, the reaction flask was detached from the vacuum line in order to inspect the interior. Rather than the powder typically produced under normal conditions, it was found that a "black flappy or spongy matter" had formed on the surface of the catalyst solution. Extracting the product using a pair of tweezers, the material was found to be "like a black rag."[87]

Shirakawa states that he still does not fully understand why Pyun's trial run failed, but he felt that the results suggested that it might be possible to synthesize polyacetylene as a thin film under the right polymerization conditions.[87] As such, he immediately initiated a series of experiments with graduate students under his guidance that investigated the effects of variable catalyst concentration, along with other conditions. Shirakawa insists, however, that Pyun was not involved in these additional experiments.

Within a couple weeks, Shirakawa and his students were able to successfully synthesize freestanding films of polyacetylene by increasing the catalyst concentration by 1000 times or greater.[87] Furthermore, it was found that thin films with a silvery metallic sheen could be obtained by first applying a concentrated catalyst solution to the glass surface of the flask, after which acetylene was polymerized onto the catalyst-coated glass. Under such conditions, the side stuck to the glass surface displayed a metallic luster, while the other side took on a black color with a matted texture. Due to the added requirements for making these silvery films, it is pointed out that Pyun's claim to have made a silver film in the initial experiment seems to be not possible.[87]

Unfortunately, Shirakawa still does not give any indication as to what led to the fortuitous error, or what led him to suspect that the catalyst concentration was the critical factor in the film formation. Rather, the most recent account stresses that "Pyun's contribution was minimal" and goes out of its way to remove any mention of films from the discussion of Pyun's initial experiment (although that still seems to be what is being described).[87] It has been pointed out, however, that this new narrative does not seem to be consistent with multiple statements Shirakawa has made in the past.[139] In addition to the previously mentioned quote from his autobiography that describes Pyun's initial product as "ragged pieces of a film,"[112] additional statements include the acknowledgment "to Messrs. H. C. Pyun and T. Ito for the preparation of poly(acetylene) films" in his 1971 paper,[107] as well as the notable acknowledgment to Pyun in his Nobel Lecture:[75] ". . . and to Dr. Hyung Chick Pyun with whom I encountered the discovery of polyacetylene film by the fortuitous error." Still, while one can argue details and semantics, it is clear that Pyun's fortuitous error was a critical event in the development of polyacetylene films and precipitated the significant research on this new form of polyacetylene that followed.

Another intriguing aspect of this discovery was the long delay in publishing the new methods for generating the polyacetylene films. Although the initial discovery was made toward the end of 1967, and Shirakawa claims that successful conditions for the synthesis of freestanding polyacetylene films were determined within a couple weeks, it was not until 1974 that Shirakawa and Ikeda published a detailed synthetic account of the film formation process.[107] Still, this did not stop them from reporting studies on the polymer films, with the first of their papers on the film's characterization submitted in November of 1970.[105] In a reflective paper published in 1996,[117] Shirakawa addressed the reasons for the six-year delay in submitting the film formation details for publication, stating that the main reason was that the film was not metallic but, rather, semiconducting. Another factor, however, was that the film formation was based on interfacial polymerization, which was viewed to be limited and thus not that important a result. To illustrate the point, he emphasized that the presentation of the film formation at Japanese meetings in 1968 generated very little interest, and that it was really only after the first reports of the chemical doping of the film in 1977,[140,141] and the associated increased conductivity, that serious interest in polyacetylene films developed.[117] While these factors could have certainly played a role, it should also be pointed out that Ikeda and Shirakawa submitted a patent application on the method of producing polyacetylene films in 1970, with the application publication published in October 1973 and the Japanese patent ultimately granted in March 1974.[142] The decision by the patent office to publish the application publication occurred in July 1973, after which their journal manuscript on the process was submitted in August 1973. The timing here could be coincidental, but it is quite realistic that they were waiting to report the details of the process until they felt that it was suitably protected by a forthcoming patent.

With a general method for the film production in hand, initial studies reported in 1971 focused on the effect of the polymerization temperature on the resulting films, similar to previous efforts by both Hatano[93] and Ikeda[100] to understand such temperature effects on polyacetylene powders. Polyacetylene films and various deuterated derivatives were thus prepared over the range of −100°C–180°C, after which the IR spectra of the resulting films were analyzed in order to gain insight into the molecular structures making up the polymer films.[105] Polymer films prepared at low temperatures (< −78°C) were described as having a "copper-like luster"[105] (Figure 5.25), with IR spectral analysis leading to the conclusion that the polyacetylene molecular structure was most consistent with an all-*cis* configuration (either *cis-transoid* or *trans-cisoid*, as illustrated in Figure 5.26).[105] Further analysis of Raman spectra in 1973 led to the conclusion that the configuration of the *cis*-polyacetylene was specifically *cis-transoid*.[106]

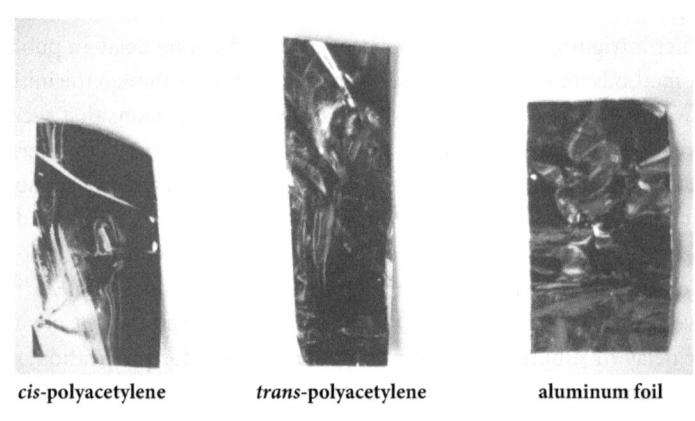

cis-polyacetylene   *trans*-polyacetylene   aluminum foil

Figure 5.25 Samples of *cis*- and *trans*-polyacetylene films compared with aluminum foil.

Reproduced courtesy of Richard Kaner, UCLA.

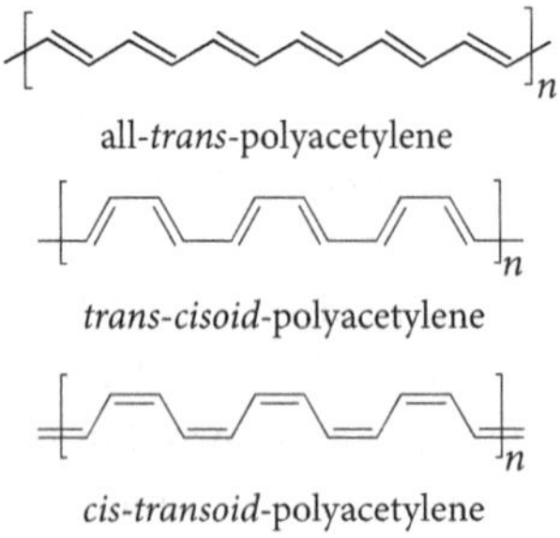

all-*trans*-polyacetylene

trans-cisoid-polyacetylene

cis-transoid-polyacetylene

Figure 5.26 Isomeric structural forms of polyacetylene.

In contrast to the low-temperature conditions, films prepared above 150°C were described as an "intense black material with a metallic luster,"[105,106] with this luster typically described as being silver in color (Figure 5.25).[74-77] Here, analysis of both the IR and Raman spectra led to the conclusion that these high-temperature materials were most consistent with an all-*trans* structure

(Figure 5.26).[105-108] More detailed results of the *cis* and *trans* content of poly-acetylene films prepared at various temperatures were then reported in 1974. As shown in Table 5.2, films prepared at –78°C were determined to consist of circa 98% *cis* content, while the *cis* content decreased to circa 60% for materials prepared at room temperature.[107]

Table 5.2 Relative *cis* versus *trans* content for polyacetylenes prepared at various temperatures

Polymerization temperature (°C)	% *cis* content	% *trans* content
–78	98.1	1.9
–18	95.4	4.6
0	78.6	21.4
18	59.3	40.7
50	32.4	67.6
100	7.5	92.5
150	0.0	100.0

Note: Values collected from Reference 107.

Finally, it was found that films prepared comprised of *cis*-polyacetylene could be converted to its *trans*-form via a *cis-trans* isomerization by heating the film at 200°C for 10 min.[108] This isomerization was found to occur at temperatures as low as 75°C, but at quite slow rates. Thermogravimetric analysis (TGA) of *cis*-rich (ca. 88% *cis* content) polyacetylene revealed an exotherm at 145°C that was not present in all-*trans* samples, which was thus assigned to the irreversible *cis*-to-*trans* isomerization of the polymer backbone. As a consequence, either the polymerization of acetylene above 150°C or heating *cis*-polyaceytlene above 150°C resulted in the isolation of the all-*trans* polyacetylene. At higher tempera-tures above 325°C, both the *cis*- and *trans*-polyacetylenes were found to undergo decomposition to give a brown solid.[108]

A study of the effect of the *cis* versus *trans* content on the electronic prop-erties of polyacetylene was then reported in 1978.[110] As shown in Figure 5.27, the resistivity underwent a near-linear decrease as the *trans* content increased, although samples with *trans* content greater than 80% did exhibit a tendency to increase in resistivity again as the *trans* content continued to approach 100%. For samples near 100% *trans* content, the resistivity was consistent for both samples polymerized at high temperature and samples thermally iso-merized. Overall, the all-*cis* samples exhibited conductivities of 10^{-9}–10^{-8} S cm^{-1}, while the analogous all-*trans* samples gave higher values of 10^{-5}–10^{-4} S cm^{-1}.[110] Interestingly, the values of the all-*trans* films were essentially the same as those previously reported by Watson[85] and Hatano[89] for highly crystalline polyacetylene powders. As Hatano had shown that the conductivity of poly-acetylene increased with the sample's crystallinity, it might be expected that the films would provide higher conductivity due to increased macroscopic order. However, this did not seem to be the case as the intrinsic electrical properties of polyacetylene in either form seemed to be nearly identical. In a similar fashion, X-ray diffraction of the polyacetylene films[107] also gave nearly identical data to that previously collected by Natta on polyacetylene powders.[73]

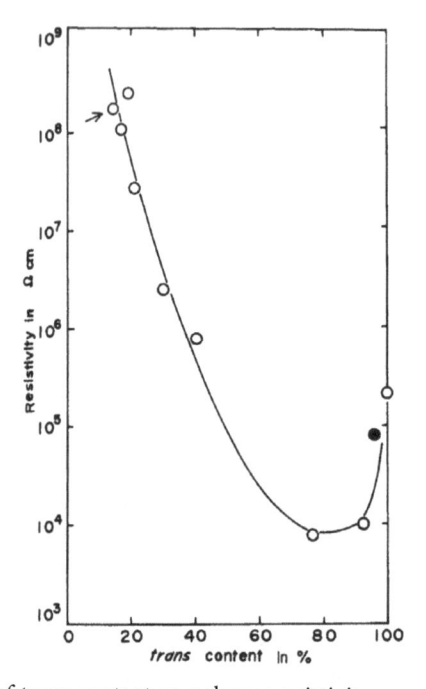

Figure 5.27 Effect of *trans*-content on polymer resistivity.

Reproduced with permission from H. Shirakawa, T. Ito, S. Ikeda. Electrical properties of polyacetylene with various *cis-trans* compositions. *Makromol. Chem.* **1978**, *179*, 1565–1573. © **1978** Hüthig & Wepf Verlag.

5.8 Berets, Smith, and effects of gases on the conductivity of polyacetylene

About the same time that the "fortuitous error" initially led to the development of polyacetylene films at Tokyo Tech, Donald J. Berets (1926–2002) and Dorian S. Smith (1933–2010) at the American Cyanamid Company in the United States were carrying out studies on the effects of gaseous additives on the conductivity of polyacetylene powders.[143] Donald Joseph Berets was born in July 1926, in New York City, and was educated at Harvard University, where he obtained a bachelor of arts (AB) degree in 1946, followed by an MA in 1947.[144] He then continued graduate studies at Harvard in physical chemistry under George Bogden Kristiakowsky (1900–1982), finishing his Ph.D. in 1949[144-146] with a dissertation titled "Studies on the Detonation of Explosive Gas Mixtures."[146] Berets then spent a year as a postdoctoral fellow at the Massachusetts Institute of Technology (MIT).[144] Following his year at MIT, Berets moved to Stamford, Connecticut, to begin a career in research and development at the American Cyanamid Company.[144-145] His work at American Cyanamid focused largely on heterogeneous catalysts for oil refining, auto emissions control, and synthetic fuels,[144] although he also had publications and patents on electronic materials and devices.[143-147] In 1986, Berets retired from American Cyanamid at the age of 60, after which he formed the Chemists Group, an affiliation of approximately 150 independent consultants in chemistry and chemical engineering.[144,145]

In addition to his work at American Cyanamid, Berets served on the Stamford Water Pollution Control Board[145] and was very active in the ACS.[144,145] Within the ACS, he served for 35 years as councilor of the Western Connecticut local ACS Section, chaired the ACS Subcommittee on Professional Standards & Ethics from 1986 to 1989, chaired ACS's Council Committee on Professional Relations in 1990, and authored the "Chemist's Code of Conduct" adopted by ACS in 1994.[144] For his long service to ACS, he received the ACS Presidential Plaque and the Western Connecticut Section's Julius Kuck Service Award in January 2002. On February 2, 2002, Berets died in Stamford at the age of 75.[144,145]

Beret's colleague, Dorian Sevcik Smith, was born in 1933 and grew up in Winthrop Harbor, Michigan.[148] He attended Illinois State Normal University (now Illinois State University),[148,149] where in addition to his coursework, he played football and was part of the university's undefeated football team in 1950.[148] To honor this achievement, the 1950 Redbirds team was inducted into the Illinois State Athletics Percy Hall of Fame in 1990.[150] Smith completed a BS in Education in 1953,[149] after which the Teachers College Board of the State of Illinois appointed him as a faculty assistant for the 1953–1954 academic year.[151] He then moved to the University of Illinois at Urbana-Champaign,[148,149] where he pursued graduate studies in chemistry under Therald Moeller (1913–1997).[149] He received an MS in 1956, after which he completed his Ph.D. with a dissertation titled "Observations on the Rare-Earths: Chemical and Electrochemical Studies in Non-Aqueous Solvents" in 1958.[149]

Smith then spent the next 10 years as an industrial chemist,[148] largely at the American Cyanamid Company, and then briefly at the Enjay Chemical Company in New York.[143] Smith then left chemistry, however, to become a financial analyst for various firms, including Donaldson, Lufkin and Jenrette (DLJ), Chemical Bank, and Yamaichi International; for the last of these he also served as their director of research.[148] After spending the majority of his working life in New York City, he retired and moved to Wilmington, North Carolina, in 1996. Smith passed away peacefully at his home on December 4, 2010, at the age of 77.[148]

In September 1967, approximately one month before the discovery of polyacetylene films, Berets and Smith submitted a paper that detailed efforts to reduce impurities in polyacetylene in order to determine the extent to which it could be made highly conducting. In the process, however, they also investigated the effects of various additives on the conductivity of polyacetylene powders.[143] The polymer samples were prepared using Natta's methods,[71] with $Et_3Al/Ti(OC_3H_7)_4$ in heptane at 70°C. The resulting black precipitate was then washed extensively using methanol saturated with HCl and finally with pure methanol. The samples prepared in this way were found to contain 0.01% titanium and 0.05% aluminum as determined by ultraviolet emission spectroscopy. The oxygen content was found to be as low as 0.7%. Samples were then compressed into 13 mm diameter pellets, which were 1–2 mm thick and possessed a silvery metallic sheen.

Efforts then began by investigation of the effect of oxygen impurities on the polyacetylene conductivity. The sample with the least oxygen content (0.7%) gave a resistivity of 7.5×10^5 Ω cm, in good agreement with polyacetylene of medium crystallinity as previously reported by Hatano.[89] Various amounts of oxygen were then introduced to the samples, which revealed that the log of the resistivity increased approximately linearly with the oxygen content (Figure 5.28). Extrapolation of the linear fit to 0% oxygen suggested a resistivity of 3×10^5 Ω cm for such a hypothetical sample. As this value is still higher than that reported by Hatano for highly crystalline samples,[89] it was concluded that the sample crystallinity was "of equal importance with chemical purity to achieve high electrical conductivity."[143]

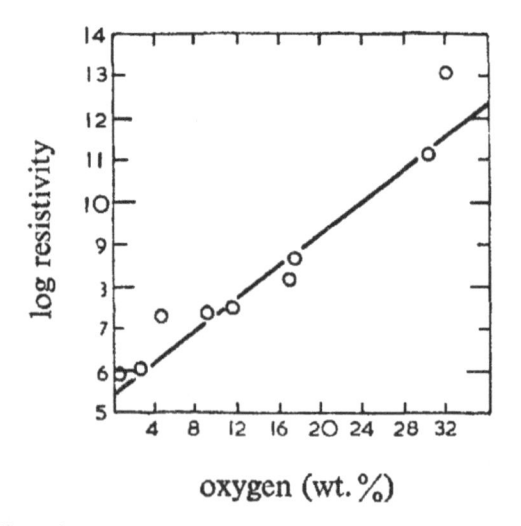

Figure 5.28 Effect of oxygen content on polyacetylene resistivity.

Adapted from D. J. Berets, D. S. Smith. Electrical properties of linear polyacetylene. *Trans. Faraday. Soc.* **1968**, *64*, 823–828, with permission of the Royal Society of Chemistry.

In the process of determining these oxygen effects, however, they observed an interesting phenomenon when oxygen initially came into contact with the polyacetylene sample:[143]

> On admission of 150 mm pressure of oxygen to the measuring apparatus (normally evacuated or under a few cm pressure of He gas), the resistivity of polyacetylene decreased by a factor of 10. If the oxygen was pumped off within a few minutes and evacuation continued at 10^{-4} mm pressure for several hours, the original electrical properties of the specimen were restored.

They went on to conclude that oxygen is first adsorbed in a reversible manner, which causes a reduction in the resistivity. Ultimately, however, oxygen reacts irreversibly with the polyacetylene, the result of which is the typically observed increase in resistivity. Of course, the initially reversible process proposed here represents the first recognition of what is now commonly referred to as oxygen doping, and it is only the enhanced reactivity of polyacetylene in comparison with other conjugated polymers that it responsible for the second irreversible process. The predominance of this second process in the more reactive polyacetylene had masked the initial process in other studies, and it had thus not been previously observed.

Berets and Smith then went on to investigate any potential effects of other gases on the polymer conductivity. The additional gases studied included BF_3, BCl_3, HCl, Cl_2, SO_2, NO_2, HCN, ethylene, NH_3, CH_3NH_2, H_2S, acetone, $(C_2H_5)_2CO$, N_2, and He.[143] While many of these gases had essentially no effect, it was found that electron acceptors (BF_3, BCl_3, Cl_2, SO_2, and NO_2), as well as the two acids (HCl and HCN), all resulted in a decrease in resistivity (i.e., an increase in conductivity). As previously observed for O_2, strongly oxidizing gases such as Cl_2 or NO_2 ultimately resulted in chemical reactions with the polymer. In contrast, electron donors such as NH_3 or CH_3NH_2 had the opposite effect, causing an increase in polymer resistivity. It was found that the most conductive samples were generated through the use of BF_3, resulting in a conductivity of ~ 0.0013 S cm^{-1}. In comparison with the initial value of circa 10^{-6} S cm^{-1}, this corresponds to an increase by three orders of magnitude, for which Berets and Smith provided the following explanation:[143]

> The effect on conductivity of the adsorbed electron-donating and electron-accepting gases is consistent with the p-type nature of the specimens. .... If holes are the dominant carriers, electron donation would be expected to compensate them and reduce conductivity; electron acceptors would be expected to increase the concentration of holes and increase conductivity; this is observed.

Although they didn't completely understand the effect of the gaseous treatments, they quite clearly state that the "electrical conductivity . . . depended on the extent of oxidation of the samples."[143] These results, however, did not seem to generate much interest, and the paper had only been cited nine times by 1977. Furthermore, Berets and Smith never followed up this investigation with any additional studies. In fact, this work was published just shortly before Smith left chemistry to become a financial analyst, and this seems to be the only paper he published after completing his Ph.D.

Although Shirakawa was well aware of the report of Berets and Smith (a third of the citations mentioned earlier came from Shirakawa and Ikeda),[106,107,110] neither Shirakawa, MacDiarmid, nor Heeger even mentioned this work as an influence contributing to the later, more successful doping of polyacetylene in 1977.[140,141] Rather, it is the related work of MacDiarmid and Heeger on poly(sulfur nitride)[152-157] that is always given as the motivation behind efforts to dope polyacetylene with halogens.[6,122,140,158]

5.9 MacDiarmid, Heeger, and poly(sulfur nitride)

Alan Graham MacDiarmid (Figure 5.29) was born on April 14, 1927 in the town of Masterton, on New Zealand's North Island.[122,123,158-160] His family was somewhat economically challenged, and they soon moved closer to Wellington, settling in the Wellington suburb of Lower Hutt, where it was believed that jobs were more plentiful at the time.[122,123,158-160] There, he was educated at the Hutt Valley High School until the age of 16, at which point his father retired and moved to the northern subtropical part of New Zealand. As a result, MacDiarmid was forced to leave high school in order take a part-time job as janitor and lab boy in the chemistry department of Victoria University College (now Victoria University of Wellington),[122,158-160] a constituent college of the University of New Zealand at the time.[158] The position consisted primarily of washing dirty labware, sweeping floors, and preparing demonstration materials for the faculty, but it also included stoking up the central heating during the night and supplying the kitchen ovens with coke.[122,158] Still, MacDiarmid was able to also find time to apply himself as a part-time student, initially taking only a course in chemistry and another in mathematics, and ultimately completed his BSc in chemistry in 1948.[122,123,158] After completing his BSc, he was then promoted to the position of demonstrator[122] and began studying the chemistry of S_4N_4 for his MSc thesis under Mr. A. D. (Bobbie) Monro (MSc 1922), the university's lecturer in chemistry.[6,122,123,158,159,161] This led to the publication of MacDiarmid's first paper in Nature in 1949,[162] after which he completed his MSc in chemistry the following year in 1950.[158,163]

Figure 5.29 Alan G. MacDiarmid (1927–2007).

Adapted from N. Hall. Twenty-five years of conducting polymers. *Chem. Commun.* **2003**, 1–4, with permission of the Royal Society of Chemistry.

That same year MacDiarmid received a Fulbright Scholarship to attend the University of Wisconsin for two semesters.[122,158,160,164] At the completion of the Fulbright Scholarship, he remained at Wisconsin to study inorganic chemistry under Norris F. Hall (1891–1962),[122,158–160] now supported by a University of Wisconsin research scholarship. Under Hall, MacDiarmid earned an MS in 1952,[158] followed by a Ph.D. in 1953[123,156,163,165] with a dissertation entitled "Isotopic Exchange in Complex Cyanide—Simple Cyanide Systems."[165] While still at Wisconsin, he obtained a Shell Oil graduate scholarship that was being offered in New Zealand, which allowed him to attend the University of Cambridge in England.[122,158–160] Thus, he left Wisconsin by ship in December 1953 to begin studies at Cambridge in 1954, where he worked under Harry J. Emeléus (1903–1993) on silicon hydrides.[122,123,158–160] While at Cambridge, MacDiarmid married Marian Matheiu,[122,158–160] whom he had met during his time in Wisconsin, and then completed his second Ph.D. in 1955,[123,158,161,164] with a dissertation entitled "The Chemistry of Some New Derivatives of the Silyl Radical."[166] He then held a temporary appointment as assistant lecturer at the University of St. Andrews in Scotland (1955–1956), while also hired to join the Department of Chemistry at the University of Pennsylvania that same year in 1955.[123,158–160,163] MacDiarmid eventually became a naturalized U.S. citizen in order to have the right to vote in U.S. elections.[122]

MacDiarmid maintained his position at the University of Pennsylvania until the end of his career. However, he also held positions at the University of Texas at Dallas, where he became the James Von Ehr Chair of Science and Technology and professor of chemistry and physics in 2002,[158-160] and at Jilin University in China, where he became professor of chemistry in 2004.[158] In addition to being awarded the 2000 Nobel Prize in Chemistry with Shirakawa and Heeger,[158-160,164] MacDiarmid also received the Frederic Stanley Kipping Award in Silicon Chemistry from the ACS in 1971,[160] the Francis J. Clamer Award from the Franklin Institute in 1993,[158] the Chemistry of Materials Award from the ACS in 1999,[158-160] the Rutherford Medal from the Royal Society of New Zealand in 2000,[158-160] and the William H. Nichols Medal in Materials Chemistry from the ACS.[159,160] In addition, he was made a member of the Order of New Zealand, the U.S. National Academy of Sciences, and the U.S. National Academy of Engineering, all in 2002.[123,158-160] The following year he was elected a Fellow of the Royal Society.[159,160] MacDiarmid continued to work until his death at the age of 79. He died on February 7, 2007,[158-160,164] after falling down the stairs in his home as he was rushing to catch a flight to New Zealand to visit with family and friends.[164]

MacDiarmid's colleague in physics, Alan Jay Heeger (Figure 5.30), was born on January 22, 1936, in Sioux City, Iowa.[124,167,168] His father's family came to the United States from Russia as Jewish immigrants in 1904, when his father was only four. While born in the United States, his mother was also a first-generation child of Jewish immigrants.[167,168] He spent his early years in nearby Akron, where his father was the manager, and later the owner, of a general store that served the local farming community, and it was in Akron that he began his primary education.[167-169] His father died in 1945 at the age of 45, when Alan was only nine.[167,168,170] As a result, the family moved to Omaha, Nebraska, in 1947, so that they could be closer to his mother's family.[167-170] There, he attended Omaha Central High School,[171] followed by undergraduate studies at the University of Nebraska.[124,167,168] His initial goal was to become an engineer, but he abandoned engineering after the first semester and completed his BS in 1957 with a dual major in physics and mathematics.[167,168,172]

Heeger then applied to number of universities for graduate school, including Cornell University, Stanford University, the University of California, Berkeley, and the California Institute of Technology (Caltech). He received acceptances from both Cornell and Berkeley and ultimately decided to attend Cornell. However, he only remained at Cornell for one year, as he did not have the funds necessary to continue his education there.[173] He then heard about a program at the Lockheed Missiles and Space Division in California that would allow him to work part-time while also continuing graduate studies, with the tuition costs covered by Lockheed. He thus applied and was accepted into the Lockheed

Figure 5.30 Alan J. Heeger (b. 1936).

Adapted from N. Hall. Twenty-five years of conducting polymers. *Chem. Commun.* **2003**, 1–4, with permission of the Royal Society of Chemistry.

program, after which he applied to the physics programs at both Stanford and Berkeley, gaining acceptance at Berkeley.[173]

As a result, Heeger then began graduate studies at Berkeley in 1958, while also working part-time for the Lockheed Missiles and Space Division in Palo Alto, California, where he worked on a silicon solar cell project for application to satellites.[167,168,173] This involved driving the 50 miles to Berkeley on Mondays, Wednesdays, and Fridays to attend morning classes, before returning to Palo Alto for work in the afternoon. This grueling schedule quickly took its toll, however, and he eventually resigned from the Lockheed program and moved to Berkeley in order to pursue full-time graduate studies.[167,173] At Berkeley, his initial goal was to do theoretical research under Charles Kittel (1916–2019) for his thesis. After approaching him, Kittel said that he was willing to take him on as a student, but he did not think that Heeger would make a first-rate theorist. As such, Kittel recommended that he should instead consider working with someone who does experimental work in close interaction with theory. Following this advice, Heeger joined the research group of Alan Portis (1926–2010).[167,168,173] Completing a dissertation entitled "Studies on the Magnetic Properties of Canted Antiferromagnets," Heeger was awarded his Ph.D. in 1961, only three years after starting at Berkeley.[173,174]

After a series of job interviews in May 1961, Heeger decided to join the Physics Department at the University of Pennsylvania (Penn). In the meantime, however, he stayed on in the Portis group for six months as a postdoctoral research fellow, while Portis was on sabbatical in Japan.[173] Thus, in late December 1961, he moved from Berkeley to Philadelphia, beginning his time as an assistant professor at the University of Pennsylvania in January of 1962.[175] At Penn, Heeger began with nuclear magnetic resonance (NMR) studies of magnetic materials. By the 1970s, however, his efforts had shifted to the metal physics of tetrathiafulvalene-tetracyanoquinodimethane (TTF-TCNQ, Figure 5.31), with specific interest in the conductivity of quasi-one-dimensional π-stacked molecular crystals.[167,168,176] This work led to serious and major controversy about the observation of potential superconducting fluctuations in an organic solid above 60 K,[156,174-177] which was ultimately determined to be flawed.[178,179] However, this work led to important new results and marked the beginnings of his interest in the conductivity of organic materials.[176] In August of 1982, Heeger left Penn to become professor of physics at the University of California, Santa Barbara (UCSB), where he currently holds the position of professor emeritus.[124,180] In addition to his faculty position as professor of physics, Heeger also served as the founding Director of UCSB's Institute for Polymers and Organic Solids (renamed as the Center for Polymers and Organic Solids in 2000).[180] In addition to being awarded the 2000 Nobel Prize in Chemistry with Shirakawa and MacDiarmid,[158,164] Heeger also received the Oliver E. Buckley Prize for Condensed Matter Physics from the American Physical Society in 1983,[176] and the Balzan Prize for the Science of New Non-Biological Materials in 1995. In addition, he was made a member of the U.S. National Academy of Sciences, and the U.S. National Academy of Engineering, both in 2002.[124]

The collaboration between the two Penn colleagues began in late 1974, after Heeger had become intrigued by reports of the inorganic species poly(sulfur nitride), $(SN)_x$, also known as polythiazyl.[6,122,124,126,158,167,168] It was first reported in 1910 as thin films exhibiting a metallic, bronze-like luster and was later shown to be conductive in the 1950s.[181] However, it was new reports[182,183] by Mortimer M. Labes (Figure 5.32) at nearby Temple University that had caught Heeger's attention. Attempting to produce more crystalline films, Labes had determined room temperature conductivities of circa 1000 S cm^{-1},[183] highly unusual for a one-dimensional polymer containing no metals. As such, Heeger was eager to study the metallic properties of the material.

Learning that MacDiarmid had some experience with sulfur nitride chemistry, Heeger approached him about collaborating on the study of this intriguing material.[6,122,157,167,168] Their efforts then began with the production of highly pure S_4N_4 and its conversion to the highly reactive S_2N_2. The spontaneous solid-state polymerization of S_2N_2 thus afforded the first reproducible preparation of analytically pure $(SN)_x$ as a lustrous golden material (Figure

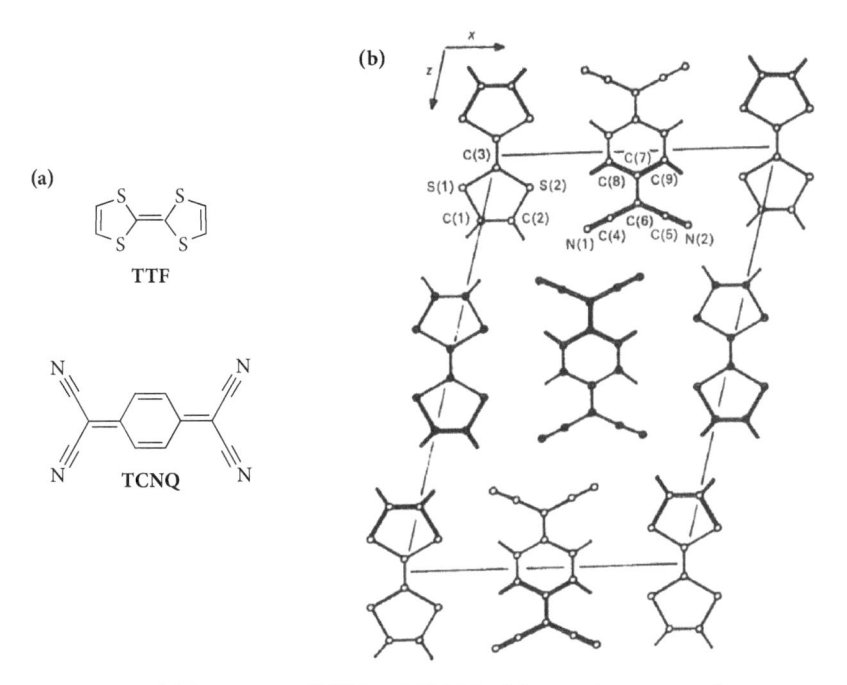

Figure 5.31 (a) Structures of TTF and TNCQ; (b) crystal structure of TTF-TCNQ showing alternating stacks of TTF and TNCQ.

Reproduced with permission of the International Union of Crystallography from T. J. Kistenmacher, T. E. Phillips, D. O. Cowan. The crystal structure of the 1:1 radical cation-radical anion salt of 2,2'-bis-1,3-dithiole {TTF} and 7,7,8,8-tetracyanoquinodimethane (TCNQ). *Acta Cryst.* **1974**, *B30*, 763–768.

5.33).[152,153] Further efforts then continued in 1976 with a detailed report of the crystal structures of both S_2N_2 and $(SN)_x$,[154] as well as the structural changes occurring during polymerization, followed by a second brief paper on the stability of $(SN)_x$ to O_2 and H_2O.[155] While the initial papers had reported conductivities up to 2500 S cm^{-1},[152] the polymer's electronic properties were reported in detail in a third 1976 paper, which was found to give room temperature conductivities of 1200–3700 S cm^{-1}, with the conductivity generally increasing with longer polymerization times.[156] These values were a marked improvement in comparison with that previously determined by Labes and coworkers for vapor-grown $(SN)_x$, which exhibited maximum conductivities of 1730 S cm^{-1}.[183] In addition, the temperature dependence of the conductivity was found to be consistent with that for ordinary metals, with values as high as 550,000 S cm^{-1} determined at 5 K.[155]

Their most significant result, however, came in a final paper in 1977 that followed up on previous reports that $(SN)_x$ reacted with halides.[157] Thus, high-purity crystals of $(SN)_x$ (ca. 1–10 mm^3) were treated with Br_2 vapor to produce $(SNBr_y)_x$ derivatives as shiny black crystals with a lustrous blue-purple tinge.

Figure 5.32 Mortimer M. Labes (1930–2023) in 1986.

Robert E. Dias, photographer, Special Collections Research Center, Temple University Libraries, Philadelphia, used with permission.

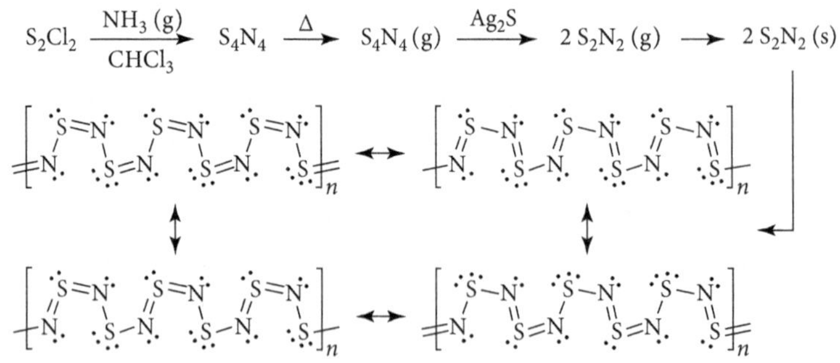

$$S_2Cl_2 \xrightarrow[\text{CHCl}_3]{\text{NH}_3\,(g)} S_4N_4 \xrightarrow{\Delta} S_4N_4\,(g) \xrightarrow{\text{Ag}_2S} 2\,S_2N_2\,(g) \longrightarrow 2\,S_2N_2\,(s)$$

Figure 5.33 Synthesis of S_4N_4 and its conversion to $(SN)_x$. Given resonance structures of $(SN)_x$ reflect proposed structures at the time.

The relative bromine content (i.e., y) was determined to be 0.38–0.42, with the bromide derivatives giving an average room temperature conductivity of 38,000 S cm^{-1}, a 10-fold increase over that found for the initial $(SN)_x$. Due to its metallic conductivity and the lack of metallic elements in its composition, poly(sulfur nitride) joined intercalated graphites and TTF-TCNQ in the growing class of materials known as *synthetic metals*.[184,185]

5.10 Doped polyacetylene films

In 1975,[116] shortly after starting the collaboration with Heeger,[126,167,168] MacDiarmid spent three-quarters of a year in Japan as a visiting professor at Kyoto University.[122,158] Toward the end of his visit in 1976,[119] he was asked to give a lecture on his silicon work at the Tokyo Institute of Technology, where he also described the ongoing $(SN)_x$ studies.[115,122,126,158] After the lecture, the head of the chemistry department invited MacDiarmid to join him for tea, and it was there that he first met Hideki Shirakawa, who had also been invited to join them.[6,122,123,126,158] As Shirakawa had not attended MacDiarmid's lecture,[126,158] MacDiarmid took out his sample of $(SN)_x$ during tea to show him its golden luster.[6,115,122,158] Upon seeing MacDiarmid's golden material, Shirakawa said that he had a similar material that was silver in color. After MacDiarmid expressed an interest in seeing it, Shirakawa returned to his lab to retrieve one of the silver-colored polyacetylene films to show MacDiarmid.[158] The polyacetylene film captured MacDiarmid's interest, and after returning to Penn, he contacted the ONR (Office of Naval Research) Program Officer Kenneth Wynne to inquire about possible supplemental funding in order to bring Shirakawa to the United States to work with them on polyacetylene.[123,158] In the end, Wynne agreed to provide $21,000 to support Shirakawa's visit,[6,123,126,158,167,168] and thus Shirakawa traveled to the United States to join MacDiarmid and Heeger as a visiting scientist in September of 1976,[87,112,115] although Heeger places his arrival a month later in October.[126]

Upon arriving at Penn, Shirakawa and MacDiarmid began by working to improve the purity of the polyacetylene films in an attempt to increase the conductivity.[122] As discussed earlier, Smith and Berets had already shown that decreasing oxygen content increased the conductivity of polyacetylene powders,[143] and thus limiting other impurities could possibly further increase the film's conductivity. Eventually, they were able to successfully produce films with purities as high as circa 98.6%.[158] Surprisingly, however, it was found that the enhanced purity did not result in greater conductivity as hoped but, rather, that the conductivity decreased with increasing purity.[122,158] In the process, the conductivity's temperature dependence was also investigated in order to determine whether the polyacetylene films exhibited the temperature profile of a metal or semiconductor.[158] As it had been previously reported in 1961 by both William Watson[85] and Masahiro Hatano[89] that the conductivity of polyacetylene powders exhibited a temperature dependence typical of semiconductors, it is unclear if they were unaware of the previous work at the time, or if it was thought that the film might somehow exhibit a different response in comparison with the powder form. However, as Shirakawa had previously referenced both studies in earlier papers,[107,108] he at least must have known the reported temperature dependence of polyacetylene powers.

Based on the observed drop in conductivity with increasing polyacetylene purity, it was proposed that the impurities in the films were perhaps acting in a similar manner to dopants in inorganic semiconductors, which thus increased the polymer conductivity.[122,158] It was thus reasoned that the addition of a halide such as bromine may result in enhanced conductivity in a manner similar to that previously observed by Heeger and MacDiarmid for $(SN)_x$.[6,122,158] This proposal was also supported by *in situ* IR measurements previously carried out by Shirakawa and Ikeda that revealed a dramatic decrease in IR transmission during the treatment of polyacetylene films with chlorine, thus suggesting that the material might have unusual electronic properties.[6,74–77] Interestingly, the previous work of Berets and Smith[143] on the use of additives to affect the polyacetylene conductivity was never mentioned as a factor, although Shirakawa was certainly aware of their report.

The landmark experiment was then performed by Shirakawa and Dr. Chwan K. Chiang, a postdoctoral fellow of Heeger,[112] on November 23, 1976.[74–77,87,112,115] The room temperature conductivity of a *trans*-polyacetylene film was measured via four-point probe while exposing the film to bromine vapor.[112,140] Upon the addition of bromine over a period of 10 min, the film conductivity increased rapidly (from 10^{-5} to 0.5 S cm^{-1}), producing a change of approximately four orders of magnitude.[140] Repeating the process using iodine in place of bromine resulted in even greater enhancements in conductivity, with values up to 38 S cm^{-1}. It was proposed that the highly conducting materials resulting from the exposure to halides were charge-transfer π-complexes similar to those believed to be formed during the halogenation of olefins. All of this was reported in an initial May 1977 communication,[140] in which the process of modulating the conductivity of polyacetylene through reactions with halides was collectively referred to as "chemical doping." This appears to be the origin of describing such organic materials as doped polymers.[1]

Further results were then reported in October of the same year, with the description "Doped Polyacetylene" now used in the title.[141] This second paper focused on the treatment of polyacetylene with both I_2 and AsF_5. In comparison with the previous iodine results, optimization of the rate of the iodine addition to *trans*-polyacetylene resulted in increased conductivities up to 160 S cm^{-1}. Better yet, the application of AsF_5 gave conductivities of 220 S cm^{-1} for the *trans* isomer, with even higher values (560 S cm^{-1}) for the *cis* isomer. It was then proposed that the residual conductivity in "pure" films is probably due to residual impurities or defects at very low levels. To illustrate this, they showed that the residual conductivity of polyacetylene can be compensated by trace amounts of a donor such as NH_3, resulting in the conductivity to fall to $<10^{-9}$ S cm^{-1} without detectable weight increase. Interestingly, no mention is made of the previous report of Berets and Smith,[143] which showed that the treatment of polyacetylene powders with NH_3 vapor dropped the conductivity to circa 10^{-14} S cm^{-1}. Lastly, it was shown that only circa 1% of the AsF_5 dopant was needed to achieve metallic conductivity, with films containing higher dopant concentrations stated to exhibit a temperature dependence consistent with a "disordered one-dimensional metal."[141] In all cases, however, the conductivity decreased as the temperature was lowered, as would be expected for a semiconducting material.

A more encompassing report of the various doped polyacetylenes produced to date, as well as their corresponding maximum conductivities (Table 5.3), was then published in early 1978.[186] Interestingly, this paper includes the first citation of Berets and Smith.[143] Still, nothing specific is mentioned about their previous work on polyacetylene, and the citation is lumped together with other examples in the following statement:[186]

> It has been known for some time that treatment of a fairly wide variety of compounds . . . with an electron-withdrawing species, usually iodine, may result in an increase of conductivity of many orders of magnitude, but the final room temperature conductivity obtained . . . is still very small.

This new study included the first reported case of n-doped polyacetylene to give a conductivity value of 80 S cm^{-1}. These Na-doped films were prepared by treating the polymers with a solution of sodium naphthalide, which reduced the polyacetylene. Conductivities were also reported for films doped with either HBr or the mixed halogen species ICl and IBr.[186] Although the conductivity of undoped *cis*-polyacetylene is considerably lower than that of the *trans* isomer (1.7×10^{-9} vs. 4.4×10^{-5} S cm^{-1}), doping of the *cis* isomer generally gave higher conductivities than doping of the *trans* isomer. However, no attempt to explain this observation was reported. The most highly conductive samples continued to be those of the AsF_5-doped films, exhibiting a maximum conductivity of 560 S cm^{-1}. By adjusting the conductivity relative to the material's density, the conductivity for these AsF_5-doped films expressed in terms of mass was determined to be roughly one-half that of copper metal (7.0×10^2 vs. 6.5×10^4 S g^{-1} cm^2).[186]

Two additional reports were then published in March of 1978,[187,188] with the first of these demonstrating that polyacetylene could be doped with

Table 5.3 Conductivities of doped polyacetylene films

Material[b]	σ (S cm^{-1})[b]	Material[a]	σ (S cm^{-1})[b]
trans-$[(CH)(HBr)_{0.04}]_x$	7×10^{-4}	trans-$[CHI_{0.20}]_x$	160
trans-$[CHCl_{0.02}]_x$	1×10^{-4}	cis-$[CH(IBr)_{0.15}]_x$	400
trans-$[CHBr_{0.05}]_x$	5×10^{-1}	trans-$[CH(IBr)_{0.12}]_x$	120
trans-$[CHBr_{0.23}]_x$	4×10^{-1}	trans-$[CH(AsF_5)_{0.03}]_x$	70
cis-$[CH(ICl)_{0.14}]_x$	50	trans-$[CH(AsF_5)_{0.10}]_x$	400
cis-$[CHI_{0.25}]_x$	360	cis-$[CH(AsF_5)_{0.14}]_x$	560
trans-$[CHI_{0.22}]_x$	30	trans-$[Na_{0.28}(CH)]_x$	80

Note: Data collected from Reference 186.
[a]The prefixes cis and trans denote the initial predoped form of the polyacetylene sample.
[b]At 25°C.

electron-donating species such as sodium and that such sodium-doped films could be compensated via oxidative doping with I_2.[187] The second report then presented a deeper study of the halogen-doping of polyacetylene, as well as probing the temperature-dependent conductivities of these materials in more detail.[187] In general, it was found that the conductivity of halogen-doped poly-acetylenes decreased with decreasing temperature, but the exact nature of this decrease (slope, deviation from linearity, etc.) depended on the specific dopant and the doping concentration. As part of these studies, the doping of cis-polyacetylene was optimized to give conductivity values above 500 S cm^{-1}. The ^{13}C NMR characterization of both the cis- and trans-isomers were then reported in third publication in April.[189]

5.11 Shirakawa, Ikeda, and continued study of polyacetylene films

Shirakawa's year at Penn concluded in the fall of 1977, and he had returned to Tokyo Institute of Technology by the start of 1978. There he continued to work on polyacetylene with Ikeda,[112] as well as publishing two additional papers as collaborations between himself and Ikeda in Japan, along with Heeger and Mac-Diarmid at Penn.[190,191] The first of these collaborative papers was published in June of 1978 and focused on optical absorption and reflection measurements carried out on stretch-oriented trans-polyacetylene films. It was concluded that polyacetylene was a direct bandgap semiconductor, with a bandgap of 1.4–1.6 eV as determined from the onset of the film absorption. Doping of the films resulted in a reduced intensity of the interband transition centered at circa 1.9 eV, concurrent with the growth of a new absorption centered around 0.8 eV.

The anisotropic reflectance from partially aligned films and the increased optical anisotropy upon doping were concluded to provide evidence for microscopic anisotropy in both the semiconducting and metallic state of polyacetylene.[190]

This was then followed in January 1979 with a computational study of the electronic structures of *cis*-polyacetylene, which was a collaboration between Shirakawa and Ikeda with researchers at Kyoto University.[192] Overall, the computational results were consistent with the *cis*-isomer adopting the *cis-transoid* structure (Figure 5.26), in good agreement with previous Raman studies.[106] However, it was pointed out that interconversion between the *cis-transoid* and *trans-cisoid* structures via appropriate vibrational modes was possible and that further vibrational analysis was warranted.

The collective study of the anisotropic nature of stretch-oriented films with Heeger and MacDiarmid was then continued in a second paper published in April of 1979, focusing on the electrical conductivity of oriented polyacetylene.[191] This study compared the electrical properties of as-grown films doped with AsF_5 to analogous samples that had been stretch oriented such that the elongated axis had been increased by a factor of either 2 or 3. As a consequence, it was found that the conductivity parallel to the elongation increased from a value of 300 S cm^{-1} for the as-grown film, to circa 1500 S cm^{-1} for the films elongated by a factor of 2, to an average value of 2150 S cm^{-1} for the most elongated films. In comparison, the conductivity perpendicular to the elongated axis showed only a minor decrease.[191]

Shirakawa and Ikeda then reported the synthesis and morphology of polyacetylene under various conditions in March of 1980, along with the detailed preparation of highly stretch-aligned films.[137] As previously discussed in Section 5.7, the polymer precipitates to form a powder when the catalyst concentration is extremely low. However, as the catalyst concentration is increased, polymerization occurs at the solution surface to form a swollen mass or gel. At sufficiently high catalyst concentrations, a film forms immediately on the surface of the catalyst solution, with the film density so high that monomer diffusion through the film becomes the rate-determining step. Stretched films were prepared from 5-mm-wide films of *cis*-polyacetylene by a conventional hand-stretching machine at room temperature, after which the prestretched films were then heated to 200°C under stress and dynamic vacuum.

A final paper was then published in late 1980 after Shirakawa had moved to the University of Tsukuba.[193] In addition to Ikeda at Tokyo Tech and Shirakawa at Tsukuba, this collaborative study also included colleagues at the University of Toyko and focused on absorption, Raman, and infrared spectroscopy in order to elucidate the structural details of doped polyacetylene. Analysis of the experimental results revealed that upon doping, vibrations in the 1600–900 cm^{-1} region of a polyene chain split into two groups. The higher-frequency group consists of vibrations of polyene parts that are not directly attacked by dopants but are perturbed along the chain, while the lower-frequency group is made up of vibrations of the positively charged polyene with the doped site at its center. Overall, it was found that the doping produces shorter segments of conjugated *trans* double bonds that are separated by positively charged units with the doped site at its center.

5.12 Continued study of doped polyacetylene films without Shirakawa

After Shirakawa returned to Japan, MacDiarmid and Heeger continued their collaboration as described earlier but also moved on with additional polyacetylene studies on their own. The first of these papers without Shirakawa was published in November of 1978, which was a new collaboration with William R. Salaneck and coworkers at the Xerox Webster Research Center in Rochester, New York. This report focused on computational modeling of the electronic structure of finite polyenes and extrapolation to polyacetylene in order to better interpret its previously reported optical specta.[194]

This was then followed by a number of papers in 1979, beginning with a report in February on the ESR characterization of neutral and doped films of *trans*-polyacetylene.[195] As with the previous paper, this too was a new collaboration, this time with the Rockwell International Science Center in California. For the undoped films, the concentration of unpaired electrons was found to be 1.44×10^{19} e$^-$/g, which is consistent with that previously found for powder samples of the low-crystalline polyacetylene by Hatano (1.1×10^{19} e$^-$/g) in 1961.[89] Such a comparison, however, was never made in this 1979 report. Upon AsF$_5$ doping, changes observed in the ESR spectra were then concluded to be indicative of the generation of free carriers, which would be consistent with the observed rise in conductivity.

A second February 1979 paper then followed on the electrochemical doping of polyacetylene films in various electrolyte solutions.[196] Of these, the use of [Bu$_4$N]ClO$_4$ resulted in a doped polyacetylene with the composition [CH(ClO$_4$)$_{0.0645}$]$_x$, which gave the highest conductivity to date, with values up to 970 S cm^{-1}. However, this was quickly exceeded by the conductivity exhibited by the stretch-oriented, AsF$_5$-doped films reported in April of that same year.[191]

A review of doped polyacetylene was then published in April of 1979, beginning with a historical overview of the material.[197] Interestingly, Natta is not mentioned at all, and the prior work of both Hatano and Berets and Smith are both presented in one to two brief statements each, with no mention whatsoever of the latter's treatment of polyacetylene with electron-deficient or electron-rich species. The total treatment of polyacetylene studies prior to the doping of films with Br_2 is given circa half of a page, followed by about nine pages of their own work.

The second collaborative report with the Xerox group was then published in May, this time focusing on the photoelectron spectra of AsF_5-doped polyacetylene.[198] From analysis of the doped and undoped samples, along with several arsenic fluoride species, it was concluded that at least a portion of the arsenic fluoride lies near the surfaces of the $(CH)_x$ fibrils. However, the results were inconclusive in determining the exact composition of the arsenic fluoride species in the final doped film.

This was then followed with a June report that detailed new methods of producing polyacetylene films of variable density via a gel intermediate.[199] It was concluded that the fibril morphology of the previous films was preserved in these materials, although with larger fibril sizes. In comparison with the higher-density films, the lower-density samples produced lower conductivities with similar doping levels. In the following months of 1979, additional papers were also reported that detailed the characterization of the magnetic susceptibility,[200] optical and IR properties,[201] and morphological studies[202] of neutral and doped polyacetylene. In a final paper published without MacDiarmid, Heeger reported a theoretical study of soliton formation in long-chain polyenes in an attempt to explain the mobile neutral defect observed in undoped $(CH)_x$.[203]

5.13 Naarmann and focused efforts to maximize conductivity

Although the 1979 collaborative study between Heeger, MacDiarmid, Shirakawa, and Ikeda on stretch-oriented polyacetylene had resulted in conductivities greater than 2000 S cm^{-1},[191] theoretical analysis suggested that conductivities greater than 20,000 S cm^{-1} could be possible for AsF_5-doped samples.[201] Efforts to achieve higher conductivities were then pursued by Herbert Naarmann at Badische Anilin- & Soda-Fabrik A. G. (BASF) (Figure 5.34) in collaboration with coworkers at the University of Montpellier. Naarmann was initially educated at the University of Münster (Westfälische Wilhelms-Universität Münster) over the period 1951–1954, where he studied organic and physical chemistry.[204] He then received his Ph.D. in 1959 at the University of Würzburg (Julius-Maximilians-Universität Würzburg) for a thesis on the identification of components of the venom of some tarantulas.[204,205] Naarmann then joined BASF in June of 1960,[204] where his work as part of the Plastics Research Laboratory focused on the structure–function relationships in organic and polymeric materials.

As early as 1982, Naarmann and coworkers at BASF were investigating the relationship between the structure and the conductivity of polyacetylene samples.[206] While some aspects of this study merely confirmed previously determined effects of crystallinity and *cis/trans* ratios on the conductivity, it also revealed a conductivity dependency on the degree of sp^3 hybridization and showed that even the polyacetylene films prepared via Shirakawa's methods contained up to 2% sp^3 content. Furthermore, it was concluded that there is a direct relationship between the sp^3 content and the polymer crystallinity.

In collaboration with colleagues at the University of Montpellier, Naarmann then applied the knowledge of these relationships in an effort to produce more crystalline and defect-free polyacetylene films, the initial results of which were reported in 1986.[207] These efforts involved the polymerization of acetylene over an aged Ti(OBu)$_4$-AlEt$_3$-silicone oil catalyst mixture, in which the silicone oil acted as a complexing solvent to generate a more active catalyst, as well as a more viscous medium for the interfacial polymerization. This combination resulted in the generation of a glossy golden polyacetylene film on the surface of the catalyst-impregnated silicone oil, which was then washed repeatedly with toluene, 6% HCl in methanol, and methanol. Doping of the purified polymer with a FeCl$_3$-nitromethane solution resulted in conductivities as high as 2340 S cm^{-1}.[207]

This was then followed by two back-to-back papers in January of the following year, both of which attempted to probe the effect of new catalyst systems and the associated polymerization conditions on the properties of the resulting polyacetylene films.[138,208] The first of these studies investigated a variant catalyst

Figure 5.34 Herbert Naarmann.

Reproduced with permission from H. Naarmann. The development of electrically conducting polymers. *Adv. Mater.* **1990**, *2*, 345–348. © **1990** Verlag GmbH & Co. KGaA, Weinheim.

mixture in which butyl lithium (BuLi) was used in place of the typical Et_3Al, with a particular focus on how changes in reaction parameters effected the morphology of the resulting films.[138] Consistent with a previous study by Shirakawa and Ikeda,[137] it was concluded that the film density was directly related to the catalyst concentration, with higher concentrations giving more dense films. Thus, very low concentrations result in powder products due to the availability of the acetylene gas to penetrate the catalyst solution. In contrast, high concentrations cause the rapid formation of a thick film that limits diffusion of acetylene to the active surface. At select intermediate concentrations, the product formed is more correctly described as a gel rather than a true film. Examination of the products via scanning electron microscope (SEM) revealed that all samples produced consisted of a fibrillar structure, which was independent of the macroscopic morphology.

The second of the two January papers then reported initial results of acetylene polymerizations utilizing a WCl_6-BuLi catalyst system.[208] While concluded to be a less reactive catalyst system than the commonly applied $Ti(OBu)_4$-AlEt₃, dilute solutions of the tungsten mixture successfully generated polyacetylene powders that were predominately comprised of the *trans* isomer (ca. 70%). While conductivities were not determined, ESR characterization of WCl_6-doped samples suggested that the material had become metallic.

Two additional papers in early 1987 then investigated the effects of various catalyst mixtures, with particular emphasis on the investigation of the previously reported $Ti(OBu)_4$-BuLi mixtures.[209,210] In all cases, the polymerizations were carried out using both toluene and silicone oil as the solvent system. Consistent with their earlier study,[207] the use of silicone oil always gave higher-quality films than the traditional use of toluene.[209] In terms of the catalyst systems, while the $Ti(OBu)_4$-BuLi mixtures produced high-quality and highly conductive polyacetylene films, the more conventional $Ti(OBu)_4$-AlEt₃ mixtures still gave the best overall results.

Another report in April of 1987 then followed that detailed the further modification of these successful approaches to high-quality polyacetylene films.[211] Here, the $Ti(OBu)_4$-AlEt₃-silicone oil medium was applied to a flat carrier consisting of a stretchable substrate (i.e., polyethylene or polypropylene). Within a glove box, the catalyst-coated substrate was then sealed underneath a specially constructed hood fitted with a gas inlet valve, after which acetylene was polymerized over the surface to generate a black, homogeneous film. The film produced in this way could be simply washed, or stretched either before or after washing, prior to removal from the supporting substrate. Upon doping with I_2, the unstretched films gave conductivities of 2000 S cm^{-1}, which could then be increased up to 18,000 S cm^{-1} upon stretch orientation before washing.

A modified catalyst system was also reported in the same paper,[211] which was found to give even higher conductivities. This new hybrid catalyst mixture combined the two previously investigated catalyst systems by adding small amounts of a BuLi solution to the commonly applied $Ti(OBu)_4$-$AlEt_3$ catalyst medium. Here, the added BuLi was believed to act as a reducing agent, thus enhancing the defect-free nature of the resulting polyacetylene films. This method was referred to as the ARA (addition of reducing agents) method and produced unstretched, I_2-doped films with conductivities of 5000 S cm^{-1}. As with the previous stretch-oriented films, this conductivity could be further increased to as high as 170,000 S cm^{-1} upon stretch orientation. When considering the measured conductivity per unit weight, such high-quality doped polyacetylene films were now exhibiting values that were greater than that of metals such as copper (i.e., ca. 150,000 S cm^{-1} g^{-1} vs. ca. 72,000 S cm^{-1} g^{-1} for Cu). Such polyacetylene samples produced over catalyst–silicone oil coated substrates then became known as a new polyacetylene type, or N-$(CH)_x$ (although it is unclear if the N was meant to represent *new* or *Naarmann*, as these samples are often referred to as Naarmann-type polyacetylene).[211]

A final 1987 paper then reported a collaborative study with Heeger, which probed the temperature- and pressure dependence of the conductivity of stretch-oriented polyacetylene films generated via the initial $Ti(OBu)_4$-$AlEt_3$-silicone oil system.[212] Consistent with the results discussed earlier, the stretch-oriented, I_2-doped material gave a room temperature conductivity above 20,000 S cm^{-1} when measured parallel to the direction of stretch orientation. The measured temperature dependence of the conductivity then revealed that it did diminish with temperature but still remained above 9000 S cm^{-1} at 0.5 K. It was thus concluded that the magnitude and temperature independence of the parallel conductivity below 1 K implied genuine metallic behavior and ruled out transport via hopping among strongly localized states. Still, the increased conductivity with higher temperatures did imply that phonon-assisted transport was involved.

Naarmann and his coworkers continued to study the electronic properties of polyacetylene films for the next few years,[213–218] but no further increases in the material conductivity were achieved. At the same time, it should be noted that other groups were not able to reproduce the results of Naarmann and his collaborators.[219,220] The closest to reproducing the results were Tsukamoto and coworkers, who used the high-boiling solvent decaline in place of silicone oil, resulting in the production of films with conductivities of 1.1×10^5 S cm^{-1} after I_2 doping.[219] Films made in this way were referred to as "ν-$(CH)_x$" in order to differentiate them from N-$(CH)_x$. It is thought that the results were highly dependent on the exact silicone oil applied,[220] which could possibly account for difficulties in reproducing the results.

Another complication, however, was that the improved conductivity of N-$(CH)_x$ was usually attributed to reduced sp^3 defects in the films, yet characterization of v-$(CH)_x$ in comparison with the original Shirakawa materials could not find any differences in sp^3 content. In the end, it was concluded that the "higher-order structure of v-$(CH)_x$ is more responsible for high conductivity than the primary structure of the polymer."[219] Both v-$(CH)_x$ and N-$(CH)_x$ were found to be more dense than the original Shirakawa materials.[220]

Shirakawa also returned in 1991 to report a comparison of his materials with N-$(CH)_x$, with the conclusion that the primary difference was the stretchability and *cis/trans* content.[221] In the end, he felt the most important factor affecting the conductivity was alignment of polyacetylene chains and the thickness of the resulting films. The conductivity of all films was found to increase with decreased thickness, with this dependence thought to be governed by a simultaneous in-plane alignment of chains when the thickness of the films is extremely thin.

5.14 Conclusions

While the polymerization of acetylene dates back to the initial investigations of Berthelot in 1866,[49,50] the successful production of the linear conjugated polyacetylene was not accomplished until the 1955 work of Natta.[71–73] As such, polyacetylene was a bit of a late addition to the study of conjugated polymers, arriving over a hundred years after the initial polymerization of aniline. Furthermore, while polyacetylene is frequently stated as the first conducting polymer, conductive forms of both polyaniline and polypyrrole predate doped polyacetylene. A significant difference between polyacetylene and these previous conducting polymers, however, is that polyacetylene is produced in its neutral form via Ziegler–Natta catalysis, while both polyaniline and polypyrrole were generated via oxidative polymerization to give the doped form as the initial product. As such, polyacetylene was then one of the first conjugated polymers in which the neutral form was purposely doped with oxidizing agents to generate the conducting form. As such, it was the extensive study of these doping processes in polyacetylene that led to much of our current understanding of the overall general doping of conjugated polymers. At the same time, while previous conductive samples of polyaniline had reached 100 S cm^{-1}, doped forms of polyacetylene were the first examples of conducting polymers capable of surpassing this threshold to give conductivity values fully in the metallic regime.[141] Continued optimization of both the polymerization of polyacetylene films and their oxidative doping ultimately led to the impressive conductivity values achieved by Naarmann in the late 1980s,[211] which still remains the highest reported conductivity for any conducting organic polymer. Thus, while the field of conjugated materials moved away from polyacetylene and the singular focus on conductivity by the early 1990s, polyacetylene will always remain a landmark conductive material.

References and Notes

1. Rasmussen, S. C. Conjugated and conducting organic polymers: The first 150 years. *ChemPlusChem* **2020**, *85*(7), 1412–1429.
2. Rasmussen, S. C. Early history of conjugated polymers: From their origins to the handbook of conducting polymers. In *Handbook of Conducting Polymers*, 4th ed. Reynolds, J. R.; Skotheim, T. A.; Thompson, B.; Eds.; CRC Press: Boca Raton, FL 2019, pp. 1–35.
3. Rasmussen, S. C. *Acetylene and Its Polymers. 150+ Years of History.* Springer Briefs in Molecular Science: History of Chemistry. Springer: Heidelberg, 2018, pp. 85–132.
4. Rasmussen, S. C. Early history of conductive organic polymers. In *Conductive Polymers: Electrical Interactions in Cell Biology and Medicine.* Zhang, Z.; Rouabhia, M.; Moulton, S.; Eds.; CRC Press: Boca Raton, FL, 2017, pp. 1–21.
5. Rasmussen, S. C. Electrically conducting plastics: Revising the history of conjugated organic polymers. In *100+ Years of Plastics. Leo Baekeland and Beyond.* E. T.Strom and S. C. Rasmussen, Eds.; ACS Symposium Series 1080, American Chemical Society: Washington, DC, 2011, pp. 147–163.
6. Hall, N. Twenty-five years of conducting polymers. *Chem. Commun.* **2003**, 1–4.
7. Saxman, A. M.; Liepins, R.; Aldissi, M. Polyacetylene: Its synthesis, doping and structure. *Prog. Polym. Sci.* **1985**, *11*, 57–89.
8. Eisch, J. J. Fifty years of Ziegler–Natta polymerization: From serendipity to science. A personal account. *Organometallics* **2012**, *31*, 4917–4932.
9. Sun, T.; Wang, T.; Dong, X.; Wang, W. An insight into the chain-propagation mechanism of propylene polymerization catalyzed by traditional Ti-based Ziegler–Natta catalysts in view of recently developed catalysts. *Des. Monomers Polym.* **2006**, *9*, 117–127.
10. Rasmussen, S. C. *Acetylene and Its Polymers. 150+ Years of History.* Springer Briefs in Molecular Science: History of Chemistry. Springer: Heidelberg, 2018, pp. 21–36.
11. Davy, E. Notice of a peculiar compound of carbon and potassium, or carburet of potassium, & c. *Br. Assoc. Adv. Sci. Rep.* **1837**, *5*, 63–64.
12. Davy, E. Notice of carburet of potassium, and of a new gaseous bi-carburet of hydrogen. *Rec. Gen. Sci.* **1836**, *4*, 321–323.
13. Davy, E. Note sur le carbure de potassium, et sur un nouveau bi-carbure d'hydrogène. *J. Pharm. Chim.* **1837**, series 2, 143–322.
14. Davy, E. Ueber Kohlenstoffkalium und einen neuen Doppelt-Kohlenwasserstoff. *Ann. Pharm.* **1837**, *23*, 144–146.
15. Davy, E. On a new gaseous compound of carbon and hydrogen. *Proc. R. Ir. Acad.* **1837**, *1*, 88–89.
16. Davy, E. On a new gaseous compound of carbon and hydrogen. *Br. Assoc. Adv. Sci. Rep.* **1838**, *6*, 50.

17. Davy, E. On a new gaseous compound of carbon and hydrogen. *Trans. R. Ir. Acad.* **1839**, *18*, 80–88.

18. Berthelot, M. Sur une nouvelle série de composés organiques, le quadricarbure d'hydrogène et ses dérivés. *C. R. Acad. Sci. Paris* **1860**, *50*, 805–808.

19. Berthelot, M. Recherches sur l'acétylène. *Ann. Chim.* **1863**, *67*, 52–77.

20. Berthelot, M. Nouvelles recherches sur la formation des carbures d'hydrogène. *C. R. Acad. Sci. Paris* **1862**, *54*, 515–519.

21. Berthelot, M. Synthèse de l'acétylène par la combinaison directe du carbone avec l'hydrogène. *C. R. Acad. Sci. Paris* **1862**, *54*, 640–644.

22. Berthelot, M. On the synthesis of acetylene by the direct combination of carbon with hydrogen. *Chem. News* **1862**, *5*, 184–185.

23. Berthelot, M. *Leçons sur les méthodes générales de synthèse en chimie organique.* Gauthier-Villars: Paris, 1864, pp. 67–84.

24. Doremus, C. G. Pierre Eugène Marcellin Berthelot. *Science* **1907**, *25*, 592–595.

25. Adloff, J. P.; Kauffman, G. B. Marcellin Berthelot (1827–1907), chemist, historian, philosopher, and statesman: A retrospective view on the centenary of his death. *Chem. Educ.* **2007**, *12*, 195–206.

26. Berthelot, M. Sur la synthése de l'acétyléne. *C. R. Acad. Sci. Paris* **1862**, *54*, 1042–1044.

27. Berthelot, M. Nouvelles contributions à l'histoire de l'acétyléne. *C. R. Acad. Sci. Paris* **1862**, *54*, 1044–1046.

28. Berthelot, M. Sur la présence et sur le role de l'acétylene dans la gaz de l'éclairage. *C. R. Acad. Sci. Paris* **1862**, *54*, 1070–1072.

29. Pratt, H. T. Thomas Leopold Willson. In *American Chemists and Chemical Engineers*. Miles, W. D., Ed.; American Chemical Society: Washington DC, 1976, pp. 512–513.

30. Nicholls, R. V. V. Gibbs, LeSueur, and Willson. Pioneers of industrial electrochemistry. In *Electrochemistry, Past and Present*. Stock, J. T.; Orna, M. V.; Eds.; ACS Symposium Series 390, American Chemical Society: Washington, DC, 1989, pp. 525–533.

31. American Chemical Society. *Discovery of the Commercial Process for Making Calcium Carbide and Acetylene.* American Chemical Society: Washington DC, 1998.

32. Paton, J. Willson, Thomas Leopold. In *Dictionary of Canadian Biography*, Vol. 14. Wilson, D. A., Ed.; University of Toronto/Université Laval, 2003; https://www.biographi.ca/en/bio/willson_thomas_leopold_14E.html (accessed December 20, 2024).

33. Willson, T. L.; Suckert, J. J. The carbides and acetylene commercially considered. *J. Franklin Inst.* **1895**, *139*, 321–341.

34. Morehead, J. T.; De Chalmot, G. The manufacture of calcium carbide. *J. Am. Chem. Soc.* **1896**, *18*, 311–331.

35. Pond, G. G. *Calcium Carbide and Acetylene*, 3rd ed. Bulletin of the Department of Chemistry, Pennsylvania State College: State College, 1917, pp. 15–16.

36. Thompson, G. F. *Acetylene Gas, Its Nature, Properties and Uses; also Calcium Carbide, Its Composition, Properties and Method of Manufacture.* Liverpool, 1898, pp 21–26.

37. Willson, T. L. Electric Reduction of Refractory Metallic Compounds. U.S. Patent 492,377, 1893.

38. Nieuwland, J. A.; Vogt, R. R. *The Chemistry of Acetylene*. American Chemical Society Monograph Series. Reinhold Publishing Corporation: New York, 1945, pp. 7–10.

39. Hare, R. Process for a fulminating powder—for the evolution of calcium and galvanic ignition of gunpowder. *Am. J. Sci.* **1839**, *27*, 268–270.

40. Hare, R. Calcium. *L'Institut* **1840**, *8*, 310–312.

41. Wöhler, F. Bildung des Acetylens durch Kohlenstoffcalcium. *Ann. Chem.* **1862**, *124*, 220.

42. Ihde, A. J. *The Development of Modern Chemistry*. Harper and Row: New York, 1964, p. 470.

43. Moissan, H. Description d'un nouveau four électrique. *C. R. Acad. Sci. Paris* **1892**, *115*, 1031–1033.

44. Moissan, H. Préparation au four électrique d'un carbure de calcium cristallisé; propriétés de ce nouveau corps. *C. R. Acad. Sci. Paris* **1894**, *118*, 501–506.

45. Böhm, L. K. Calcium carbide and acetylene. *Min. Ind.* **1900**, *9*, 63–74.

46. Morris, P. J. T. The industrial history of acetylene: The rise and fall of a chemical feedstock. *Chem. Ind.* **1983**, *47*, 710–715.

47. Rasmussen, S. C. Cuprene: A historical curiosity along the path to polyacetylene. *Bull. Hist. Chem.* **2017**, *42*, 63–78.

48. Rasmussen, S. C. *Acetylene and Its Polymers. 150+ Years of History*. Springer Briefs in Molecular Science: History of Chemistry. Springer: Heidelberg, 2018, pp. 37–65.

49. Berthelot, M. Action de la chaleur sur quelques carbures d'hydrogène. (Première partie). *C. R. Acad. Sci. Paris* **1866**, *62*, 905–910.

50. Berthelot, M. Les polymères de l'acétylène. Première partie: synthèse de la benzine. *C. R. Acad. Sci. Paris* **1866**, *63*, 479–484.

51. Blyth, J.; Hoffman, A. W. Ueber das Styrol und einige seiner Zersetzungsproducte. *Ann. Chem. Pharm.* **1845**, *53*, 289–329.

52. Rasmussen, S. C. Revisiting the early history of synthetic polymers: Critiques and new insights. *Ambix* **2018**, *65*, 356–372.

53. Rasmussen, S. C. From polymer to macromolecule: Origins and historical evolution of polymer terminology. *Bull. Hist. Chem.* **2020**, *45*, 91–100.

54. Thenard, P.; Thenard, A. Acétylène liquéfié et solidifié sous l'influence de l'effluve électrique. *C. R. Acad. Sci. Paris* **1874**, *78*, 219.

55. Kolotyrkin, V. M.; Gil'man, A. B.; Tsapuk, A. K. Production of organic surface films by the action of electrons, ultraviolet radiation, and the glow discharge. *Russ. Chem. Rev.* **1967**, *36*, 579–591.

56. v. Wilde, M. P. Vermischte Mittheilungen. *Ber. Dtsch. Chem. Ges.* **1874**, *7*, 352–357.

57. Berthelot, M. Quatrième mémoire. Absorption de l'hydrogène libre par l'influence de l'effluve. *Ann. Chim. Phys.* **1877**, *10*, 66–69.

58. Willson, T. L.; Suckert, J. J. The carbides and acetylene commercially considered. *J. Franklin Inst.* **1895**, *139*, 321–341.

59. Erdmann, H.; Köthner, P. Einige Beobachtungen über Acetylen und dessen Derivate. *Z. Anorg. Chem.* **1898**, *18*, 48–58.

60. Hanriot, M. Extrait des Procès-verbaux des séances. *Bull. Soc. Chim. Fr.* **1899**, *21*, 529–531.

61. Sabatier, P.; Senderens, J. B. Action du cuivre sur l'acétylène: formation d'un hydrocarbure très condensé, le cuprène. *C. R. Acad. Sci. Paris* **1900**, *130*, 250–252.

62. Natta, G. Giulio Natta - biographical. In *Nobel Lectures, Chemistry 1963–1970*. Elsevier Publishing Company: Amsterdam, 1972, pp. 79–81.

63. Bawn, C. E. H. Giulio Natta, 1903–1979. *Nature* **1979**, *280*, 707.

64. Anon. Nobelist Giulio Natta dies in Italy. Chem. *Eng. News* **1979**, *57*(20), 48.

65. Bamford, C. H. Giulio Natta—an appreciation. *Chem. Brit.* **1981**, *17*, 298–300.

66. Morris, P. J. T. *Polymer Pioneers*. Center for the History of Chemistry: Phiadelphia, 1986, pp. 81–83.

67. Seymour, F. B. Giulio Natta – a Pioneer in Polypylene. In *Pioneers in Polymer Science*. Kluwer Academic Publishers: Dordrecht, the Netherlands, 1989, pp. 207–212.

68. Hargittai, I.; Comotti, A.; Hargittai, M. Giulio Natta. *Chem. Eng. News* **2003**, *81*(6), 26–28.

69. Porri, L. Giulio Natta—his life and scientific achievements. *Macromol. Symp.* **2004**, *213*, 1–5.

70. As previously detailed in Chapter 2, *dottore* was the terminal degree in Italy prior to the mid-1980s, consisting of a total of four to five years of study and a thesis. In many respects, it would be similar to a modern master's degree.

71. Natta, G.; Pino, P.; Mazzanti, G. Polimeri ad elevato peso molecolore degli idrocarburi acetilenici e procedimento per la loro preparozione. Italian Patent 530,753, July 15, 1955; *Chem. Abst.* **1958**, *52*, 15128b.

72. Natta, G.; Mazzanti, G.; Pino, P. Hochpolymere von Acetylen-Kohlenwasserstoffen, erhalten mittels Organometall-Komplexen von Zwischenschalenelementen als Katalysatoren. *Angew. Chem.* **1957**, *69*, 685–686.

73. Natta, G.; Mazzanti, G.; Corradini, P. Polimerizzazione stereospecifica dell'acetilene. *Atti Accad. Naz. Lincei Rend. Cl. Sci. Fis. Mat. Nat.* **1958**, *25*, 3–12.

74. Shirakawa, H. The discovery of polyacetylene film: The dawning of an era of conducting polymers. In *Les Prix Nobel. The Nobel Prizes 2000*. Frängsmyr, T., Ed.; Nobel Foundation: Stockholm, 2001, pp. 217–226.

75. Shirakawa, H. The discovery of polyacetylene film: The dawning of an era of conducting polymers (Nobel Lecture). *Angew. Chem. Int. Ed.* **2001**, *40*, 2574–2580.

76. Shirakawa, H. Nobel Lecture: The discovery of polyacetylene film—the dawning of an era of conducting polymers. *Rev. Mod. Phys.* **2001**, *73*, 713–718.

77. Shirakawa, H. The discovery of polyacetylene film. The dawning of an era of conducting polymers. *Synth. Met.* **2002**, *125*, 3–10.

78. Natta, G.; Mazzanti, G.; Pregaglia, G.; Peraldo, M. Preparazione e struttura di alti polimeri lineari dell'esino-1. *Gazz. Chim. Ital.* **1959**, *89*, 465–494.

79. Luttinger, L. B. A new catalyst system for the polymerization of acetylenic compounds. *Chem. Ind.* **1960**, 1135.

80. Luttinger, L. B. Hydridic reducing agent-group VIII metal compound. A new catalyst system for the polymerization of acetylenes and related compounds. I. *J. Org. Chem.* **1962**, *27*, 1591–1596.

81. Luttinger, L. B.; Colthup, E. C. Hydridic reducing agent-group VIII metal compound. A new catalyst system for the polymerization of acetylenes and related compounds. II. *J. Org. Chem.* **1962**, *27*, 3752–3756.

82. Daniels, W. E. The polymerization of acetylenes by nickel halide-tertiary phosphine complexes. *J. Org. Chem.* **1962**, *29*, 2936–2938.

83. Stille, J. K. Some new polymers from unsaturated monomers and their derivatives. *Prepr. ACS Div. Petrol. Chem.* **1961**, *6*, 89–95.

84. Stille J. K.; Frey, D. A. Polymerization of non-conjugated diynes by complex metal catalysts. *J. Am. Chem. Soc.* **1961**, *83*, 1696–1701.

85. Watson, W. H., Jr.; McMordie, W. C., Jr.; Lands, L. G. Polymerization of alkynes by Ziegler-type catalyst. *J. Polym. Sci.* **1961**, *55*, 137–144.

86. William H. Watson, Obituary. https://www.tributearchive.com/obituaries/24260178/william-h-watson (accessed May 15, 2022).

87. Shirakawa, H. Path to the synthesis of polyacetylene films with metallic luster: In response to Rasmussen's article. *Substantia* **2022**, *6*, 121–127.

88. Tohoku University, Tohoku University Fact Book 2014. http://www.tohoku.ac.jp/en/about/images/factbook2014.pdf (accessed May 17, 2022)

89. Hatano, M.; Kanbara, S.; Okamoto, S. Paramagnetic and electric properties of polyacetylene. *J. Polym. Sci.* **1961**, *51*, S26–S29.

90. Kanbara, S.; Hatano, M.; Hosoe, T. 遷移金属アセチルアセトナート‐トリエチルアルミニウム系によるアセチレンの重合 (Polymerization of acetylene by transition metal acetylacetonate-triethylaluminum system). *J. Soc. Chem. Ind. Jpn.* **1962**, *65*, 720–723.

91. Ikeda, S.; Hatano, M. Polymerization reaction by organic transition metal complexes. *Kōgyō kagaku zasshi* **1963**, *66*(8), 1032–1037.

92. Shimamura, K.; Hatano, M.; Kanbara, S.; Nakada, I. Electrical conduction of polyacetylene under high pressure. *J. Phys. Soc. Jpn.* **1967**, *23*, 578–581.

93. Hatano, M. アセチレン重合体の構造と電気的性質 (Structures and electrical properties of acetylene polymers). *Tanso* **1967**, *26*–31.

94. Successive Director, Laboratory for Chemistry and Life Science Institute of Innovative Research, Tokyo Institute of Technology. http://www.res.titech.ac.jp/english/about/director.html (accessed July 24, 2022).

95. History, Laboratory for Chemistry and Life Science Institute of Innovative Research, Tokyo Institute of Technology. http://www.res.titech.ac.jp/english/about/history.html (accessed July 24, 2022).

96. Ikeda, S. Recent development in polymer synthesis. *J. Synth. Org. Chem. Jpn.* **1962**, *20*, 76–86.

97. Ikeda, S.; Tamaki, A. Syntheses of benzene-$^{14}C_6$ and benzene-$^2H_6$ using a Ziegler-catalyst. *Radioisotopes* **1963**, *12*, 368–372.

98. Ikeda, S.; Tamaki, A.; Yoshida, K. Measurement of C_2 component in acetylene-^{14}C polymerization system by Ziegler catalyst by radio gas chromatography. *Radioisotopes* **1964**, *13*, 415–417.

99. Ikeda, S.; Tamaki, A. On the mechanism of the cyclization reaction of acetylene polymerization. *J. Polym. Sci. B Polym. Lett. Ed.* **1966**, *4*, 605–607.

100. Ikeda, S. チグラー触媒によるエチレンおよびアセチレン重合の立体化学 (Stereochemistry of ethylene and acetylene polymerization by Ziegler catalyst). *J. Soc. Chem. Ind. Jpn.* **1967**, *70*, 1880–1886.

101. Yamamoto, A.; Morifuji, K.; Ikeda, S.; Saito, T.; Uchida, Y.; Misono, A. Butadiene polymerization catalysts. Diethylbis(dipyridyl)iron and diethyldipyridylnickel. *J. Am. Chem. Soc.* **1965**, *87*, 4652–4653.

102. Saito, T.; Uchida, Y.; Misono, A.; Yamamoto, A.; Morifuji, K.; Ikeda, S. Diethyldipyridylnickel. Preparation, characterization, and reactions. *J. Am. Chem. Soc.* **1966**, *88*, 5198–5201.

103. Yamamoto, A.; Ikeda, S. A coordination compound of diethyldipyridylnickel with acrylonitrile. A polymerization catalyst of acrylonitrile. *J. Am. Chem. Soc.* **1967**, *89*, 5989–5990.

104. Yamamoto, A.; Morifuji, K.; Ikeda, S.; Saito, T.; Uchida, Y.; Misono, A. Diethyl-bis(dipyridyl)iron. A butadiene cyclodimerizaton catalyst. *J. Am. Chem. Soc.* **1968**, *90*, 1878–1883.

105. Shirakawa, H.; Ikeda, S. Infrared spectra of poly(acetylene). *Polym. J.* **1971**, *2*, 231–244.

106. Shirakawa, H.; Ito, T.; Ikeda, S. Raman scattering and electronic spectra of poly(acetylene). *Polym. J.* **1973**, *4*, 460–462.

107. Ito, T.; Shirakawa, H.; Ikeda, S. Simultaneous polymerization and formation of polyactylene film on the surface of concentrated soluble Ziegler-type catalyst solution. *J. Polym. Sci. Polym. Chem. Ed.* **1974**, *12*, 11–20.

108. Ito, T.; Shirakawa, H.; Ikeda, S. Thermal *cis-trans* isomerization and decomposition of polyacetylene. *J. Polym. Sci. Polym. Chem. Ed.* **1975**, *13*, 1943–1950.

109. Ito, T.; Shirakawa, H.; Ikeda, S. ポリアセチレンのシス-トランス組成と固体構造 (*Cis-trans* composition and solid structure of polyacetylene). *Kobunshi Ronbunshu* **1976**, *33*, 339–345.

110. Shirakawa, H.; Ito, T.; Ikeda, S. Electrical properties of polyacetylene with various cis-trans compositions. *Makromol. Chem.* **1978**, *179*, 1565–1573.

111. Rasmussen, S. C. The path to conducting polyacetylene. *Bull. Hist. Chem.* **2014**, *39*, 64–72.

112. Shirakawa, H. Hideki Shirakawa. In *Les Prix Nobel. The Nobel Prizes 2000*. Frängsmyr, T., Ed.; Nobel Foundation: Stockholm, 2001; pp. 213–216.

113. Shirakawa, H. Another role of grants-in-aid: Feeding research fruits into society. *Kakenhi Essay Series* **2010**, *15*, 1–5.

114. Tokyo Institute of Technology Library's Thesis Database. http://tdl.libra.titech.ac.jp/hkshi/en/recordID/dissertation.bib/TT00000879 (accessed August 16, 2013).

115. Hargittai, I. Risking reputation: Conducting polymers. In *Drive and Curiosity: What Fuels the Passion for Science*. Prometheus Books: Amherst, NY, 2011, pp. 173–190.

116. Rasmussen, S. C. New insight into the "fortuitous error" that led to the 2000 Nobel Prize in Chemistry. *Substantia* **2021**, *5*(1), 91–97.

117. Shirakawa, H. Reflections on "Simultaneous Polymerization and Formation of Polyacetylene Film on the Surface of Concentrated Soluble Ziegler-Type Catalyst Solution," by Takeo Ito, Hideki Shirakawa, and Sakuji Ikeda, *J. Polym. Sci.: Polym. Chem. Ed.*, 12, 11 (1974). *J. Polym. Sci. A Polym. Chem.* **1996**, *34*, 2529–2530.

118. Chien, J. C. W. *Polyacetylene. Chemistry, Physics, and Material Science*. Academic Press: Orlando, FL, 1984; pp. xi–xiv.

119. Kaner, R. B.; MacDiarmid, A. G. Plastics that conduct electricity. *Sci. Am.* **1988**, *258*(2), 106–111.

120. Miller, J. S. The 2000 Nobel Prize in Chemistry—a personal accolade. *ChemPhysChem* **2000**, *1*, 229–230.

121. Brennan, M. Electrifying plastics. *Chem. Eng. News* **2000**, *78*(42), 4–5.

122. MacDiarmid, A. G. Alan G. MacDiarmid. In *Les Prix Nobel, The Nobel Prizes 2000*; Frängsmyr, T., Ed.; Nobel Foundation: Stockholm, 2001; pp. 183–190.

123. Hargittai, B.; Hargittai, I. *Candid Science V: Conversations with Famous Scientists*. Imperial College Press: London, 2005, pp. 401–409.

124. Hargittai, B.; Hargittai, I. *Candid Science V: Conversations with Famous Scientists.* Imperial College Press, London, 2005, pp. 411–427.
125. Scott, C. History of conductive polymers. In *Nanostructured Conductive Polymers.* Eftekhari, A., Ed.; John Wiley & Sons: Chichester, UK, 2010; pp. 3–17.
126. Heeger, A. J. *Never Lose Your Nerve!* World Scientific Publishing: Singapore, 2016; pp. 131–150.
127. Byun Hyung Jik. The Chosun Ilbo (chosun.com). http://focus.chosun.com/ people/people-01.jsp?id=20494 (accessed October 6, 2017).
128. Byun, J. Personal communication with the son of Pyun, 2018.
129. Established in 1959, the name of the institute changed to the Korea Advanced Energy Research Institute in 1980, before being restored to the original name in 1989. Although Pyun's papers list the institute's location as Seoul, it is really situated in Daejeon, ca. 100 miles outside of Seoul.
130. Pyun, H. C. Pre-equilibrium in the Schmidt reaction of benzhydrols. *J. Korean Chem. Soc.* **1964**, *8*, 25–29.
131. Pyun, H. C.; Kim, J. R. Isotopic exchange 5-bromouracil-Br[82]. *J. Korean Chem. Soc.* **1964**, *8*, 39–42.
132. Byun, H. J. What is a Nobel Prize? 2013 (Pyun's unpublished account).
133. Byun, H. J. I can't doubt the facts. . . *KAERI Magazine* **2002**, *10*, 7–8.
134. Pyun, H. C. Pre-equilibrium in the Schmidt reaction of benzhydrols. Ph.D. dissertation, Seoul National University, 1970.
135. Pyun, H. C.; Kim, J.; Lee, W.-M. Copolymerization of phenyl acetylene with stryrene. *J. Korean Chem. Soc.* **1969**, *13*, 387–393.
136. Wnek, G. E. Comments on "Simultaneous Polymerization and Formation of Polyacetylene Film on the Surface of Concentrated Soluble Zeigler-Type Catalyst Solution," by Takeo Ito, Hideki Shirakawa, and Sakuji Ikeda, *J. Polym. Sci.: Polym. Chem. Ed.*, 12, 11 (1974). *J. Polym. Sci. A Polym. Chem.* **1996**, *34*, 2531–2532.
137. Shirakawa, H.; Ikeda, S. Preparation and morphology of as-prepared and highly stretch-aligned polyacetylene. *Synth. Met.* **1980**, *1*, 175–184.
138. Munardi, A.; Aznar, R.; Theophilou, N.; Sledz, J.; Schue, F.; Naarnann, H. Morphology of polyacetylene produced in the presence of the soluble catalyst $Ti(OnBu)_4$-*n*-BuLi. *Eur. Polym. J.* **1987**, *23*, 11–14.
139. Rasmussen, S. C. Comments on Shirakawa's response. *Substantia* **2022**, *6*, 129–131.
140. Shirakawa, H.; Louis, E. J.; MacDiarmid, A. G.; Chiang, C. K.; Heeger, A. J. Synthesis of electrically conducting organic polymers: Halogen derivatives of polyacetylene, $(CH)_x$. *J. Chem. Soc., Chem. Commun.* **1977**, 578–580.
141. Chiang, C. K.; Fincher, C. R., Jr.; Park, Y. W.; Heeger, A. J.; Shirakawa, H.; Louis, B. J.; Gau, S. C.; MacDiarmid, A. G. Electrical conductivity in doped polyacetylene. *Phys. Rev. Lett.* **1977**, *39*, 1098–1101.
142. Ikeda, S.; Shirakawa, H. Method for Producing Film-like and Fibrous Acetylene High Polymer. Japanese Patent Application Publication 48-032581, October 9, 1973; Japanese Patent 0722516, March 18, 1974.
143. Berets, D. J.; Smith, D. S. Electrical properties of linear polyacetylene. *Trans. Faraday. Soc.* **1968**, *64*, 823–828.
144. Morrissey, S. Obituaries. *Chem. Eng. News* **2002**, *80*(20), 56,
145. Donald Joseph Berets. Harvard Magazine. May–June 2002, https://www.harvardmagazine.com/2002/05.

146. Berets, D. J. Studies on the detonation of explosive gas mixtures. Ph.D. dissertation, Harvard University, 1949.

147. Berets, D. J.; Castellion, G. A.; Haacke, G. Electrochromic Information Displays. U.S. Patent 3,968,639, July 13, 1976.

148. Dorian Sevcik Smith Obituary. *Wilmington Star-News*, December 15, 2010.

149. Smith, D. S. Observations on the rare-earths: Chemical and electrochemical studies in non-aqueous solvents. Ph.D. dissertation, University of Illinois, 1958.

150. Illinois State Athletics Percy Hall of Fame (1990), 1950 Football. http://goredbirds.com/hof.aspx?hof=281 (Accessed September 25, 2022).

151. State of Illinois. *Proceedings of the Teachers College Board of the State of Illinois*, July 1, 1953–June 30, 1954, pp. 155–156.

152. MacDiarmid, A. G.; Mikulski, C. M.; Russo, P. J.; Saran, M. S.; Garito, A. F.; Heeger, A. J. Synthesis and structure of the polymeric metal, $(SN)_x$, and its precursor, S_2N_2. *J. Chem. Soc. Chem. Commun.* **1975**, 476–477.

153. Mikulski, C. M.; Russo, P. J.; Saran, M. S.; MacDiarmid, A. G.; Garito, A. F.; Heeger, A. J. Synthesis and structure of metallic polymeric sulfur nitride, $(SN)_x$, and its precursor, disulfur dinitride, S_2N_2. *J. Am. Chem. Soc.* **1975**, *97*, 6358–6363.

154. Cohen, M. J.; Garito, A. F.; Heeger, A. J.; MacDiarmid, A. G.; Mikulski, C. M.; Saran, M. S.; Kleppinger, J. Solid state polymerization of S_2N_2 to $(SN)_x$. *J. Am. Chem. Soc.* **1976**, *98*, 3844–3848.

155. Mikulski, C. M.; MacDiarmid, A. G.; Garito, A. F.; Heeger, A. J. Stability of polymeric sulfur nitride, $(SN)_x$, to air, oxygen, and water vapor. *Inorg. Chem.* **1976**, *15*, 2943–2945.

156. Chiang, C. K.; Cohen, M. J.; Garito, A. F.; Heeger, A. J.; Mikulski, C. M.; MacDiarmid, A. G. Electrical conductivity of $(SN)_x$. *Solid State Commun.* **1976**, *18*, 1451–1455.

157. Chiang, C. K.; Cohen, M. J.; Peebles, D. L.; Heeger, A. J.; Akhtar, M.; Kleppinger, J.; MacDiarmid, A. G.; Milliken, J.; Moran, M. J. Transport and optical properties of polythiazyl bromides: $(SNBr_{0.4})_x$. *Solid State Commun.* **1977**, *23*, 607–612.

158. MacDiarmid, A. G. Oral history interview by Cyrus Mody at University of Pennsylvania, Philadelphia, Pennsylvania. Oral History Transcript #0325. Chemical Heritage Foundation, Philadelphia, 2005.

159. Holmes, A. B.; Klein, M. L.; Baughman, R. H. Alan Graham MacDiarmid. 14 April 1927–7 February 2007. *Biogr. Mems Fell. R. Soc.* **2023**, *74*, 259–282.

160. Kaner, R. B.; Holmes, A. B.; MacDiarmid, A. G. April 14, 1927–February 7, 2007. Biographical Memoirs, National Academy of Sciences, Washington, DC, 2024.

161. Halton, B. *Chemistry at Victoria - the Wellington University. A Personalized Account of the Hundred Years from 1899*. 3rd ed. Victoria University of Wellington: Wellington, 2018, pp. 28–29.

162. MacDiarmid, A. G. Preparation of mono-halogen substituted compounds of sulphur nitride. *Nature* **1949**, *164*, 1131–1132.

163. MacDiarmid, A. G. "Synthetic metals": A novel role for organic polymers (Nobel Lecture). *Angew. Chem. Int. Ed.* **2001**, *40*, 2581–2590.

164. Halford, B. Alan MacDiarmid dies at 79. *Chem. Eng. News* **2007**, *85*(7), 16.

165. MacDiarmid, A. G. Isotopic exchange in complex cyanide—simple cyanide systems. Ph.D. dissertation, University of Wisconsin-Madison, 1953.

166. MacDiarmid, A. G. The chemistry of some new derivatives of the silyl radical. Ph.D. dissertation, University of Cambridge, 1955.

167. Heeger, A. J. Alan J. Heeger. In *Les Prix Nobel, The Nobel Prizes 2000*; Frängsmyr, T., Ed.; Nobel Foundation: Stockholm, 2001; pp. 139–143.

168. Heeger, A. J. Semiconducting and metallic polymers: The fourth generation of polymeric materials (Nobel Lecture). *Angew. Chem. Int. Ed.* **2001**, *40*, 2591–2611.

169. Heeger, A. J. *Never Lose Your Nerve!* World Scientific Publishing: Singapore, 2016, pp. 23–26.

170. Heeger, A. J. *Never Lose Your Nerve!* World Scientific Publishing: Singapore, 2016, pp. 34–39.

171. Heeger, A. J. *Never Lose Your Nerve!* World Scientific Publishing: Singapore, 2016, p. 55.

172. Heeger, A. J. *Never Lose Your Nerve!* World Scientific Publishing: Singapore, 2016, pp. 68–71.

173. Heeger, A. J. *Never Lose Your Nerve!* World Scientific Publishing: Singapore, 2016, pp. 77–91.

174. Heeger, A. J. Studies on the Magnetic Properties of Canted Antiferromagnets. Ph.D. dissertation, University of California, Berkeley, 1961.

175. Heeger, A. J. *Never Lose Your Nerve!* World Scientific Publishing: Singapore, 2016, pp. 93–98.

176. Heeger, A. J. *Never Lose Your Nerve!* World Scientific Publishing: Singapore, 2016, pp. 45–54.

177. Lubkin, G. B. Attempts to explain and observe TCNQ behavior at 60 K. *Phys. Today* **1973**, *26*, 17, 19–20.

178. Thomas, G. A.; Schafer, D. E.; Wudl, F.; Horn, P. M.; Rimai, D.; Cook, J. W.; Glocker, D. A.; Skove, M. J.; Chu, C. W.; Groff, R. P.; Gillson, J. L.; Wheland, R. C.; Melby, L. R.; Salamon, M. B.; Craven, R. A.; De Pasquali, G.; Bloch, A. N.; Cowan, D. O.; Walatka, V. V.; Pyle, R. E.; Gemmer, R.; Poehler, T. O.; Johnson, G. R.; Miles, M. G.; Wilson, J. D.; Ferraris, J. P.; Finnegan, T. F.; Warmack, R. J.; Raaen, V. F.; Jerome, D. Electrical conductivity of tetrathiafulvalenium-tetracyanoquinodimethanide (TTF-TCNQ). *Phys. Rev. B* **1976**, *13*, 5105–5110.

179. Malozemoff, A. P.; Gallagher, W. J.; Greene, R. L.; Laibowitz, R. B.; Tsuei, C. C. Superconductivity at IBM - a centennial review: Part II – Materials and physics. *IEEE/CSC SNF* **2012**, *6* (21), RN28-2(1-25).

180. Heeger, A. J. *Never Lose Your Nerve!* World Scientific Publishing: Singapore, 2016, pp. 169–180.

181. Labes, M. M.; Love, P.; Nichols, L. F. Polysulfur nitride-a metallic, superconducting polymer. *Chem. Rev.* **1979**, *79*, 1–15.

182. Walatka, V. V.; Labes, M. M.; Perlstein, J. H. Polysulfur nitride—a one-dimensional chain with a metallic ground state. *Phys. Rev. Lett.* **1973**, *31*, 1139–1142.

183. Hsu, C.; Labes M. M. Electrical conductivity of polysulfur nitride. *J. Chem. Phys.* **1974**, *61*, 4640–4645.

184. Rasmussen, S. C. On the origin of "synthetic metals." *Mater. Today* **2016**, *19*, 244–245.

185. Rasmussen, S. C. On the origin of "synthetic metals": Herbert McCoy, Alfred Ubbelohde, and the development of metals from nonmetallic elements. *Bull. Hist. Chem.* **2016**, *41*, 64–73.

186. Chiang, C. K.; Druy, M. A.; Gau, S. C.; Heeger, A. J.; Louis, E. J.; MacDiarmid, A. G.; Park, Y. W.; Shirakawa, H. Synthesis of highly conducting films of derivatives of polyacetylene, $(CH)_x$. *J. Am. Chem. Soc.* **1978**, *100*, 1013–1015.

187. Chiang, C. K.; Gau, S. C.; Fincher, C. R., Jr.; Park, Y. W.; MacDiarmid, A. G.; Heeger, A. J. Polyacetylene, $(CH)_x$: n-type and p-type doping and compensation. *App. Phys. Lett.* **1978**, *33*, 18–20.

188. Chiang, C. K.; Park, Y. W.; Heeger, A. J.; Shirakawa, H.; Louis, E. J.; MacDiarmid, A. G. Conducting polymers: Halogen doped polyacetylene. *J. Chem. Phys.* **1978**, *69*, 5098–5104.

189. Maricq, M. M.; Waugh, J. S.; MacDiarmid, A. G.; Shirakawa, H.; Heeger, A. J. Carbon-13 nuclear magnetic resonance of *cis*- and *trans*-polyacetylenes. *J. Am. Chem. Soc.* **1978**, *100*, 7729–7730.

190. Fincher, C. R., Jr.; Peebles, D. L.; Heeger, A. J.; Druy, M. A.; Matsumura, Y.; MacDiarmid, A. G.; Shirakawa, H.; Ikeda, S. Anisotropic optical properties of pure and doped polyacetylene. *Solid State Commun.* **1978**, *27*, 489–494.

191. Park, Y. W.; Druy, M. A.; Chiang, C. K.; MacDiarmid, A. G.; Heeger, A. J.; Shirakawa, H.; Ikeda, S. Anisotropic electrical conductivity of partially oriented polyacetylene. *J. Polym. Sci. Polym. Lett. Ed.* **1979**, *17*, 195–201.

192. Yamabe, T.; Tanaka, K.; Terama-e, H.; Fukui, K.; Shirakawa, H.; Ikeda, S. The electronic structures of cis-polyacetylene. *Solid State Commun.* **1979**, *29*, 329–333.

193. Harada, I.; Furukawa, Y.; Tasumi, M.; Shirakawa, H.; Ikeda, S. Spectroscopic studies on doped polyacetylene and β-carotene. *J. Chem. Phys.* **1980**, *73*, 4746–4757.

194. Duke, C. B.; Paton, A.; Salaneck, W. R.; Thomas, H. R.; Plummer, E. W.; Heeger, A. J.; MacDiarmid, A. G. Electronic structure of polyenes and polyacetylene. *Chem. Phys. Lett.* **1978**, *59*, 146–150.

195. Goldberg, I. B.; Crowe, H. R.; Newman, P. R.; Heeger, A. J.; MacDiarmid, A. G. Electron spin resonance of polyacetylene and AsF5-doped polyacetylene. *J. Chem. Phys.* **1979**, *70*, 1132–1136.

196. Nigrey, P. J.; MacDiarmid, A. G.; Heeger, A. J. Electrochemistry of polyacetylene, $(CH)_x$: Electrochemical doping of $(CH)_x$ films to the metallic state. *J. Chem. Soc. Chem. Commun.* **1979**, 594–595.

197. Chiang, C. K.; Heeger, A. J.; Macdiarmid, A. G. Synthesis, structure, and electrical properties of doped polyacetylene. *Ber. Bunsenges. Phys. Chem.* **1979**, *83*, 407–417.

198. Salaneck, W. R.; Thomas, H. R.; Duke, C. B.; Paton, A.; Plummer, E. W.; Heeger, A. J.; MacDiarmid, A. G. Photoelectron spectra of AsF_5-doped polyacetylenes. *J. Chem. Phys.* **1979**, *71*, 2044–2050.

199. Wnek, G, E.; Chien, J. C. W.; Karasz, F. E.; Druy, M. A.; Park, Y. W.; MacDiarmid, A. G.; Heeger, A. J. Variable-density conducting polymers: Conductivity and thermopower studies of a new form of polyacetylene: $(CH)_x$. *J. Polym. Sci. Polym. Lett. Ed.* **1979**, *17*, 779–786.

200. Weinberger, B. R.; Kaufer, J.; Heeger, A. J.; Pron, A.; MacDiarmid, A. G. Magnetic susceptibility of doped polyacetylene. *Phys. Rev. B* **1979**, *20*, 223–230.

201. Fincher, C. E., Jr.; Ozaki, M.; Tanaka, M.; Peebles, D.; Lauchlan, L.; Heeger, A. J.; MacDiarmid, A. G. Electronic structure of polyacetylene: Optical and infrared studies of undoped semiconducting $(CH)_x$ and heavily doped metallic $(CH)_x$. *Phys. Rev. B* **1979**, *20*, 1589–1601.

202. Karasz, F. E.; Chien, J. C. W.; Galkiewicz, R.; Wnek, G. E.; Heeger, A. J.; MacDiarmid, A. G. Nascent morphology of polyacetylene. *Nature* **1979**, *282*, 286–288.

203. Su, W. P.; Schrieffer, J. R.; Heeger, A. J. Solitons in polyacetylene. *Phys. Rev. Lett.* **1979**, *42*, 1698–1701.

204. Naarmann, H. The development of electrically conducting polymers. *Adv. Mater.* **1990**, *2*, 345–348.

205. Naarmann H. Über die Identifizierung von Inhaltsstoffen des Giftes einiger Vogel-spinnen: über Amidierungen in vivo et vitro, insbesondere mit Trimethylen-diamin, zur Biogenese des Trimethylendiamins. Ph.D. dissertation, Julius-Maximilians-Universität Würzburg, 1959.

206. Haberkorn, H.; Naarmann, H.; Penzien, K.; Schlag, J.; Simak, P. Structure and conductivity of poly(acetylene). *Synth. Met.* **1982**, *5*, 51–71.

207. Theophilou, N.; Aznar, R.; Munardi, A.; Sledz, J.; Schue, F.; Naarnann, H. E.S.R. Study of the Ti(OBu)$_4$ catalyst mixture in silicone oil with regard to the synthesis of homogeneous and highly conducting (CH)$_x$. *Synth. Met.* **1986**, *16*, 337–342.

208. Theophilou, N.; Munardi, A.; Aznar, R.; Sledz, J.; Schué, F.; Naarnann, H. Poly-merization of acetylene in the presence of a tungsten-based catalyst, and geometric structure, electrical properties and morphology of the polymer. *Eur. Polym. J.* **1987**, *23*, 15–20.

209. Theophilou, N.; Aznar, R.; Munardi, A.; Sledz, J.; Schue, F.; Naarnann, H. Polymerization of acetylene with new catalytic systems and optimization of the properties of the polymers. *J. Macromol. Sci. Chem. A* **1987**, *24*, 797–812.

210. Munardi, A.; Theophilou, N.; Aznar, R.; Sledz, J.; Schue, F.; Naarnann, H. Poly-merization of acetylene with Ti(OC$_4$H$_9$)$_4$/butyllithium as catalyst system and silicone oil as reaction medium. *Makromol. Chem.* **1987**, *188*, 395–399.

211. Naarmann, H.; Theophilou, N. New process for the production of metal-like, stable polyacetylene. *Synth. Met.* **1987**, *22*, 1–8.

212. Basescu, N.; Liu, Z.-X.; Moses, D.; Heeger, A. J.; Naarmann, H.; Theophilou, N. High electrical conductivity in doped polyacetylene. *Nature* **1987**, *327*, 403–405.

213. Schimmel, T.; Rieß, W.; Gmeiner, J.; Denninger, G.; Schwoerer, M.; Naarmann, H.; Theophilou, N. DC-conductivity on a new type of highly conducting poly-acetylene, N-(CH)$_x$. *Solid State Commun.* **1988**, *65*, 1311–1315.

214. Theophilou, N.; Naarmann, H. Influences of the catalyst system on the morphol-ogy, structure and conductivity of a new type of polyacetylene. *Makromol. Chem. Macromol. Symp.* **1989**, *24*, 115–128.

215. Schimmel, T.; Denninger, G.; Riess, W.; Voit, J.; Schwoerer, M.; Schoepe, W.; Naarmann, H. High-σ polyacetylene: DC conductivity between 14 mK and 300 K. *Synth. Met.* **1989**, *28*, D11–D18.

216. Winter, H.; Sachs, G.; Dormann, E.; Cosmo, R.; Naarmann, H. Magnetic proper-ties of spin-labelled polyacetylene. *Synth. Met.* **1990**, *36*, 353–365.

217. Schimmel, T.; Schwoerer, M.; Naarmann, H. Mechanisms limiting the D.C. con-ductivity of high-conductivity polyacetylene. *Synth. Met.* **1990**, *37*, 1–6.

218. Schimmel, T.; Glaser, M.; Schwoerer, M.; Naarmann, H. Conductivity barriers and transmission electron microscopy on highly conducting polyacetylene. *Synth. Met.* **1991**, *41*, 19–25.

219. Tsukamoto, J.; Takahashi, A.; Kawasaki, K. Structure and electrical properties of polyacetylene yielding a conductivity of 10^5 S/cm. *Jpn. J. Appl. Phys.* **1990**, *29*, 125–130.

220. Arbuckle, G. A.; Buecheler, N. M.; Valentine, K. G. Characterization of "highly conducting" polyacetylene. *Chem. Mater.* **1994**, *6*, 569–572.

221. Shirakawa, H.; Zhang, Y.-X.; Mochizuki, K.; Akagi, K. S-(CH)$_x$ and N-(CH)$_x$— what is the difference? *Synth. Met.* **1991**, *41*, 13–18.

Chapter 6
Polythiophene

6.1 Introduction

The final parent conjugated polymer covered in the current volume, polythiophene (Figure 6.1), stands out to occupy a special place in the history of these materials.[1,2] To begin with, unlike all of the other polymers discussed so far, the history of polythiophene dates back less than 50 years and is thus the only example discussed here that does not predate the seminal polyacetylene work of the 1970s. This alone is curious, as polythiophene can be synthesized via oxidative polymerization in the same way as polyaniline and polypyrrole.[3,4] In fact, the mechanism for the oxidative polymerization of thiophene is commonly held to be identical to that of pyrrole (Figure 6.1; see Chapter 3 for detailed discussion).[5-7] As such, the only primary difference is that thiophene is not as susceptible to oxidation than either pyrrole or aniline and thus requires the application of either stronger chemical oxidizing agents or more anodic potentials.[3] Still, the treatment of thiophene with oxidizing agents was not studied during the early periods in the same way as aniline and pyrrole, and thus the first significant report of polythiophene was not of materials generated via oxidative polymerization.

Rather than oxidative polymerization, early polythiophene was initially produced via metal-mediated cross-coupling methods as an extension of methods previously applied to polyphenylene.[1,2,4] This may have been at least partially due to the fact that since its very discovery, thiophene was closely associated with benzene, with both aromatic species exhibiting similar chemistry. As a consequence of its developmental paths, polythiophene can be viewed to combine the historical path of polyphenylene with the much older historical evolution of oxidatively generated conjugated polymers. Thus, like polyaniline and polypyrrole, conductive polythiophene films could be easily produced via electropolymerization. More critically, however, the application of metal-mediated cross-coupling methods,[8-11] combined with the introduction of functionalized thiophenes, has led to the development of a wealth of neutral, solution-processible polythiophenes. Furthermore, like polyphenylene, neutral polythiophenes are much more environmentally stable than neutral polypyrroles. It is perhaps this broad versatility that has led to the eventual

The Origins and Early History of Conjugated Organic Polymers. Seth C. Rasmussen, Oxford University Press.
© Oxford University Press (2025). DOI: 10.1093/9780197638194.003.0006

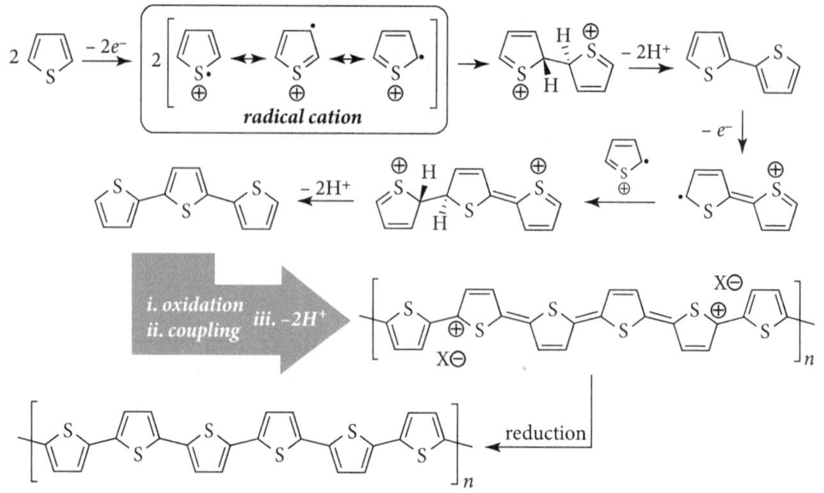

Figure 6.1 Modern oxidative polymerization mechanism of thiophene.

dominance of thiophene-based materials in the modern era. In order to attempt to understand the late introduction of polythiophene, as well as place all of its history in context, we will again begin with a review of the history of monomeric thiophene.

6.2 A brief history of thiophene

The history of thiophene began in the autumn of 1882, when the German chemist Victor Meyer (1848–1897, Figure 6.2) was demonstrating the decarboxylation of benzoic acid during a lecture at Zurich Polytechnic (now ETH Zurich).[12] In order to identify the benzene product, he applied the indophenine test for benzene (Figure 6.3) developed by Adolph von Baeyer (1835–1917) a few years earlier in 1879.[13] To his surprise, the test was negative, although he had successfully practiced the same indophenine test that morning before his lecture. As pointed out by his lecture assistant, Traugott Sandmeyer (1854–1922), the only significant difference is that Meyer had used a sample of coal tar benzene during the morning practice, rather than the synthetic benzene generated during the demonstration. Intrigued, Meyer began investigating this unexpected negative test the same day.[12]

By the end of November, Meyer had found that the purest benzene samples isolated from coal tar always gave a positive indophenine reaction, while the same samples first shaken repeatedly with sulfuric acid did not, nor did

Figure 6.2 Viktor Meyer (1848–1897).

Reproduced from R. Meyer. Victor Meyer. 1848–1897. *Ber. Dtsch. Chem. Ges.* **1908**, *41*, 4504–4718.

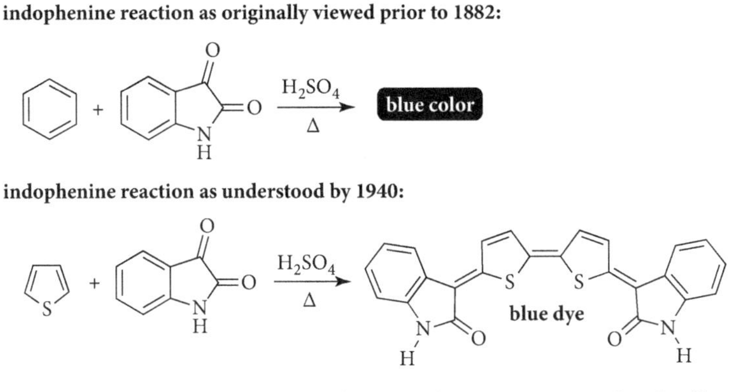

indophenine reaction as originally viewed prior to 1882:

indophenine reaction as understood by 1940:

Figure 6.3 Indophenine test as originally viewed prior to 1882 and as finally understood by 1940.

benzene samples prepared from benzene derivatives. He reported this in an initial communication, in which he proposed three hypotheses to account for these observations. He presented the last of these as follows:[14]

Benzene from coal tar could contain two substances that are physically and chemically very similar, and which differ in that one is more reactive than the other; the more reactive would connect to the isatin; and on treating it with sulfuric acid it would first be converted into sulphonic acid, leaving the less reactive body. The latter would also be present in benzene from benzoic acid.

Meyer then followed this up with a full paper in June, 1883, in which he detailed the successful isolation of the second species present in benzene from coal tar.[15] Meyer began by shaking the initial benzene samples with H_2SO_4 for several hours. The benzene layer was then removed; the remaining black acid layer diluted with water and then treated with lead carbonate ($PbCO_3$) to generate a lead sulfonate species. This lead salt was then mixed with ammonium chloride and dry distilled to give a volatile oil that boiled at 84°C. Meyer was able to confirm that it was this new oil that was responsible for the blue color of a positive indophenine test, and it was determined that the purest commercially available benzene from coal tar contained only about 0.5% of this second species. Further analysis of the oil ultimately gave a formula of C_4H_4S, with Meyer settling on the name *thiophene* for this second component, stating:[15] "The name thiophene might be appropriate for the latter, which expresses on the one hand the sulfur content of the substance and on the other hand the great similarity of it and its derivatives with the phenyl compounds." In the same paper, Meyer prepared the bromo and dibromo derivatives and considered a number of possible structures for the new thiophene, most of which he quickly discarded. In the end, he concluded that thiophene was most likely the sulfur analog of furan, and it was this structure that he presented as a preliminary formulation.[14] Of course, this structure was quickly adopted and has since been confirmed.

In August of that same year, Meyer and Sandmeyer published an attempted synthesis of thiophene by passing ethylene or acetylene through boiling sulfur.[16] While such efforts produced primarily a mixture of charcoal, hydrogen sulfide (H_2S), and carbon disulfide (CS_2), a small quantity of an oil was isolated, which gave tests and reactions consistent with thiophene. This report, however, was extremely short, with little detail. In a later 1885 report,[17] Meyer stated that this attempt had been followed up with a number of additional syntheses of thiophene, but that none of them had been published, as they had not proved to be effective methods of its production. One such example of additional syntheses was a direct follow-up to the original reported attempt, in which ethylene or illuminating gas was passed over heated pyrite. Again, however, no real details were reported, and the yield of thiophene produced in this way was never given.

The first effective synthesis of a thiophene was then reported by Carl Paal (1860–1935) of the University at Erlangen in February of 1885.[18] In an extension of previous work on the synthesis of furans, Paal found that heating acetophenoneacetone with P_2S_5 resulted in the generation of phenylmethylthiophene in good yield (Figure 6.4). Paal then followed this up with a second February report utilizing mucic acid rather than acetophenoneacetone.[19] Here, heating with barium sulfide resulted in the generation of α-thiophenecarboxylic acid (Figure 6.4) in low yield. The carboxylic acid was then converted to its calcium salt, mixed with $Ca(OH)_2$, and distilled to give the parent thiophene.

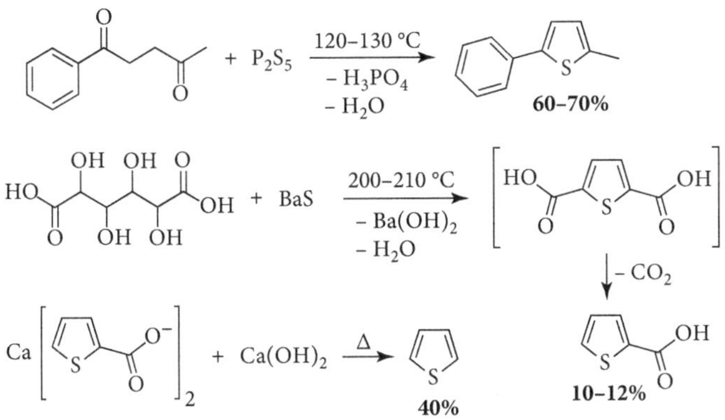

Figure 6.4 Paal's 1885 synthesis of thiophenes.

Another related thiophene synthesis was published at the same time as Paal's second paper, this time by the German chemist Jacob Volhard (1834–1910, Figure 6.5) and his student Hugo Erdmann (1862–1910) of the University of Halle.[20] Volhard's methodology began with the reaction of succinic anhydride and P_2S_5 to give the unfunctionalized parent thiophene (Figure 6.6). Although the efficiency of this reaction is not given, Volhard states that better yields are obtained through the use of sodium succinate and P_2S_3. While this improvement gave only modest yields of circa 40%, this became the most effective method for the synthesis of thiophene for the next couple of decades.

The next significant synthetic advancement then came in 1914, when the German chemist Wilhelm Steinkopf (1879–1949) revisited the synthesis of thiophene from acetylene.[21] In a similar approach to Meyer's use of pyrite, Steinkopf found that passing acetylene over heated pyrite successfully produced thiophene, with control of the temperature key to effective yields. While it was found that the reaction occurred at temperatures as low as 260°C, systematic investigations showed the optimum temperature to be circa 300°C. This process resulted in a brown liquid that consisted of about 40%–50% thiophene, which could be isolated via distillation to give circa 95% pure thiophene. To obtain completely pure material, the thiophene was first converted to its mercury oxyacetate compound, from which thiophene could be liberated by treatment with HCl.

Figure 6.5 Jacob Volhard (1834–1910).

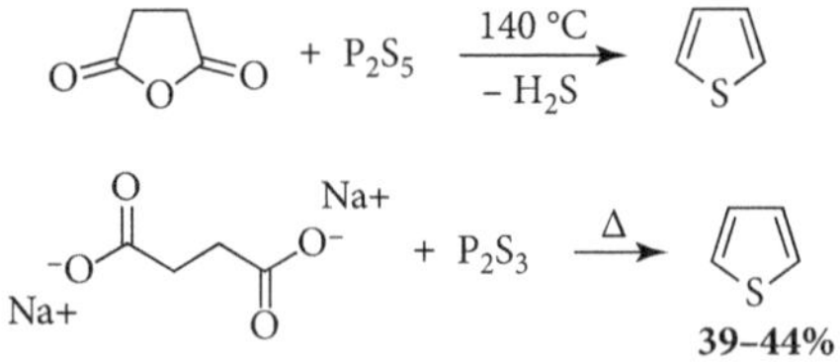

Figure 6.6 Volhard's 1885 synthesis of thiophene from succinic acid.

Finally, in 1946, the vapor phase synthesis of thiophene from butane and sulfur was reported by the industrial chemist Herbert E. Rasmussen (b. 1915) and coworkers of the Socony-Vacuum Oil Company.[22] In this process, commercial butane of approximately 95% purity and molten sulfur were first both preheated separately to about 600°C. The two reactant vapors were then introduced rapidly through a mixing nozzle into a reactor tube at 600°C, allowing reaction for only a fraction of a second, after which the product stream was rapidly quenched, and crude thiophene separated from other products such as butylene and butadiene. The nonthiophene species were then recycled back to contribute to the next pass, with this process capable of continuing for long periods without interruption. In this way, thiophene could be produced from butane in overall yields of circa 50%. The Socony-Vacuum process thus made thiophene available in commercial quantities and it is variants of this general methodology that are still the primary route for the commercial production of thiophene.

6.3 Early thiophene polymerizations via acids

As early as 1883, Meyer had reported that the treatment of thiophene with H_2SO_4 resulted in the formation of a black resin.[15] By the 1940s, numerous authors had noted that thiophene can be polymerized by acidic materials, typically resulting in amorphous, highly insoluble products.[23-27] Acids utilized in these cases included H_2SO_4, FSO_3H, p-toluenesulfonic acid (CH_3-C_6H_4-SO_3H), HF, dihydroxyfluoboric acid ($H[BF_2(OH)_2]$), and $BF_3 \cdot Et_2O$.[23,24] In addition, polymerization was promoted at increased reaction temperatures.[24]

The first attempts to elucidate potential structures came in 1950, when Howard D. Hartough (1913–1992) and coworkers at the Socony-Vacuum Oil Company found that more moderate acids such as 100% H_3PO_4 resulted in the generation of liquid polymers of thiophene and alkylthiophenes.[26] Although a small amount of insoluble polymer was produced, the remaining product could be fractionated via distillation to show that it consisted predominantly of trimers, with smaller amounts of pentamers and a trace of dimeric material. Analysis of the trimer fraction led to the conclusion that it consisted of 2,4-bis(2-thienyl)thiolane, for which a possible mechanism was proposed (Figure 6.7). The possible structure of the pentamer fraction was not determined.

A related study was then reported in 1967 by Albert Wassermann (1901–1971) and coworkers at University College, London.[27] As with the Hartough study, Wassermann utilized acids of reduced strength, in this case the organic species trichloro- and trifluoro-acetic acid, in order to study the polymerization of pyrrole, furan, and thiophene. In addition to the specific choice of acids utilized, they also carefully controlled the reaction temperature, concentration, and time in order to limit the formation of insoluble products. In the case of thiophene, a soluble polymer was isolated in the form of a yellowish amorphous powder. Based on molecular weight analysis, it was determined that the sample was primarily composed of tetramers. Further analysis via nuclear magnetic resonance (NMR) and infrared (IR) spectroscopy led to the conclusion that the thiophene units had become partially saturated in the process of polymerization, with potential repeat units shown in Figure 6.8. These potential repeat units were designated as either Type II or Type III, and while it was believed that the thiophene polymer contained both classes of units, it was thought that the structure consisted of a majority of Type II units.

The Hartough study was then revisited by R. F. Curtis and coworkers at the University College of Swansea in 1971.[28] Repeating the methods of Hartough

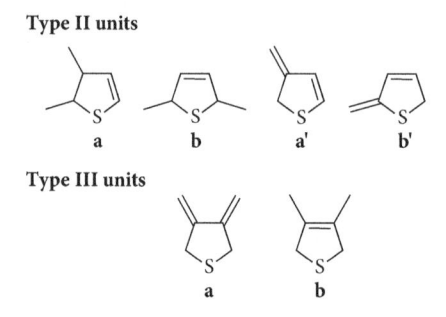

Figure 6.7 Proposed mechanism for the acid-mediated production of 2,4-bis(2-thienyl)thiolane.

Type II units

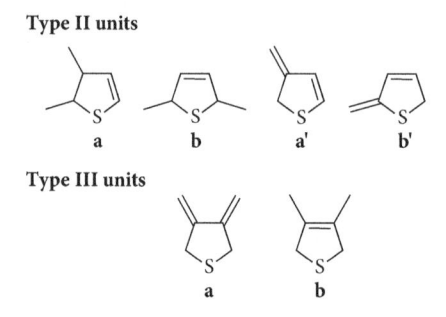

a   b   a'   b'

Type III units

a   b

Figure 6.8 Proposed repeat units of oligothiophenes via acid-mediated polymerization.

and coworkers,[26] Curtis was able to provide some additional detail. Subjecting the trimer isolated by Hartough to repeated chromatography revealed it to be a mixture of two crystalline products, both of identical molecular weight. Further analysis by NMR led to the conclusion that these species were conformational isomers, with the two terminal thiophenes adopting either a *cis-* or *trans-* orientation (Figure 6.9). Chromatography of the previously assigned pentamer fraction revealed a compound with the formula $C_{16}H_{14}S_3$, which was believed to be derived from a tetramer with the loss of H_2S. Further analysis led to the conclusion that this was comprised of a central 1,2,3,4-tetrahydrodibenzo[*b*]thiophene with two terminal 2-thienyl groups, as shown in Figure 6.9.

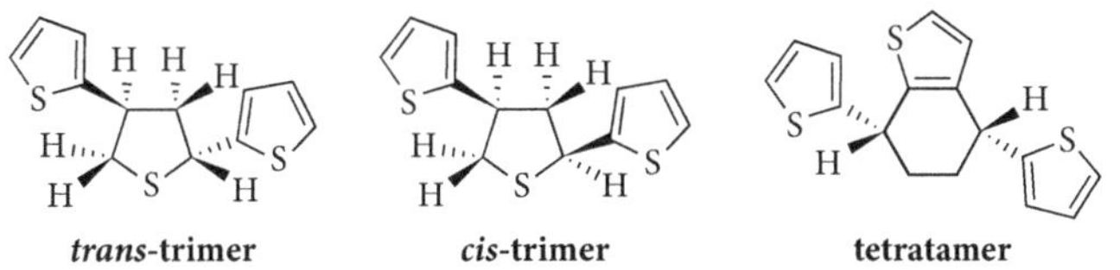

trans-trimer   cis-trimer   tetratamer

Figure 6.9 Proposed structures of the *cis* and *trans* isomers of the trimer, as well as the larger tetramer.

All of these studies confirmed that the polymers derived from the acid-mediated polymerization of thiophene contained considerable saturation defects and limited conjugation. As such, these polymers were not early examples of the fully conjugated poly(α,α'-thiophene), as implied by some in the literature.[4] The earliest possible example of what could be considered to be the parent conjugated polymer polythiophene can be found in a 1971 U.S. patent granted to James J. Louvar of the Universal Oil Products Company for the electropolymerization of heterocycles (including thiophene) in aqueous CH_3CO_2H-KOH mixtures.[29] However, the patent included no product characterization beyond a claim that IR spectroscopy and elemental analysis confirmed the product to be polythiophene. As such, it wasn't until 1980 that the first reliable reports of polythiophene appeared, with a number of studies appearing nearly simultaneously. The earliest of these reports brings us back to Tokyo Tech with the work of Takakazu Yamamoto (1944–2014), who was initially introduced in Chapter 4.

6.4 Yamamoto and polythiophene via Kumada coupling

Takakazu Yamamoto was born on March 31, 1944, in Toyama, Japan, approximately 190 miles northwest of Tokyo.[30,31] He was educated at the Tokyo Institute of Technology (Tokyo Tech), where he completed his graduate diploma in 1966 and a master's degree in 1968.[31,32] He then continued his doctoral studies at Tokyo Tech under Sakuji Ikeda (see Chapter 5), in collaboration with A. Yamamoto at the University of Tokyo.[30] He completed his Doctor of Engineering (Dr. Eng.) in 1971, with a dissertation on diorgano(dipyridyl)metal complexes (Fe, Ni, and Co).[30-33] He then joined Tokyo Tech's Chemical Resources Laboratory as a research associate from 1971 to 1976.[30-32] During the period of 1972–1973, he also spent time as a postdoctoral fellow with Lee J. Todd

(1936–2011) at Indiana University.[30,34,35] Yamamoto was promoted to associate professor in 1976 and then full professor in 1986.[30-32] He officially retired from Tokyo Tech in 2009, although he continued research activity in the energy conversion materials division of the Chemical Resources Laboratory.[31]

During his career, Yamamoto received numerous awards and honors, including the Scientific Award from the Ichimura Foundation (1988), the Award of the Society of Polymer Science, Japan (1998), the Hoko Award of the Hattori Hokokai Foundation (1988), the Seiichi Tejima Invention Award (1990), the Inoue Prize for Science (1993), the Japanese Government's Medal of Honor, purple ribbon (2004), and the Award for Outstanding Achievement in Polymer Science and Technology of the Society of Polymer Science, Japan (2009).[30,31] Yamamoto passed away suddenly on October 29, 2014, at the age of 70.[31]

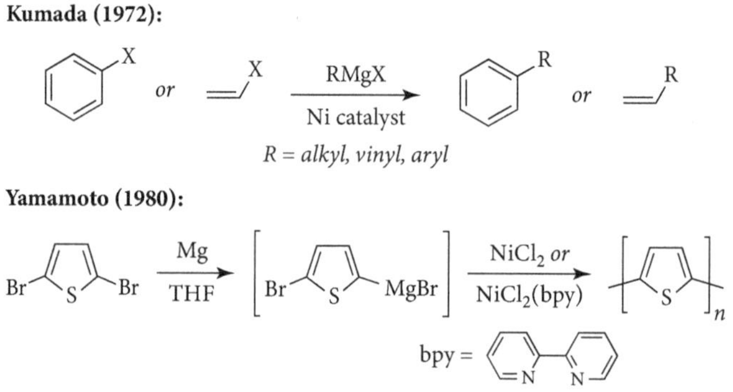

Figure 6.10 Kumada cross-coupling and its application to the polymerization of 2,5-dibromothiophene.

After applying nickel-catalyzed polycondensation to the production of polyphenylene and polymethylene in the late 1970s (see Chapter 4),[36,37] Yamamoto turned to the polymerization of 2,5-dibromothiophene in 1979, initial results of which were published in January of 1980.[38] As outlined in Figure 6.10, the 2,5-dibromothiophene was first reacted with a single equivalent of magnesium in dry tetrahydrofuran (THF), thus resulting in the formation of an intermediate species containing both bromo- and bromomagnesium functionalities. The addition of a Ni catalyst then allowed growth of the polymer via polycondensation with elimination of $MgBr_2$. The catalytic cross-coupling process has come to be known as Kumada (or Kumada–Corriu) cross-coupling, as its application to C–C bond formation was independently discovered by Makoto Kumada (1920–2007)[39] and Robert Corriu (1934–2016) in 1972. Yamamoto, however, was the first to apply this metal-catalyzed process to the polymerization of poly(arylene)s.[36,37]

This polycondensation gave a black precipitate that was then purified via sequential Soxhlet extractions with hot methanol and hot $CHCl_3$ in order to remove low-molecular-weight material. The remaining product represented 78% of the initially isolated polymer and was described as a black, insoluble powder that did not melt below 360°C.[38] Characterization via IR spectroscopy gave a single sharp band at 788 cm^{-1}, the simplicity of which led to the conclusion that the backbone was comprised of a single repeating unit linked together at the 2- and 5-positions. Finally, the electric properties of the polymer were investigated to give a conductivity of 5.3×10^{-11} S cm^{-1} for the neutral polymer. The polymer was then treated with I_2 vapor, increasing the conductivity to 3.4×10^{-4} S cm^{-1}. The extent of doping was then determined by the increase in polymer weight, giving a value of 0.32 iodine atom equivalents per thiophene unit.

This initial report was then followed by a second polythiophene paper the following year.[40] In addition to providing new characterization data for the previously reported polythiophene, the polycondensation of the isomeric 2,4-dibromothiophene was also reported, thus allowing direct comparison between the two polymeric materials. While both polymers were obtained as insoluble, amorphous powders, the poly(α,β-thiophene) exhibited reduced conjugation in comparison with the poly(α,α'-thiophene). This could be seen in both the color of the respective materials—black versus light brown—and the fact that the α,α'-material exhibited significantly higher reactivity with I_2.

While the original report had just focused on the insoluble fraction, this second study also characterized the $CHCl_3$-soluble fraction via select solution measurements, beginning with determination of the molecular weights by vapor pressure osmometry.[40] Although both soluble fractions were of low molecular weight, the α,β-material (MW = 2050) was nearly twice that of the α,α'-material (MW = 1100–1400), likely due to enhanced solubility in the less conjugated material. Further solution characterization included both absorption and fluorescence spectroscopy, with the α,α'-material exhibiting an absorption maximum at 410 nm and strong fluorescence. In contrast, the α,β-material exhibited blue-shifted absorbance (280 nm) and no fluorescence.

Lastly, the electrical conductivity of both polymers was studied, both as the undoped neutral materials and doped with either I_2 or SO_3.[40] While the conductivities of the neutral materials were fairly comparable (5×10^{-11} S cm^{-1} vs. 5×10^{-13} S cm^{-1} for the α,α'- and α,β-materials, respectively), I_2 doping resulted in little enhancement of the α,β-material (7×10^{-11} S cm^{-1}). In comparison, optimization of the I_2 doping of poly(α,α'-thiophene) increased its conductivity to 4×10^{-2} S cm^{-1}. Doping with SO_3 resulted in enhanced conductivity in both materials, with surprisingly similar results (1×10^{-4} S cm^{-1} vs. 9×10^{-5} S cm^{-1}).

The effects of thiophene connectivity were further investigated in February of 1982 with a series of copolymers generated from mixtures of 2,5-dibromothiophene and 2,4-dibromothiophene (Figure 6.11).[41] As expected, the polymer absorption was dependent on the ratio of the two repeat units, with the maximum of the lowest-energy transition shifting to longer wavelength with an increase in the content of thiophene-2,5-diyl units. All of the polymers studied exhibited similar conductivities for the neutral species (10^{-11}–10^{-14} S cm^{-1}), yet the effect of doping with I_2 was much more significant for the polymers of greater thiophene-2,5-diyl content.

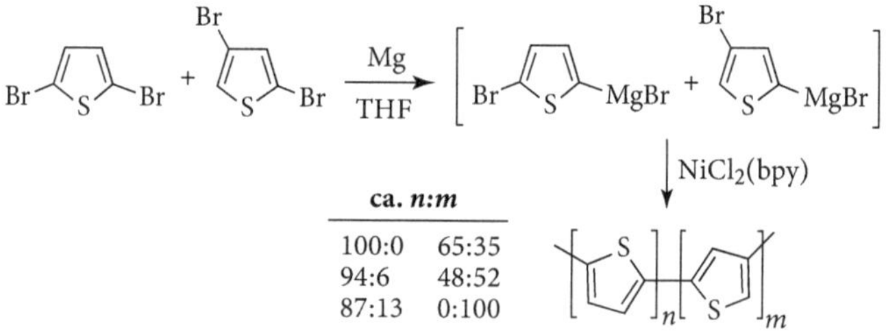

Figure 6.11 Synthesis of random copolymers from mixtures of 2,5-dibromo- and 2,4-dibromo-thiophene.

Also in February of 1982, Yamamoto extended these methods to the polymerization of 3-methylthiophene (Figure 6.12) in an attempt to produce more soluble analogs.[42] The crude polymer was washed repeatedly with methanol, dried, and then separated into a CHCl$_3$-soluble fraction (95%) and an insoluble residue (5%). Besides increased CHCl$_3$ solubility, the overall properties were quite similar to those of the unfunctionalized polythiophene. The CHCl$_3$-soluble fraction was found to have an M_n of 2400 (n = ca. 25) and its UV-vis spectrum exhibited an absorption maximum at 420 nm. After doping with I_2, the CHCl$_3$-soluble fraction was found to have a conductivity of 2.8×10^{-2} S cm^{-1}, while the insoluble fraction gave a value of 4.5×10^{-1} S cm^{-1}.

A paper compiling many of the previous results discussed earlier was then reported the following year, which also included some new details on various points.[43] These new details included a study of the Grignard intermediates formed during the polymerization. The product of the 1:1 reaction of Mg and 2,5-dibromothiophene was hydrolyzed and analyzed by gas chromatography to reveal not a single product, but a mixture of the unreacted 2,5-dibromothiophene and its mono- and di-Grignard products in a 1:2:1 ratio (Figure 6.12). Also included was a deeper analysis of the NMR spectrum of poly(3-methylthiophene), which showed a single aromatic proton, but two distinct CH$_3$ signals. The two CH$_3$ signals were proposed to be due to either the

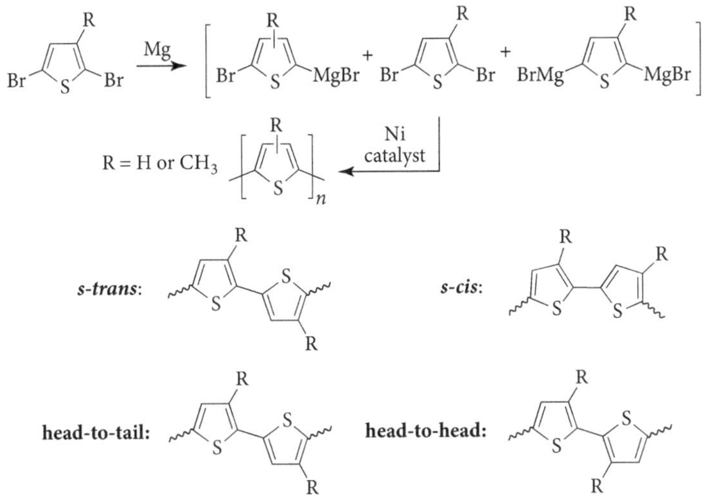

Figure 6.12 Detailed general polymerization of dibromothiophenes and potential conformational and regiochemical units in poly(3-methylthiophene).

presence of both s-*trans* and s-*cis* conformations in solution, or different regio-coupled units, referred to as head-to-tail and head-to-head (Figure 6.12), terms still used to describe the regiochemistry of modern polyalkylthiophenes.

A second 1983 paper then compiled the various studies on the electrical con-ductivities of the polythiophenes produced to date, focusing on the doping of these materials with both I_2 and SO_3.[44] The maximum conductivity achieved for I_2-doped samples of the parent polythiophene now reached circa 5×10^{-2} S cm^{-1}, while the SO_3-doped materials gave values of circa 1×10^{-3} S cm^{-1}. Fur-thermore, the mechanism of conductivity was proposed to be due to the removal of electrons from the polythiophene to result in the formation of cationic cen-ters that could then migrate along the π-conjugated backbone. Two additional reports were then published in 1984 and 1985 on the application of I_2-doped polythiophenes as active materials for positive electrodes in galvanic cells.[45,46]

A second 1985 paper then reported Raman and ^{129}I Mössbauer studies of I_2-doped polythiophenes.[47] The Raman spectrum exhibited a band at 108 cm^{-1}, which was assigned to the symmetric stretching vibration of a linear triiodide anion (I_3^-). A relatively weak secondary band at 145 cm^{-1} was then assigned to the antisymmetric vibration of a distorted I_3^- ion, while a strong band at 170 cm^{-1} was concluded to be due to an I_2 molecule coordinated to a I_3^- ion. All of this led to the overall conclusion that the iodine in I_2-doped polythiophene exists predominately as a polyiodide made up of alternating I_3^- and I_2 units. This was further supported by the Mössbauer spectra, which clearly ruled out the presence of isolated I^- ions.

While Yamamoto dominated the initial preparation of polythiophene via polycondensation, John W.-P. Lin and Lesley P. Dudek at Xerox had also reported a related metal-catalyzed polymerization study, about five months after Yamamoto's first 1980 paper.[48] The methods reported were very similar to those of Yamamoto, but with a focus on various metal acetylacetonate (acac) catalysts for the cross-coupling of the bromothienylmagnesium bromide intermediate. Although the polymerization was successfully catalyzed by nickel, iron, and cobalt complexes, the best overall results were achieved with $Ni(acac)_2$. The polythiophene products exhibited properties that were nearly identical to Yamamoto's previous reports, although higher conductivities of up to 0.1 S cm^{-1} were achieved for I_2-doped samples. After this report, Yamamoto revisited the conductivities of their materials and stated that it was found that the commercial 2,5-dibromothiophene used in the early Toyko Tech studies was impure and contained about 6% 2,4-dibromothiophene.[41] This then led to α,β-defects in the polymers and the observed lower conductivities. Once pure 2,5-dibromothiophene reagents were utilized, Yamamoto was able to achieve similar conductivities to those reported by Lin and Dudek. The Xerox study, however, was also able to show that thermal annealing could further increase the sample conductivities up to 6 S cm^{-1}. This initial report from Lin and Dudek was never followed up with any additional papers.

6.5 Polythiophene via electropolymerization

Considering that the electropolymerization of aniline and pyrrole were both known by 1979, it is not surprising that the electropolymerization of thiophene soon followed. At the time, the leader of the method was Arthur Diaz at IBM, who was working extensively on its application to the production of polypyrrole films (see Chapter 3). As such, it was only a matter of time before Diaz started looking at the electropolymerization of thiophene. However, the first published report of polythiophene via electrochemical methods came not from Diaz, but from V. L. Afanas'ev and coworkers at the Moscow Institute of Chemical Physics of the Russian Academy of Sciences.[49] In a very brief communication in late 1980, they reported the electropolymerization of thiophene via experimental conditions previously developed by Diaz and coworkers for pyrrole.[50] Using tetrabutylammonium tetrafluoroborate ($Bu_4N^+BF_4^-$) as the electrolyte, they successfully generated polythiophene-BF_4 films with a reported composition of $(C_{40}H_{2.5}S)(BF_4)_{0.6}$. Measurement of the film conductivities gave values of 10^{-3} S cm^{-1}, but few other details were reported.

In early 1981, Diaz then published a paper in *Chemical Scripta* that consisted of the proceedings of a talk he had previously given at the International

Conference on Low Dimensional Synthetic Metals in August of 1980.[51] This paper was an overview of his work on polypyrrole but also included a number of other polymers prepared by electropolymerization, with polythiophene among these materials. In a table of collected data, polythiophene is listed as having a switching potential of 700 mV and a conductivity of 10^{-3}–10^{-1} S cm^{-1}, the low end of which is in good agreement with the value reported by Afanas'ev. As briefly stated by Diaz:[51] "We have also prepared a thiophene polymer by the electrochemical route and find that the resulting polymer has a more anodic oxidation potential and a lower room temperature conductivity than polypyrrole." Also in early 1981,[52] Diaz and coworkers at IBM published a preliminary note that expanded on the previous polypyrrole work at IBM to provide correlations between the potential of oxidation and the number of repeat units in oligomers and polymers of pyrroles, phenylenes, and thiophenes. Although the preparation of the thiophene materials was not included in the report, it was stated that they had been previously prepared in an unpublished 1980 study. In terms of the polymer of thiophene, a potential of 700 mV is given for the film oxidation, with the neutral film exhibiting an absorption maximum at 410 nm. This energy is in excellent agreement with that previously reported by Yamamoto for polythiophene via polycondensation.[40]

A more extensive report was then published in the spring of 1983 that provided the full electropolymerization details for thiophene and various thiophene derivatives.[53] Along with Diaz, this paper included his IBM coworkers Robert J. Waltman and Joachim Bargon as coauthors, with Waltman given as corresponding author, and all three authors were constants on all of the polythiophene papers from the IBM group. Bargon received his MS and Ph.D. in physics from the Technical University in Darmstadt and came to the United States in 1969 as a postdoctoral fellow in organic chemistry at the California Institute of Technology in Pasadena.[54] He joined IBM in 1970 as a postdoctoral fellow at the Thomas J. Watson Research Center in Yorktown Heights, New York, before transferring to the San Jose Research Laboratory in 1971. At San Jose, Bargon had served as manager of the Chemical Dynamics Department since 1976. Waltman received a BS in radiochemistry and an MS in chemistry from San Jose State University and had worked in the Physical Science Department since joining the San Jose Research Laboratory in 1980.[54]

The conditions detailed were nearly identical to those previously reported for polypyrrole to give polythiophene films compensated with counterions of HSO_4^-, BF_4^-, or PF_6^-. In addition to the electropolymerization of thiophene and 2,2'-bithiophene, polymers were successfully generated from the derivatives 3-methylthiophene, 3-bromothiophene, 3-thiopheneacetonitrile, and 3,4-dibromothiophene. Freestanding films, however, could only be obtained from thiophene, 2,2'-bithiophene, and 3-methylthiophene. Typical thicknesses of the freestanding films were 10^{-4}–10^{-3} cm and exhibited conductivities of 0.02–1 S cm^{-1}, with poly(3-methylthiophene)-PF$_6$$^-$ giving the highest conductivity.[53] While these values were considerably higher than those previously reported,[49,51] they were still far below those previously achieved for electropolymerized polypyrroles.[50,51,53] Waltman and Bargon republished much of this data as part of an internal IBM publication later that same year, although interestingly without Diaz as a coauthor.[54]

This was followed by a 1984 paper that focused on the surface characteristics of various polymer films, including both electropolymerized polythiophene and poly(3-methylthiophene).[55] Analysis of the surface by scanning electron micrograph (SEM) revealed that electrochemically generated films can have a wide variation in the surface topology. As the electropolymerized films are viewed as ionic composites in which the polymer is cationic and compensated by amounts of inorganic anion, electron spectroscopy for chemical analysis (ESCA) was used to determine the anion content by weight in these films. For polythiophene-ClO$_4$$^-$, this was determined to be 17%, with a value of only 11% for poly(3-methylthiophene)-ClO$_4$$^-$. Lastly, the wetting characteristics of both the neutral and oxidized films were measured via contact angle measurements, with contact angles of 86°–90° for the neutral thiophene films. Surprisingly, the electropolymerized polythiophene films show only a marginal change upon oxidation.

Two additional 1984 papers then followed that focused primarily on the effects of substituents on the potentials of oxidation for various thiophene monomers and their corresponding polymers.[56,57] Analysis of this data showed a linear correlation between the potentials of oxidation for the monomers and the respective Hammett substituent constants (σ_p^+). It was suggested that one possible interpretation of these results is that there is an optimum potential range (i.e., reactivity of the radical cation intermediate), which favors radical cation coupling in the follow-up reaction to yield polymeric films. This conclusion was then used to explain why some thiophene derivatives successfully generated electropolymerized materials, while others did not. Lastly, for those monomers that did form polymers, a linear correlation was found between the potentials for the oxidation of the monomers and those of the corresponding polymers.

With the popularity of the electropolymerization methods developed at IBM for polypyrrole, it was not surprising that others also began applying these methods to thiophene. Beyond the initial report by Afanas'ev,[49] the first detailed efforts outside of IBM were published early in 1982 by Gérard Tourillon and Francis Garnier in France.[58] This study reported a variety of electropolymerized materials, including those from pyrrole, indole, azulene, thiophene, and furan. Similar to the previous methods of Diaz and coworkers,[50] the polythiophene films were electropolymerized onto Pt from solutions of thiophene and Bu_4NClO_4 in CH_3CN. However, the application of other solvents (THF, CH_2Cl_2), electrolytes (Bu_4NBF_4, $LiClO_4$), and electrode substrates (Au, SnO_2, In_2O_3) was also investigated.[58] Via such methods, polythiophene films were generated, analysis of which indicated the presence of ClO_4^- ions and a polymer structure consisting of four thiophene units to each counterion. This proposed structure was consistent with that previously concluded by Diaz and coworkers for electropolymerized polypyrrole,[50] although it corresponded to a higher level of anion content than previously determined for polythiophene-ClO_4^- films.[55] SEM of the polymer films revealed a surface that was regular and homogeneous yet became more rough and irregular as the film thickness was increased. Conductivity measurements via four-point probe gave values of 10–100 S cm^{-1}, considerably higher than previous values for polythiophenes and the first polythiophene values that compared well with previous polypyrrole results. Although the authors attributed the increased conductivity to reduced impurities in the electropolymerized films, the potentially higher doping levels could have also played a considerable role.

The effect of the dopant (i.e., counterion) on the properties of the electropolymerized films was then the focus of a second paper in 1983,[59] which also expanded the scope to include the functionalized derivatives generated from 3-methyl-, 3,4-dimethyl- and 3,4-diethyl-thiophene. The initially electropolymerized films could also be electrochemically reduced to allow characterization of the neutral films. Both polythiophene and poly(3-methylthiophene) exhibited an absorption maximum at 480 nm, with corresponding absorption coefficients of 1–5×10^5 cm^{-1}. Using a band model, the absorption onset was used to calculate a bandgap of circa 2–2.5 eV. In contrast, the polydialkylthiophenes gave considerably blue-shifted maxima (280–330 nm), which was attributed to steric interactions caused by the alkyl groups that resulted in a twist between neighboring thiophenes. Electrochemical oxidation resulted in a loss of the visible absorption and the growth of a new, very broad band extending from 650 to 2000 nm, which was attributed to the existence of free carriers. In terms of conductivity, although some differences were observed with the choice of dopant, it was generally concluded that the macroscopic conductivity increased with the doping level, but that the maximum conductivity was limited by the film morphology. As with the previous study, measured conductivities of doped polythiophenes ranged from 10 to 100 S cm^{-1}, with the highest values achieved from poly(3-methylthiophene)-$CF_3SO_3^-$ films. Fully dedoped polythiophene films exhibited conductivities of circa 10^{-7} S cm^{-1}.

A second 1983 paper then reported the electrochromic properties of electrochemically prepared films of polypyrrole, polythiophene, poly(3-methylthiophene), and poly(3,4-dimethylthiophene) on semitransparent Pt electrodes on glass.[60] As explained, the contrast and perception of color were related to the position and intensities of the absorption profiles of the neutral and oxidized states. For polythiophene and its 3-methyl derivative, the films were red in the neutral state and blue in the oxidized state, while the dimethyl derivative switched between pale blue and deep blue. Switching to the oxidized state was fast, occurring in a few milliseconds, while switching back to the neutral state was much slower due to some amount of structural reorganization. Samples of poly(3-methylthiophene) were found to exhibit high stability, with 80% of its initial activity observed after 1.2×10^5 cycles.

Another 1983 report then focused on polythiophene's enhanced stability in comparison with other conjugated materials such as polyacetylene, polyphenylene, and polypyrrole.[61] Beginning with poly(3-methylthiophene), it was shown that both the doped and undoped forms were relatively unaffected by ambient air, with neither water nor oxygen detectable in undoped samples after storage in air for eight months. In a similar manner, the electrochemically doped samples containing $CF_3SO_3^-$ exhibited only minor reduction in conductivity after eight months in air. The stability of poly(3-methylthiophene) was then evaluated under electrochemical cycling between the oxidized and neutral states, with the sample exhibiting reproducible I–V curves through 20 cycles. Overall, polythiophenes exhibited enhanced environmental stability and more stable redox cycling in comparison with other conjugated polymers.

Finally, two additional 1984 papers detailed the application of electrochemically grown polythiophene films to photovoltaic[62] and photoelectrochemical[63] cells. In the first case, the photovoltaic cells were comprised of neutral polymer films sandwiched between a gold electrode and a 10 nm semitransparent Al overlayer. While promising power efficiencies were achieved at low illumination ($P_i < 1 \, \mu W \, cm^{-2}$), these decreased rapidly to circa 0.01% at higher light intensities. The electrochemistry of polymer films on either Au or Pt electrodes were then studied under illumination to probe the parameters that impacted the resulting photocurrent. They followed this with a 1986 report in which poly(3-methylthiophene) was electrochemically grown onto an electrode comprised of a thin platinum or gold layer (15 nm) deposited onto a glass substrate.[64] The polymer film utilized was either neutral or doped at a level of 5%. A semitransparent (10 nm) Al electrode was then evaporated onto the top of the polymer layer. At a light power of $10 \, \mu W \, cm^{-2}$, an I_{SC} of $12 \, \mu A/cm^2$, a V_{OC} of 0.35 V, and the fill factor of 0.25 were achieved. These reports appear to be the first examples of organic photovoltaic cells utilizing thiophene materials, which eventually became the primary and most successful materials for such applications.[65]

Additional reports of electropolymerized polythiophenes based on the IBM methods also came later in 1982 from Yoshio Inuishi (1921–1994) and Katsumi Yoshino (b. 1941) at Osaka University.[66] Born in 1921,[67] Yoshio Inuishi attended

Osaka Imperial University (renamed Osaka University in 1947), receiving a bachelor's degree in electrical engineering in 1944.[66,67] He then continued his graduate education at Osaka under the direction of Prof. Tokuo Suita. From 1955 to 1956, he also worked with Arthur R. von Hippel (1898–2003) at MIT as a Fulbright research fellow.[67,68] He completed his Dr. Eng. in 1959, with a dissertation on electrical insulation breakdown.[69] Remaining at Osaka as part of the faculty of the Department of Electrical Engineering, Inuishi founded the University's Research Center for Superconducting Materials and Electronics and became the first director. He passed away on October 26,1994.[67,68]

Katsumi Yoshino was born on December 10, 1941, in Shimane, Japan.[70,71] Attending Osaka University, he received a bachelor's in engineering (BE) from the Department of Electrical Engineering in 1964, followed by a master's in engineering (ME) in 1966.[70,71] He then completed his Dr. Eng. under Inuishi in 1969,[71] after which he became a research associate in 1969.[70,71] He served as a visiting scientist at the Hahn-Meitner Institute for Nuclear Research in Berlin from 1974 to 1975[70,71] and became an associate professor of the Department of Electrical Engineering at Osaka in 1978.[71] He was promoted to professor in 1988[70,71] and retired from Osaka as a professor emeritus in 2005.[70] He served as the vice president of the Institute Electrical Engineers Japan (IEEJ) from 1996 to 1997, and as president of the Japanese Liquid Crystal Society (JLCS) from 1999 to 2000.[71] During his career, he received various awards, including the Japan Society of Applied Physics (JSAP) Award in 1984, the Osaka Science Prize in 1990, the IEEJ Book of the Year Award in 1997, IEEJ Outstanding Achievement Award in 1998, the JLSC Outstanding Achievement Award in 2003, and the Outstanding Achievement Award of the Society of Polymer Science, Japan, in 2007.[70]

The conditions utilized by Yoshino and Inuishi incorporated a few modifications from the general methods discussed, including the use of $AgClO_4$ as the electrolyte and a working electrode of indium-tin oxide (ITO).[66] Using these methods, the application of a potential of 2.4 V to the working electrode resulted in the production of a dark greenish film that could be easily removed from the electrode. Chemical analysis determined the films to have a composition of $[C_{3.91}H_{2.91}S_{1.00}(ClO_4)_{0.26}]_x$, which led to the view that the positive charge on the thiophene was delocalized over four thiophene units, similar to previous conclusions by Tourillon and Garnier.[58] The electropolymerized films were determined to have a thickness of circa 10 μm and a room temperature sheet conductivity of 0.6 S cm^{-1}. In contrast, the conductivity perpendicular to the film surface was determined to be only 10^{-4} S cm^{-1}, which seemed to indicate highly anisotropic conductivity. Lastly, the positive dependence of the conductivity on temperature indicated a classical semiconducting relationship.

A brief communication the following year[72] then reported optimization of the polymerization conditions to give conductivities of 20–106 S cm^{-1}, similar to those reported by Tourillon and Garnier.[58,59] The most conductive films were grown from LiBF$_4$/PhCN solutions, which were determined to have doping levels of circa 30 mol% of BF$_4^-$ per thiophene molecule, a bit higher than that determined for the original films of the 1982 study. Finally, neutral films were successfully obtained via electrochemical reduction to give materials with an optical bandgap of 2.0 eV and a conductivity of 2×10^{-8} S cm^{-1}.

This was then followed quickly by a second 1983 communication that provided some additional details of the optimized materials.[73] For measurement of the electrical properties, films of 10–20 μm that had been stripped from the ITO electrode were used, while thin films of circa 0.3 μm thickness that remained on the ITO substrate were used for optical measurements. As with the samples reported in the original 1982 paper,[66] the conductivities of the optimized samples exhibited a maximum value of 106 S cm^{-1} determined at 290 K. This conductivity decreased monotonously with temperature down to 10 K, after which it plateaued to a nearly constant value of 2×10^{-2} S cm^{-1} (Figure 6.13). The absorption spectrum of undoped samples exhibited maxima at circa 260 nm, with an absorption edge at about 2.0 eV, both of which agrees reasonably well with the previous report of Tourillon and Francis Garnier.[59] The absorption spectrum of BF$_4^-$ doped film exhibited a decrease of transmission below the energy of 2.3 eV, which was considered to be due to free carriers associated with the high conductivity of the sample.

A review of some of the early results of electropolymerized polythiophenes was then reported that same year as part of an internal Osaka University publication,[74] after which Yoshino continued on independently with an additional study on the chemical doping of neutral polythiophene in 1984.[75] These efforts focused on both the luminescence and photoinduced absorption of polythiophene films. Excitation at 4 K with 488 nm light gave rise to a broad luminescence band with a maximum at circa 653 nm and a radiative lifetime of $\tau \leq 10$ ps. Photoinduced absorption revealed peaks at 0.45 and 1.35 eV using 90 mW of 488 nm light for excitation at 4 K. While it is stated that positively charged excitonic polaron might be expected to give two peaks of comparable intensity at 0.4 and 1.4 eV, the authors point out that care should be taken in coming to any firm conclusions. An additional absorption is found in doped films, proposed to be due to the growth of bipolarons, which could in principle give rise to two observable transitions in the bandgap.

Additional reports of electropolymerized polythiophenes also came from Shu Hotta and coworkers from Matsushita Electric Industrial Co. starting in early 1983.[76] Hotta received his Dr. Eng. degree in 1988 from Kyoto University, while working at Matsushita Electric Industrial Co., Ltd. (now Panasonic

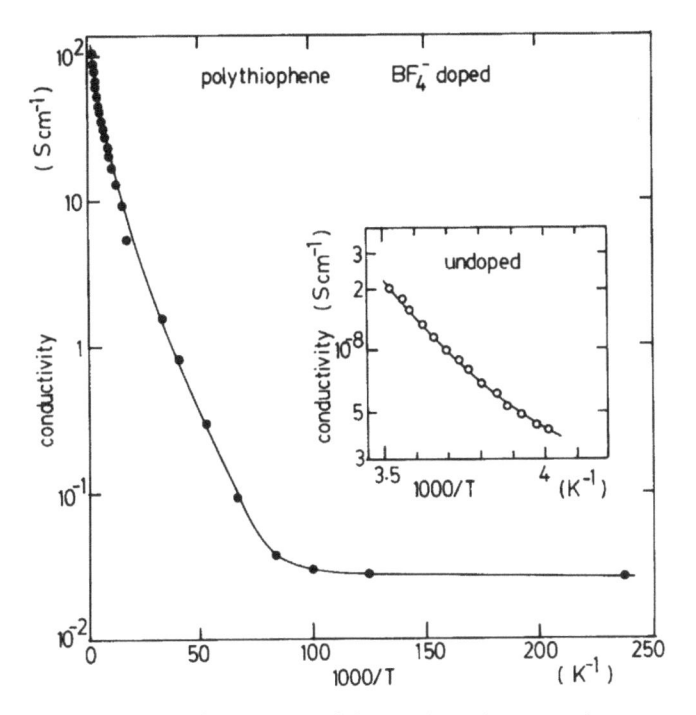

Figure 6.13 Temperature dependence of electrical conductivity of polythiophene-BF$_4$$^-$. The temperature dependence of the undoped film treated with gaseous ammonia is shown in the inset.

Reprinted from K. Kaneto, K. Yoshino, Y. Inuishi. Electrical and optical properties of polythiophene prepared by electrochemical polymerization. *Solid State Commun.* **1983**, 46, 389–391, with permission from Elsevier.

Corporation).[77] He then became professor of the Department of Polymer Science and Engineering at Kyoto Institute of Technology in 2003. It was at Matsushita, and prior to his doctoral work, that Hotta, Tomiharu Hosaka, and Wataru Shimotsuma reported the electropolymerization of thiophene onto ITO substrates using various solvent–electrolyte solutions.[76] The solvents applied included nitrobenzene, dimethyl sulfate, and diethyl sulfate, while the electrolytes included BuN_4BF_4, Et_4NClO_4, H_2SO_4, and LiTCNQ. The prepared films were 5–10 μm thick, green-black in color, and found to be amorphous by X-ray diffraction. Under all conditions, polythiophene-$ClO_4$$^-$ films were found to be the most conductive, with values as high as circa 16 S cm^{-1}. The conductivity was found to be dependent on both the electropolymerization temperature, with growth at lower temperatures giving higher conductivities, and whether the measurements were carried out in air or under vacuum. In the latter case, conductivities measured at the reduced pressure were about twice as large as those measured in air.

This was then followed by a second paper on the electropolymerization of 3-methylthiophene in early 1984.[78] Here, electropolymerization was carried out at 5°C from solutions of Bu_4NClO_4 in nitrobenzene to give 8-μm-thick poly(3-methylthiophene)-ClO_4^- films deposited onto an ITO substrate. The as-deposited films were then electrochemically reduced by applying a constant current density of 1 mA cm^{-2}. Elemental analysis indicated that the neutral film still contained small amounts of chloride, but that the ClO_4^- content had been almost entirely eliminated. The neutral films were then doped by soaking the films in methanol solutions of I_2 at room temperature. The conductivity was found to increase with dopant concentration, with the highest room temperature conductivity found to be 5.3 S cm^{-1}. The dopant molar ratio for the most conductive samples was estimated to be circa 30% per thiophene unit. While the conductivities here are about 10 times that reported by Yamamoto for the iodine doping of poly(3-methylthiophene) produced via Kumada coupling,[44] they are far lower than that found by Tourillon and Garnier for poly(3-methylthiophene)-$CF_3SO_3^-$ films.[59]

6.6 Kossmehl and polythiophenes via chemical oxidation

Following the initial papers on electrochemical polymerization of thiophene, Gerhard Kossmehl (or Koßmehl, see Chapter 4) and Georg Chatzitheodorou of the Free University of Berlin reported the first example of thiophene polymerizations via chemical oxidation in the fall of 1981.[79] With the goal of investigating the doping of polythiophene, polymer samples were prepared via the cross-coupling methods previously reported by Yamamoto,[38] after which they were doped via treatment with AsF_5. The measured room temperature conductivities were found to increase with the pressure of AsF_5 applied, with a maximum value of 1.4×10^{-3} S cm^{-1}.

More interestingly, however, it was also found that conducting polythiophenes could be produced by treating either thiophene or 2,2'-bithiophene with AsF_5, giving conductivities up to 0.021 S cm^{-1}. In an effort to characterize the polymers produced in this way, the initially formed conductive materials were dedoped with 25% aqueous NH_3 to give insoluble polymers with a silver-like appearance.[79] The measured conductivities of the dedoped materials were similar to the neutral polythiophenes generated via Kumada coupling, and analysis by mass spectroscopy suggested that the neutral materials were made up of thiophene oligomers with a maximum number of 9–10 units. Such products were believed to be due to an oxidative polymerization coupled with both the reduction of AsF_5 and the formation of AsF_6^- anions:

$$AsF_5 + 2H^+ + 2e^- \rightarrow AsF_3 + 2HF$$

$$2AsF_5 + 2HF \rightarrow 2AsF_6^- + 2H^+$$

A mechanism for this polymerization was also proposed, starting with the generation of a thiophene radical cation as shown in Figure 6.14. While aspects differ from the modern accepted mechanism given in Figure 6.1, this seems to be the first attempt to explain the oxidative polymerization of thiophene.

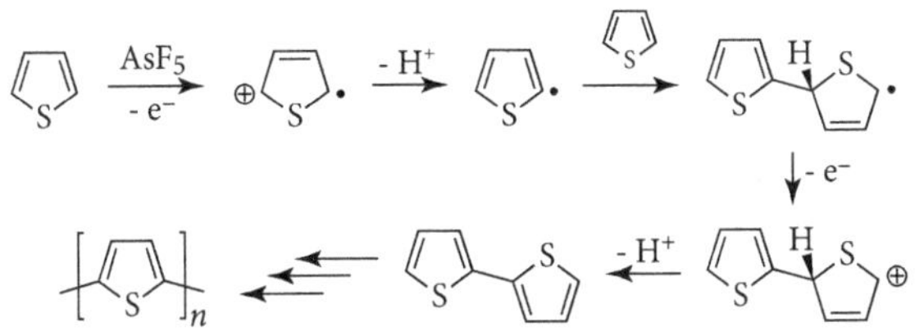

Figure 6.14 Proposed mechanism for the oxidative polymerization of thiophene by AsF_5.

A second paper on polythiophene-AsF_6^- materials then followed in 1982, which expanded on the results given in the initial publication.[80] Again, the polymerization of thiophene or 2,2'-bithiophene with AsF_5 is described, in which thin black films (ca. 0.04 µm) were formed on the walls of the reaction vessel. Treatment of these films with aqueous NH_3 then gave insoluble and infusible films with a silver-like appearance. Analysis of the initial black films by IR spectroscopy led to the view that these were comprised of oxidized polythiophene with cationic centers that were balanced by AsF_6^- counterions. The formation of these materials was again explained by the same reactions and mechanisms proposed in the initial publication. Finally, the study of the doping of these materials with AsF_5 under various conditions was reported. As before, the conductivity increased with AsF_5 pressure, with a maximum value of 0.021 S cm^{-1} achieved via 450 torr of AsF_5. However, the conductivities' values began to decrease again with the application of AsF_5 at pressures above 450 torr.

A third paper in 1983 then expanded the scope to include the oxidants $NOSbF_6$, $NOBF_4$, and $NOPF_6$.[81] Thiophene or 2,2'-bithiophene was reacted with various oxidants in either dry CH_2Cl_2 or 1,2-dichloroethane to give black precipitates that were viewed to be oxidized polythiophenes, the cationic centers of which were balanced by SbF_6^-, BF_4^-, or PF_6^- counterions. These materials exhibited conductivities as high as 0.07 S cm^{-1}, and analysis by mass spectroscopy suggested these materials to be higher molecular weight than the previous materials via AsF_5, with molecules consisting of up to 16 units.

6.7 Heeger, Wudl, and the further study of polythiophene

Although most frequently associated with polyacetylene, Alan Heeger (see Chapter 5) expanded the scope of his research after his move to the University of California, Santa Barbara (UCSB), in 1982. As part of this move, Heeger also helped establish UCSB's Institute for Polymers and Organic Solids and served as its founding director.[82] Another critical member of this new Institute was the organic chemist Fred Wudl (b. 1941), who we will revisit in more detail in Chapter 7. At the Institute, Heeger and Wudl then began a long-standing collaboration, beginning with the first of many reports on polythiophene in 1984.[83]

This initial paper reported optimization of the polycondensation of dihalothiophenes via Kumada coupling.[83] These efforts focused on the use of various dihalothiophenes (dibromo, bromo/chloro, or diiodo) as the monomer, as well as a range of solvents as the reaction media. The best results were obtained using 2,5-diiodothiophene and ether as a solvent for the formation of the Grignard intermediate, after which the ether was removed and replaced with anisole as solvent for the polycondensation step. The catalyst applied here was phosphine-chelated dichloro(1,3-diphenylphosphinopropane)nickel(II), commonly abbreviated as $NiCl_2(dppe)$. After polymerization, the samples were purified by sequential Soxhlet extractions with methanol, THF, $CHCl_3$, and chlorobenzene. Such methods resulted in a crystalline product of high purity, with an estimated molecular weight of about 4000 (46 thiophene units), significantly higher than that previously obtained by Yamamoto.[40,42]

Characterization of the polymer revealed an absorption maximum at circa 2.7 eV (460 nm), with a band edge at circa 2.0 eV, both of which are in good agreement with previous reports on electropolymerized polythiophene.[59,73] Electron spin resonance (ESR) spectroscopy revealed a symmetric line centered at a g value of 2.0026, indicating that the species responsible for the spin resonance was a defect in the "*cis* polyene" backbone. Calibration of the absolute intensity indicated a room temperature spin susceptibility of 8.4×10^{-8} emu mol^{-1}. Finally, the room temperature resistivity of pressed pellets was found to be $\sim 10^9$ Ω cm, which lowered to $\sim 10^2 - 10^1$ Ω cm (0.01–0.1 S cm^{-1}) after doping with iodine vapor. The iodine doping also resulted in an increase in the absorption throughout the infrared, indicative of free carrier generation. Further study with doping via AsF_5 showed an increase in conductivity with dopant concentration in a manner consistent with previous studies, with conductivity up to 14 S cm^{-1} at a dopant concentration of 24 mol%. While this conductivity was a marked improvement in comparison with previous polythiophenes generated through Kumada coupling,[38,40-44] this was still well below the conductivities achieved via electropolymerization.[57,58,73]

A second 1984 paper then shifted to electropolymerized polythiophene, with a focus on its application as an electrode in an electrochemical cell.[84] Using fairly conventional electropolymerization methods, doped polythiophene films compensated with ClO_4^- were deposited onto Pt foil substrates. This polymer electrode and a Li counterelectrode were inserted into rectangular glass tubing, with the two electrodes separated by glass filter paper. A 1 M solution of $LiClO_4$ in propylene carbonate was used as the electrolyte. Once assembled, the evacuated rectangular glass was sealed twice across the electrode leads. The internal resistance of the electrochemical cell was ~100 Ω. After charging to an open-circuit voltage of 3.8 V, short-circuit currents of about 20 mA mg^{-1} were obtained, with a maximum power density of 2.5×10^4 W kg^{-1}. It was pointed out that the high voltages and excellent stability obtained with polythiophene as a cathode (vs. Li) suggest the possibility of the construction of an all-polymer battery consisting of a polythiophene cathode and a polyacetylene anode.

Two additional 1984 papers then focused on elucidation of the nature of the doped state of polythiophene.[85,86] In the first of these, the polythiophene utilized was generated via the electropolymerization of 2,2'-bithiophene, which allowed its growth at lower applied potentials.[82] These electropolymerized films were then utilized to carry out an *in situ* study of the absorption spectrum during electrochemical doping. The initially produced film was dedoped through the application of a reducing potential, which resulted in an absorbance spectrum essentially identical to that previously obtained from neutral samples produced via Kumada coupling. A series of absorption spectra were then taken at different applied voltages, ranging from 3.6 V to 4.3 V (vs. Li), as shown in Figure 6.15. As the doping proceeded, the intensity of the visible transition decreased continuously, with a shift of the absorption maximum to higher energy. In addition, two new absorption features appeared below the gap edge with intensities that increased as the dopant level increased. The maximum of the lower-energy transition remained at a constant energy (−0.65 eV), while the higher-energy transition shifted to higher energy as the dopant level increased.

Analysis of the results led to a model in which doping results in the generation of polarons and bipolarons, rather than the solitons of polyacetylene. Furthermore, the experimental evidence indicated an energy level structure as illustrated in Figure 6.16, in which $\hbar\omega_1 \approx 0.60$–0.65 eV and $\hbar\omega_2 \approx 1.4$–1.45 eV. The value for the interband absorption ($\hbar\omega_I$) could be estimated from the data to be circa 2.1 eV, which is in reasonable agreement with previous studies.

The second paper reported photoinduced absorption and photoinduced ESR measurements on neutral polythiophene synthesized by their previously reported optimization of polycondensation via Kumada coupling.[86] The observation of relatively sharp photoinduced mid-IR peaks (at 1020, 1120, 1200, and 1320 cm^{-1}) was concluded to demonstrate the formation of localized

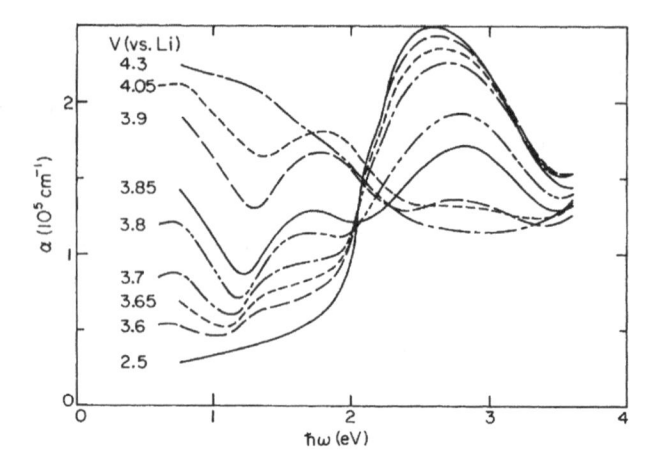

Figure 6.15 *In situ* absorption curves for polythiophene during electrochemical doping with ClO_4^-.

Reprinted with permission from T.-C. Chung, J. H. Kaufman, A. J. Heeger, F. Wudl. Charge storage in doped poly(thiophene): Optical and electrochemical studies. *Phys. Rev. B* **1984**, *30*, 702–710. © 1984 by the American Physical Society.

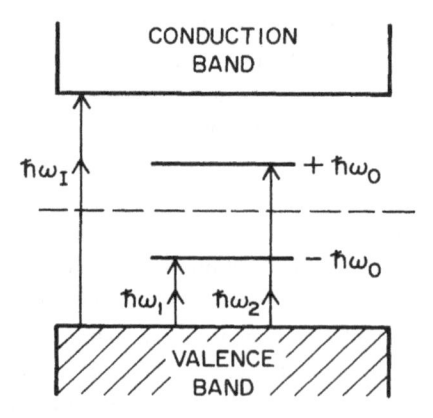

Figure 6.16 Energy-level diagram for poly(thiophene) at dilute doping concentrations.

Reprinted with permission from T.-C. Chung, J. H. Kaufman, A. J. Heeger, F. Wudl. Charge storage in doped poly(thiophene): Optical and electrochemical studies. *Phys. Rev. B* **1984**, *30*, 702–710. © 1984 by the American Physical Society.

structural distortions, consistent with photogeneration of polarons or bipolarons. In a similar manner, the observation of photoinduced spins via ESR was concluded to correspond to the photoproduction of polarons. In addition, a broad photoinduced absorption peak at 3600 cm^{-1} (0.45 eV) was attributed to the lowest-energy electronic excitation of the photogenerated polarons. All of this then led to the tentative conclusion that charge in the doped polymer was stored in charged bipolarons, whereas at least some of the charged photoexcitations are polarons. In the case of photoexcitation, the neutral, spin-singlet biopolaron would have a short lifetime and weak IR activity and thus would not contribute to the photoinduced properties.

This was then continued with a follow-up ESR study reported in early 1985.[87] As with the previous paper, this utilized polythiophene generated via Kumada coupling, which was then doped with either AsF_5 or I_2 for the ESR measurements. At high concentration of AsF_5 (ca. 24 mol%), a temperature-independent Pauli spin susceptibility of 3×10^{-5} emu/mole of thiophene was found, indicative of a metallic density of states. In contrast, measurements on moderately doped samples via iodine (<10 mol% I_3^-) indicated Curie spin concentrations independent of the doping level, but dependent on temperature and significantly larger than in undoped polythiophene. It was concluded that the features observed in these iodine-doped samples were due to polaron and bipolaron formation with one kinetically metastable spin-1/2 polaron per two polythiophene chains, consistent with their previous report.

A second 1985 paper then reported electron spin echo measurements on neutral polythiophene, which exhibited clear modulation of the spin-echo decay.[88] Analysis of the modulation and its Fourier transformed power spectrum indicated that the wave function of the spin defect is extended over ~25 proton nuclei. The functional behavior of the longitudinal relaxation, T_1, versus temperature was very similar to that observed previously for polyacetylene, although the polythiophene values were significantly longer.

A final 1985 paper then reported X-ray scattering measurements to determine the crystallinity and crystal structure of neutral polythiophene samples produced via Kumada coupling as previously described.[89] The as-synthesized material was found to be circa 35% crystalline, with the crystallinity increasing to circa 55% after annealing at 300°C. These crystalline regions were imperfect, however, with mean square deviation from the perfect lattice sites of approximately 0.3 Å.

Although the crystallographic data were incomplete, an initial model of the polythiophene structure was presented. The observed d spacings implied a similar crystal structure and chain packing to that of polyphenylene, which implied an individual chain structure in which all of the thiophene units adopted an s-*trans* conformation. The results were consistent with either an orthorhombic unit cell with lattice constants of $a = 7.80$ Å, $b = 5.55$ Å, and $c = 8.03$ Å, or a monoclinic unit cell with $a = 7.83$ Å, $b = 5.55$ Å, $c = 8.20$ Å, and $\beta = 96°$. In either case, the polymer axis is along c, with two polymer chains per unit cell.

Several additional papers in 1986 and 1987 then continued efforts to characterize the formation of polarons and bipolarons in doped polythiophenes, as well as photoexcited carriers in neutral polythiophenes.[90-93] While these added some additional experimental evidence and helped to refine the previously reported models, these were only minor extensions of their previous studies. However, it was at this same time that efforts moved to the study of polyalkylthiophenes, not only by Heeger and Wudl, but also a number of other research groups as well.

6.8 Polyalkylthiophenes

Although several reports had included the study of poly(3-methylthiophene),[42,43,53,55,59,60] the bulk of polythiophene studies from 1980 to 1985 were all focused on the unfunctionalized parent. While Yamamoto had included some solution measurements of both polythiophene[40] and poly(3-methylthiophene)[42] in a couple reports, this was largely limited to lower-molecular-weight fractions (MW = 1100–2500) of limited solubility, and most analysis was limited to solid-state materials. This changed, however, with the production of poly(3-alkylthiophene)s containing longer side chains, thus resulting in more soluble materials. The first of these reports was an initial preprint from Ron L. Elsenbaumer in 1985,[94] which was then followed up with a communication in the spring of 1986.[95]

The desired poly(3-alkylthiophene)s for these studies were produced by first generating the necessary 3-alkylthiophene (alkyl = methyl, ethyl, butyl, octyl) precursors from 3-bromothiophene via Kumada coupling with the appropriate alkylmagnesium bromide (Figure 6.17).[94,95] These 3-alkylthiophenes were then di-iodinated, and polycondensation was carried out using similar conditions

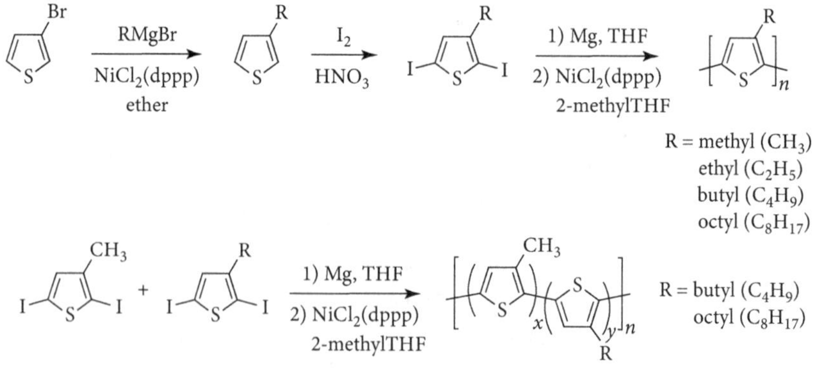

Figure 6.17 Elsenbaumer's synthesis of poly(3-alkylthiophene)s and analogous copolymers via Kumada coupling.

to that previously reported by Heeger and Wudl,[80] except that THF and 2-methyltetrahydrofuran (2-methylTHF) were used as the reaction solvents.

The resulting polythiophenes containing butyl or octyl side chains were found to be readily soluble at room temperature in common organic solvents such as THF, 2-methylTHF, nitropropane, toluene, xylene, methylene chloride, anisole, nitrobenzene, and benzonitrile.[95] The polymers were deep red in color, with an absorption maximum at 460 nm. Average molecular weights were determined to be in the range of 3000–8000, and analysis of poly(3-butylthiophene) indicated exclusive α,α'-coupling with a random mixture of head-to-head, tail-to-tail, and head-to-tail regiocoupling (Figure 6.12). X-ray diffraction found films cast from toluene to be completely amorphous.

Doping of polymer films with I_2, $NOBF_4$, $NOPF_6$, or $NOSbF_6$ resulted in conductivities of 3–4 S cm^{-1}. Surprisingly, the conductivity was found to be relatively insensitive to both the nature of the alkyl substituent and the chosen dopant. It was found that polymers with better film-forming properties and higher conductivities could be produced by random copolymerization of 2,5-diiodo-3-methylthiophene and 3-butyl-2,5-diidothiophene (Figure 6.18). These materials were found to have lower room temperature solubility, but a higher glass transition temperature (T_g) than the homopolymers of the longer alkyl side chains, with doping by I_2, $FeCl_3$, or $NOSbF_6$ giving conductivities of 10–50 S cm^{-1}.

Other early contributions on poly(3-alkylthiophene)s also came from Masaaki Sato, Susumu Tanaka, and Kyoji Kaeriyama of the Research Institute for Polymers and Textiles in Tsukuba, Japan. This began with a brief 1985 report that optimized the electropolymerization of 3-methylthiophene.[96] The resulting films of poly(3-methylthiophene) compensated by various dopants gave conductivities of 450–510 S cm^{-1}, five times that previously achieved by Tourillon and Garnier.[59] In the same study, they also applied these optimized conditions to produce a poly(3-ethylthiophene) film with a conductivity of 270 S cm^{-1}.

These efforts were then extended to longer alkyl analogs in an early 1986 report.[97] As with the work of Elsenbaumer, the various 3-alkylthiophenes (alkyl = hexyl, octyl, dodecyl, octadecyl, icosyl) were first prepared from 3-bromothiophene via Kumada cross-coupling (Figure 6.18), after which the corresponding polymeric materials were generated via electropolymerization. In order to adequately solubilize the long-chain 3-alkylthiophenes, nitrobenzene was used as the electropolymerization solvent, with tetraethylammonium hexafluorophosphate (Et_4NPF_6) as the supporting electrolyte. Polymerization was carried out at 5°C, after which the electrodeposited films were peeled off the anode and washed with hexane.

The conductivities of the oxidized films prepared in this way were then determined by four-point probe to give values ranging from 11 to 95 S

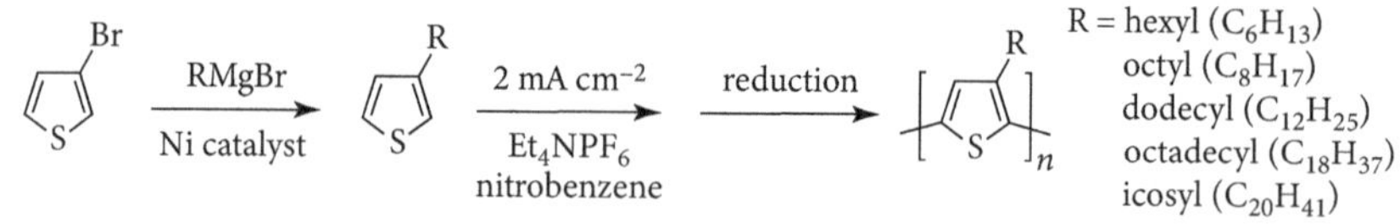

Figure 6.18 Kaeriyama's synthesis of long-chain poly(3-alkylthiophene)s via electropolymerization.

cm^{-1}.[97] The highest conductivity of 95 S cm^{-1} was determined for the poly(3-hexylthiophene)-PF_6^- film. While Elsenbaumer had observed no real effect on conductivity with the length of the side chains,[95] the conductivity of the materials here decreased as the length of the alkyl side chain was increased, with the largest drop found by the change from dodecyl (67 S cm^{-1}) to octadecyl (17 S cm^{-1}). This observed effect of chain length was likely due to the fact that the alkyl side chains applied here were significantly larger than those studied by Elsenbaumer.

Neutral films were then obtained by reduction of the initially deposited films. Of the films prepared, neutral poly(3-hexylthiophene), poly(3-octylthiophene), and poly(3-dodecylthiophene) were soluble in chloroform, benzene, and tetrahydronaphthalene, with the dodecyl analog exhibiting the greatest solubility. The final two long-chain derivatives were less soluble. The degrees of polymerization for the hexyl, octyl, and dodecyl analogs were determined to be 230, 140, and 90, respectively, with the shorter alkyl chains giving the higher molecular weights. Finally, films of poly(3-dodecylthiophene) were cast from tetrahydronaphthalene solution, which gave the same IR spectrum as initial neutral films obtained by electrochemical reduction of the as-grown film. In a similar manner, the cyclic voltammogram of cast films were found to be the same as that of as-grown films, all of this indicating that no chemical changes took place during dissolving, casting, and drying of the films.

A third example was reported in the fall of 1986 by Yoshino and coworkers, who presented the first production of a poly(3-alkylthiophene) via chemical oxidation.[98] Unlike the previous reports of Koßmehl on chemical oxidation via AsF_5 or various NO^+ salts,[79-81] the approach here utilized the addition of 3-hexylthiophene to a $CHCl_3$ solution of anhydrous $FeCl_3$ under nitrogen. After allowing this mixture to stir for two hours at 30°C, the reaction was added to methanol. The methanol precipitated the resulting polymer, while also solubilizing the various salt species in the reaction mixture. In addition, the reducing nature of methanol dedoped the polymer product, thus producing in its neutral form. After isolation of the polymer by filtration, further washing with H_2O and methanol ensured dedoping and removal of any remaining salts. In the reported oxidative polymerization of 3-hexylthiophene here, the product was isolated in a yield of 85%, with both elemental analysis and IR spectroscopy confirming its identity as poly(3-hexylthiophene).

Polymer thin films could be cast from CH_2Cl_2 solutions, which gave optical spectra with a maximum at circa 460 nm and an onset at 2.0 eV. The conductivity of such neutral films was determined to be 10^{-10} S cm^{-1}, similar to that previously reported for neutral polythiophene. Oxidation of the films with I_2 vapor then raised the conductivity of poly(3-hexylthiophene) to circa 10 S cm^{-1}. It was also stated that poly(3-alkylthiophenes) could be obtained via the use of $RuCl_3$ or $MoCl_5$ rather than $FeCl_3$. However, the oxidative polymerization of thiophene species via $FeCl_3$ went on to become the most common and preferred method for the generation of neutral, soluble polymers.

Yoshino and coworkers then followed this with a second report in the spring of 1987 that continued the study of poly(3-alkylthiophene)s generated via oxidative polymerization with $FeCl_3$.[99] The polymers were generated as previously reported for poly(3-hexylthiophene), with the focus here on the octyl, dodecyl, and docosyl analogs. The polymers were characterized by gel permeation chromatography to give M_n values ranging from 20,600 to 28,600, and M_w values ranging from 108,800 to 146,500. Of course, the large differences in the M_n and M_w values gave quite high polydispersion indexes (ca. 5–6). The most novel aspect of this study, however, was the first report of fusible conjugated polymers, with differential scanning calorimetry (DSC) of poly(3-octylthiophene) demonstrating a peak at 156°C that was attributed to fusing of the sample. Fusing of the material was also confirmed visually, with heating to this temperature causing it to become soft and fluid. As a result, powder samples obtained via oxidative polymerization were shown to form films by hot press treatments, with the polymer melting point found to decrease with increasing alkyl chain length. The electrical conductivity of neutral poly(3-octylthiophene) films prepared by this method was found to be 10^{-9} S cm^{-1}, with conductivities as high as 11 S cm^{-1} after doping with iodine.

Another report then followed in August of 1987 that focused on the solution properties of same polymers of the previous paper.[100] In general, it was found that poly(3-alkylthiophene)s with an alkyl chain length greater than 4 were found to be soluble in solvents such as $CHCl_3$, THF, and CH_2Cl_2. Absorption spectra of poly(3-octylthiophene) in CH_2Cl_2 exhibited a similar shape to that of cast films, but blueshifted by about 0.37 eV. This energetic shift was proposed to be due to reduced intermolecular interactions in solution in comparison with thin films, thus resulting in a shift to the shorter wavelengths in solution. Furthermore, transmission intensity was found to be dependent on both concentration and temperature, suggesting that the side alkyl chains played an important role in the characteristics of polymer solution. Lastly, relatively strong light scattering was also observed at high concentrations, although this behavior was found to be strongly dependent on the solvent.

A more extensive study in late 1987 then focused on solvent and temperature effects on the absorption spectra of these materials.[101] In the process, the solubility of the polymers was characterized in a much larger range of solvents. Even within those solvents that suitably solubilized the polymers, it was found that the color of the solution ranged from yellow to orange to red, with the color dependent on the solvent applied.

The absorption spectrum of all three polymers in CH_2Cl_2 exhibited a clear temperature dependence, in which a large red shift of the absorption edge was clearly seen with decreasing temperature. The shift of the absorption edge was proposed to reflect a change of the conjugation length, which, for conjugated polymers consisting of aromatic rings, should be dependent on the planarity of these rings. In addition, it is probable that the planarity of the neighboring rings changes with either temperature or solvent. The origin of this effect was then proposed to be as follows: At high temperature, the polymer backbone adopts a random coil conformation in which neighboring thiophene rings form a torsion angle, effectively resulting in a shorter conjugation length and thus absorption at higher energy. At low temperature, the interaction between polymer chains becomes stronger, resulting in a rod-like conformation of linear chains. In such cases, the conjugation length would be effectively longer compared with that at high temperature, which should result in the lower absorption onset.

A 1988 report then shifted the focus to the thin film properties of poly(3-pentylthiophene) and poly(3-dodecylthiophene).[102] Both polymers were again produced via oxidative polymerization with $FeCl_3$ and then cast to generate polymer films with a thickness of 0.1–1 µm. As shown in Figure 6.19, the absorption spectra of poly(3-dodecylthiophene) show a temperature dependence, with the maximum of 512 nm observed at low temperatures shifting to higher energy with increasing temperature. At the same time, the film color changed from violet to yellow with increased temperature. Similar trends can also be seen for the pentyl analog.

These results were interpreted to suggest that even in films, a rod-coil-type transition can be induced by changing temperature. As with the previous solution studies, the film at low temperature is comprised of a linear alignment of polymer chains, with a rodlike structure of the polymer backbone. At high temperatures, however, the polymers adopt the coil-like conformation, for which the interband absorption shifts to a higher energy due to a reduced degree of planarity, thus resulting in decreased effective conjugation and a larger bandgap. Furthermore, it was proposed that at higher temperatures, polymer chains tend to be independent of each other due to the steric hindrance of side chains.

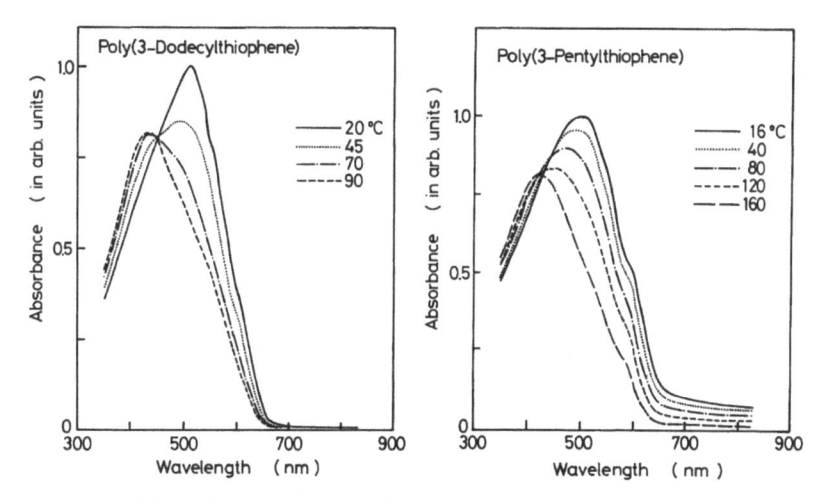

Figure 6.19 Effect of temperature on the absorption spectra of
poly(3-pentylthiophene) and poly(3-dodecylthiophene) films.

Reprinted with permission from K. Yoshino, S. Nakajima, D. H. Park, R. Sugimoto.
Thermochromism, photochromism and anomalous temperature dependence of luminescence in
poly(3-alkylthiophene) film. *Jpn. J. Appl. Phys.* **1988**, *27*, L716–L718. © 1988 The Physical Society
of Japan and The Japan Society of Applied Physics.

It was after these initial reports of poly(3-alkylthiophene)s that Heeger and
Wudl also turned to such functionalized derivatives. These studies were all
in collaboration with Shu Hotta, with the first report appearing in 1987.[103]
As with previous studies, 3-butyl- and 3-hexyl-thiophene were prepared from
3-bromothiophene via Kumada couplings, after which the polymers were prepared via a modification of the electropolymerization methods previously developed by Hotta.[77] The as-polymerized films of circa 10 μm thickness gave conductivities of 40 and 30 S cm^{-1} for the butyl and hexyl derivatives, respectively.
The polymers were then electrochemically reduced and washed via sequential
Soxhlet extractions with methanol and acetone. Polymer solutions were prepared by dissolving the neutral films in CHCl$_3$, CH$_2$Cl$_2$, toluene, or THF at circa
60°C. The weight-average molecular weight (M$_w$) of the poly(3-hexylthiophene)
was determined to be 48,000, with a corresponding polydispersity of circa
2. While the M$_w$ was considerably smaller than that achieved via oxidative
polymerization with FeCl$_3$,[99] the polydispersity was much more uniform.

The absorption spectra of poly(3-hexylthiophene) in THF exhibited a maximum at 435 nm, with a corresponding absorption onset of circa 530 nm. In
comparison, solution-cast films gave an absorption maximum at circa 490 nm,
with an absorption onset of about 640 nm corresponding to a bandgap of circa
2 eV. The significant blueshift of both the maximum and onset in solution

was proposed to be due to disordered conformations of the polymer chains in solution. In comparison, solution-cast solid films result in an ordered backbone with more extensive delocalization of the π-electrons. The fact that solution-cast films of the poly(3-alkylthiophene)s were essentially equivalent to that of highly crystalline polythiophene led to the conclusion that this implied weak interchain electronic transfer interactions, such that the electronic structure is highly anisotropic or quasi-one-dimensional.

This was then followed up with two additional 1987 papers that both focused on the doping of poly(3-hexylthiophene) in solution.[104,105] Interestingly, Wudl is not included as an author on one of the two papers, although both papers published largely the same material. For these studies, the polymer was prepared as outlined in the previous paper and then dissolved in $CHCl_3$. Doped polymer solutions were then prepared by adding the appropriate amount of $NOPF_6$ in acetonitrile to the polymer solution. Solutions with various doping levels were then characterized by both ESR and absorption spectroscopy, with an example of changes in absorption given in Figure 6.20.

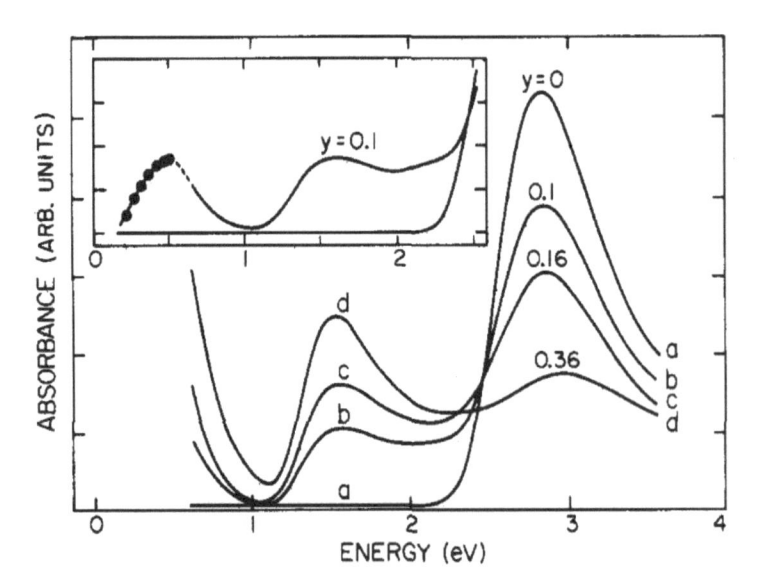

Figure 6.20 Absorption spectra of neutral (y = 0) and doped (y ≈ 0.1, 0.16, 0.36) poly(3-hexylthiophene) in $CHCl_3$.

Reprinted with permission from M. J. Nowak, S. D. D. V. Rughooputh, S. Hotta, A. J. Heeger. Polarons and bipolarons on a conducting polymer in solution. *Macromolecules* **1987**, *20*, 965–968. © 1987 American Chemical Society.

It was observed that increased doping resulted in a depletion of the π-π* transition in the visible region and the growth of two principal subgap features in the infrared. The two principal subgap absorption bands with maxima $\hbar\omega_1$ ≈

1.55 eV and $\hbar\omega_2 \approx 0.5$ eV in the doped polymer were concluded to be consistent with charge storage predominantly in bipolarons and agreed well with the previously presented energy-level diagram given in Figure 6.16. The conclusion that bipolarons were the lowest energy charge-storage configuration, combined with the results of detailed analysis of the absorption spectra, led to the view that electron–electron Coulomb correlations were relatively weak, even for isolated conducting polymer chains in dilute solution.

Hotta then reported an independent study in 1988 with a distinct shift in emphasis from the previous collaborations with Wudl and Heeger.[106] The methods applied to the generation of the poly(3-alkylthiophene)s were essentially the $FeCl_3$-based methods of Yoshino, although with much longer reaction times (two days in comparison with two hours). In addition, it was shown that 3-alkyl-2-halothiophenes could also be polymerized in the same way and that copolymers could be synthesized from solutions containing two or more thiophene monomers. Here, the copolymer of 3-hexyl- and 3-methyl-thiophene was prepared as an illustrative example.

Poly(3-hexylthiophene) was characterized by GPC to give M_w values of circa 250,000, with a polydispersity index of 5.5. The M_w value here was nearly twice that previously found by Yoshino, although with a nearly identical polydispersity index.[99] In comparison, the copolymer gave an M_w of only 76,000, although with an even greater polydispersity index of 9.2. The absorption spectra of a solution-cast film of poly(3-hexylthiophene) exhibited a maximum at 505 nm, with two noticeable shoulders at circa 550 and 600 nm.

Finally, the effect of stretching of the solution-cast films on the resulting conductivity was investigated. After doping with iodine, unstretched films of poly(3-hexylthiophene) gave a conductivity of 27 S cm^{-1}. Stretching resulted in significant enhancement of conductivity with values as high as 200 S cm^{-1} after stretching the film by a factor of 5. The increase in conductivity was concluded to be due to the presence of an oriented structure in the stretched films, with the presence of a highly oriented structure directly demonstrated by infrared dichroic measurements.

6.9 Conclusions

Although the history of polythiophene does not date back as far as the other conjugated polymers discussed in the previous chapters, a significant amount of work was reported quickly following its introduction in 1980. The exact reason for its rather late introduction is unclear, as reasonable synthetic routes to thiophene date back to the second decade of the 20th century, and its polymerization can be carried out via oxidative polymerization in a manner similar to that of both polypyrrole and polyaniline. The fact that higher potentials are required for the oxidation of thiophene in comparison with aniline or pyrrole certainly played a role, as this required the application of stronger oxidizing agents, but a number of suitably strong oxidizing agents had been previously applied to aniline and pyrrole. Another contributing factor could be the strong association between thiophene and benzene in the early literature, resulting in the application of various known reactions of benzene to thiophene, rather than a focus on extending the chemistry of pyrrole to its sulfur analog. This effect is also illustrated by the fact that the first report of polythiophene was via polycondensation of the dihalide, a process previously applied for the production of polyphenylene. Lastly, early attempts to polymerize thiophene focused on various acid-promoted methods, which resulted in materials with various saturation defects rather than the strongly conjugated material reported in later studies.

Still, since its first report in 1980, polythiophene has held a special place in the history of conjugated polymers. Not only did it tie together the historical paths of benzene-based polymers such as polyphenylene with the previous oxidatively polymerized materials, but the synthetic diversity of thiophene led to the production of polyalkylthiophenes, thus providing soluble materials. As a result, this allowed the first real detailed studies of conjugated polymers in the solution state. More importantly, this also resulted in either melt- or solution-processible materials, which greatly expanded the potential applications of these materials and began the shift away from maximizing conductivity to the more general optical and electronic properties of these materials. As such, it is not surprising that so much of the current field is dominated by thiophene-based materials.

References and Notes

1. Rasmussen, S. C. Early history of conjugated polymers: From their origins to the handbook of conducting polymers. In *Handbook of Conducting Polymers*, 4th ed. Reynolds, J. R.; Skotheim, T. A.; Thompson, B., Eds.; CRC Press: Boca Raton, FL, 2019, pp. 1–35.
2. Rasmussen, S. C. Conjugated and conducting organic polymers: The first 150 years. *ChemPlusChem* **2020**, *85*, 1412–1429.
3. Toshima, N.; Hara, S. Direct synthesis of conducting polymers from simple monomers. *Prog. Polym. Sci.* **1995**, *20*, 155–183.
4. Feast, W. J.; Tsibouklis, J.; Pouwer, K. L.; Groenendaal, L.; Meijer, E. W. Synthesis, processing and material properties of conjugated polymers. *Polymer* **1996**, *37*, 5017–5047.
5. Waltman, R. J.; Bargon, J. Electrically conducting polymers: A review of the electropolymerization reaction, of the effects of chemical structure on polymer film properties, and of applications towards technology. *Can. J. Chem.* **1986**, *64*, 76–95.

6. John, R.; Wallace, G. G. The use of microelectrodes to probe the electropolymerization mechanism of heterocyclic conducting polymers. *J. Electroanal. Chem.* **1991**, *306*, 157–167.

7. Smith, J. R.; Cox, P. A.; Campbell, S. A.; Ratcliffe, N. M. Application of density functional theory in the synthesis of electroactive polymers. *J. Chem. Soc. Faraday Trans.* **1995**, *91*, 2331–2338.

8. Babudri, F.; Farinola, G. M.; Naso, F. Synthesis of conjugated oligomers and polymers: The organometallic way. *J. Mater. Chem.* **2004**, *14*, 11–34.

9. Carsten, B.; He, F.; Son, H. J.; Xu, T.; Yu, L. Stille polycondensation for synthesis of functional materials. *Chem. Rev.* **2011**, *111*, 1493–1528.

10. Leone, A. K.; Mueller, E. A.; McNeil, A. J. The history of palladium-catalyzed cross-couplings should inspire the future of catalyst-transfer polymerization. *J. Am. Chem. Soc.* **2018**, *140*, 15126–15139.

11. Baker, M. A.; Tsai, C.-H.; Noonan, K. J. T. Diversifying cross-coupling strategies, catalysts and monomers for the controlled synthesis of conjugated polymers. *Chem. Eur. J.* **2018**, *24*, 13078–13088.

12. Cameron, M. D. Victor Meyer and the thiophene compounds. *J. Chem. Educ.* **1949**, *26*, 521–524.

13. Sumpter, W. C. The chemistry of Isatin. *Chem. Rev.* **1944**, *34*, 393–434.

14. Meyer, V. Ueber Benzole verschiedenen Ursprungs. *Ber. Dtsch. Chem. Ges.* **1882**, *15*, 2893–2894.

15. Meyer, V. Ueber den Begleiter des Benzols im Steinkohlentheer. *Ber. Dtsch. Chem. Ges.* **1883**, *16*, 1465–1478.

16. Meyer, V.; Sandmeyer, T. Künstliche Bildung des Thiophens. *Ber. Dtsch. Chem. Ges.* **1883**, *16*, 2176.

17. Meyer, V. Synthesen des Thiophens. *Ber. Dtsch. Chem. Ges.* **1885**, *18*, 217–218.

18. Paal, C. Synthese von Thiophen- und Pyrrolderivaten. *Ber. Dtsch. Chem. Ges.* **1885**, *18*, 367–371.

19. Paal, C.; Tafel J. Thiophen aus Schleimsäure. *Ber. Dtsch. Chem. Ges.* **1885**, *18*, 456–460.

20. Volhard, J.; Erdmann, H. Synthetische Darstellung von Thiophen. *Ber. Dtsch. Chem. Ges.* **1885**, *18*, 454–455.

21. Steinkopf, G. W.; Kirchhoff, G. I. Die Darstellung von Thiophen aus Acetylen. *Justus Liebigs Ann. Chem.* **1914**, *403*, 1–11.

22. Rasmussen, H. E.; Hansford, R. C.; Sachanen, A. N. Reactions of aliphatic hydrocarbons with sulfur. Production of olefins, diolefins, and thiophene. *Ind. & Eng. Chem.* **1946**, *38*, 376–382.

23. Kutz, W. M.; Corson, B. B. Alkylation of thiophene by olefins and alcohols. *J. Am. Chem. Soc.* **1946**, *68*, 1477–1479.

24. Hartough, H. D.; Kosak, A. I. Acylation studies in the thiophene and furan series. IV. Strong inorganic oxyacids as catalysts. *J. Am. Chem. Soc.* **1947**, *69*, 3093–3096.

25. Johnson, D. H. Thiophen removal from benzene with aluminium chloride. *J. Chem. Soc. C* **1967**, 2275–2281.

26. Meisel, S. L.; Johnson, G. C.; Hartough, H. D. Polymerization of thiophene and alkylthiophenes. *J. Am. Chem. Soc.* **1950**, *72*, 1910–1912.

27. Armour, M.; Davies, A. G.; Upadhyay, J.; Wassermann, A. Colored electrically conducting polymers from furan, pyrrole, and thiophene. *J. Polym. Sci. A* **1967**, *5*, 1527–1538.

28. Curtis, R. F.; Jones, D. M.; Thomas, W. A. The "trimer" and "pentamer" from the polymerisation of thiophen by polyphosphoric acid. *J. Chem. Soc. C* **1971**, 234–238.

29. Louvar, J. J. Polymerization of heterocyclic compounds. U. S. Patent US 3,574,072, April 6, 1971.

30. Nishihara, H. Biography of Professor Takakazu Yamamoto. *J. Inorg. Organomet. Polym.* **2009**, *19*, 1–2.

31. Osakada, K.; Tezuka, Y. Obituary: Takakazu Yamamoto, 1944–2014. *React. Funct. Polym.* **2015**, *86*, 282.

32. Prabook. Takakazu Yamamoto. https://prabook.com/web/takakazu.yamamoto/ 352256 (accessed June 4, 2022).

33. Yamamoto, T. *Reactions of alkyl dipyridyl iron, nickel, and cobalt complexes with olefins.* Ph.D. dissertation, Tokyo Institute of Technology, 1971.

34. Yamamoto, T.; Todd, L. J. σ-Bonded complexes of some heteroatom boranes with iron and molybdenum derivatives. *J. Organometallic Chem.* **1974**, *67*, 75–80.

35. Yamamoto, T.; Garber, A. R.; Wilkinson, J. R.; Boss, C. B.; Streib, W. E.; Todd, L. J. *J. Chem. Soc. Chem. Comm.* **1974**, 354–356.

36. Yamamoto, T. Electrically conducting and thermally stable p-conjugated poly(arylene)s prepared by organometallic processes. *Prog. Polym. Sci.* **1992**, *17*, 1153–1205.

37. Yamamoto, T. Cross-coupling reactions for preparation of p-conjugated polymers. *J. Organometallic Chem.* **2002**, *653*, 195–199.

38. Yamamoto, T.; Sanechika, K.; Yamamoto, A. Preparation of thermostable and electric-conducting poly(2,5-thienylene). *J. Polym. Sci. Polym. Lett. Ed.* **1980**, *18*, 9–12.

39. Tamao, K.; Sumitani, K.; Kumada, M. Selective carbon-carbon bond formation by cross-coupling of Grignard reagents with organic halides. Catalysis by nickel-phosphine complexes. *J. Am. Chem. Soc.* **1972**, *94*, 4374–4376.

40. Yamamoto, T.; Sanechika, K.; Yamamoto, A. Preparation of poly(2,4-thienylene) and comparison of its optical and electrical properties with those of poly(2,5-thienylene). *Chem. Lett.* **1981**, *10*, 1079–1082.

41. Sanechika, K.; Yamamoto, T.; Yamamoto, A. Preparation of copolymers composed of 2,5-thienylene and 2,4-thienylene units. Effect of copolymer composition on electronic spectrum, electric conductivity, and chemical properties. *J. Polym. Sci. Polym. Lett. Ed.* **1982**, *20*, 365–371.

42. Yamamoto, T.; Sanechika, K. Preparation and properties of π-conjugated poly(3-methyl-2,5-thienylene). *Chem. Ind.* **1982**, *9*, 301–302.

43. Yamamoto, T.; Sanechika, K.; Yamamoto, A. Preparation and characterization of poly(thienylene)s. *Bull. Chem. Soc. Jpn.* **1983**, *56*, 1497–1502.

44. Yamamoto, T.; Sanechika, K.; Yamamoto, A. Formation of adducts of poly(thienylene)s with electron acceptors and electric conductivities of the adducts. *Bull. Chem. Soc. Jpn.* **1983**, *56*, 1503–1507.

45. Yamamoto, T.; Masanobu, Z.; Yamamoto, A. Li|LiI|Iodine galvanic cells using iodine-poly(2,5-thienylene) adducts as active materials of positive electrodes. *Chem. Lett.* **1984**, *13*, 1577–1580.

46. Yamamoto, T.; Masanobu, Z.; Yamamoto, A. Secondary cells using poly(2,5-thienylene)s and poly(2,5-pyrrolylene)s as materials for positive electrodes. $Zn|ZnI_2|I_2$ secondary cell. *Chem. Lett.* **1985**, *14*, 563–566.

47. Sakai, H.; Mizota, M.; Maeda, Y.; Yamamoto, T.; Yamamoto, A. Resonance Raman and ^{129}I Mössbauer spectroscopy study of iodine doped in poly(thienylene)s. *Bull. Chem. Soc. Jpn.* **1985**, *58*, 926–931.

48. Lin, J. W.-P.; Dudek, L. P. Synthesis and properties of poly(2,5-thienylene). *J. Polym. Sci. Polym. Chem. Ed.* **1980**, *18*, 2869–2873.

49. Afanas'ev, V. L.; Nazarova, I. B.; Khidekel, M. L. Polythiophene - an electrically conductive organic compound. *Izv. Akad. Nauk SSSR Ser. Khim.* **1980**, (7), 1687–1688.

50. Diaz, A. F.; Kanazawa, K. K.; Gardini, G. P. Electrochemical polymerization of pyrrole. *J. Chem. Soc. Chem. Commun.* **1979**, 635–636.

51. Diaz, A. Electrochemical preparation and characterization of conducting polymers. *Chemica Scripta* **1981**, *17*, 145–148.

52. Diaz, A. F.; Crowley, J.; Bargon, J.; Gardini, G. P.; Torrance, J. B. Electrooxidation of aromatic oligomers and conducting polymers. *J. Electroanal. Chem.* **1981**, *121*, 355–361.

53. Waltman, R. J.; Bargon, J.; Diaz, A. F. Electrochemical studies of some conducting polythiophene films. *J. Phys. Chem.* **1983**, *87*, 1459–1463.

54. Bargon, J.; Mohmand, S.; Waltman, R. J. Electrochemical synthesis of electrically conducting polymers from aromatic compounds. *IBM J. Res. Develop.* **1983**, *27*, 330–341.

55. Hernandez, R.; Diaz, A. F.; Waltman, R.; Bargon, J. Surface characteristics of thin films prepared by plasma and electrochemical polymerizations. *J. Phys. Chem.* **1984**, *88*, 3333–3337.

56. Waltman, R. J.; Diaz, A. F.; Bargon, J. Electroactive properties of polyaromatic molecules. *J. Electrochem. Soc.* **1984**, *131*, 740–744.

57. Waltman, R. J.; Diaz, A. F.; Bargon, J. Electroactive properties of polyaromatic molecules. *J. Electrochem. Soc.* **1984**, *131*, 1452–1456.

58. Tourillon, G.; Garnier, F. New electrochemically generated organic conducting polymers. *J. Electroanal. Chem.* **1982**, *135*, 173–178.

59. Tourillon, G.; Garnier, F. Effect of dopant on the physicochemical and electrical properties of organic conducting polymers. *J. Phys. Chem.* **1983**, *87*, 2289–2292.

60. Garnier, F.; Tourillon, G.; Gazard, M.; Dubois, J. C. Organic conducting polymers derived from substituted thiophenes as electrochromic material. *J. Electroanal. Chem.* **1983**, *148*, 299–303.

61. Tourillon, G.; Garnier, F. Stability of conducting polythiophene and derivatives. *J. Electrochem. Soc.* **1983**, *130*, 2042–2044.

62. Glenis, S.; Horowitz, G.; Tourillon, G.; Garnier, F. Electrochemically grown polythiophene and poly(3-methylthiophene) organic photovoltaic cells. *Thin Solid Films* **1984**, *111*, 93–103.

63. Glenis, S.; Tourillon, G.; Garnier, F. Photoelectrochemical properties of thin films of polythiophene and derivatives: Doping level and structure effects. *Thin Solid Films* **1984**, *122*, 9–17.

64. Glenis, S.; Tourillon, G.; Garnier, F. Influence of the doping on the photovoltaic properties of thin films of poly-3-methylthiophene. *Thin Solid Films* **1986**, *139*, 221–231.

65. Dastoor, P. C.; Belcher, W. J. How the West was won? A history of organic photovoltaics. *Substantia* **2019**, *3*(2) Suppl. 1, 99–110.

66. Kaneto, K.; Yoshino, K.; Inuishi, Y. Electrical properties of conducting polymer, poly-thiophene, prepared by electrochemical polymerization. *Japan J. Appl. Phys.* **1982**, *21*, L567–L568.

67. Ieda, M. To the memory of Professor Yoshio Inuishi. *Proceedings of 1995 International Symposium on Electrical Insulating Materials*, **1995**, 1–6.

68. Ieda, M. Remembering Professor Yoshio Inuishi. *IEEE Transactions on Dielectrics and Electrical Insulation* **1996**, *3*, 327–330.

69. Inuishi, Y. Study on electrical insulation breakdown in solids and liquids. Ph.D. dissertation, Osaka University, 1959.

70. Katsumi Yoshino. IEEE Xplore. https://ieeexplore.ieee.org/author/37291839800 (accessed June 7, 2022).

71. Katsumi Yoshino. Prabook. https://prabook.com/web/katsumi.yoshino/1423053 (accessed June 7, 2022).

72. Kaneto, K.; Kohno, Y.; Yoshino, K.; Inuishi, Y. Electrochemical preparation of a metallic polythiophene film. *J. Chem. Soc. Chem. Commun.* **1983**, 382–383.

73. Kaneto, K.; Yoshino, K.; Inuishi, Y. Electrical and optical properties of polythiophene prepared by electrochemical polymerization. *Solid State Commun.* **1983**, *46*, 389–391.

74. Kaneto, K.; Yoshino, K.; Inuishi, Y. Electrochemically prepared conducting polymers, poly-thiophene and poly-pyrrole. *Technology Reports of the Osaka University* **1983**, *33*, 75–79.

75. Hattorii, T.; Hayes, W.; Wong, K.; Kaneto, K.; Yoshino, K. Optical properties of photoexcited and chemically doped polythiophene. *J. Phys. C: Solid State Phys.* **1984**, *17*, L803–L807.

76. Hotta, S.; Hosaka, T.; Shimotsuma, W. Electrochemically prepared polythienylene films. *Synth. Met.* **1983**, *6*, 69–71.

77. Hotta, S.; Yamao, T. The thiophene/phenylene co-oligomers: Exotic molecular semiconductors integrating high-performance electronic and optical functionalities. *J. Mater. Chem.* **2011**, *21*, 1295–1304.

78. Hotta, S.; Hosaka, T.; Soga, M.; Shimotsuma, W. Electrochemically prepared poly(3-methylthiophene) films doped with iodine. *Synth. Met.* **1984**, *9*, 381–387.

79. Koßmehl, G.; Chatzitheodorou, G. Electrical conductivity of poly(2,5-thiophenediyl)-AsF$_5$-complexes. *Makromol. Chem. Rapid Commun.* **1981**, *2*, 551–555.

80. Koßmehl, G.; Chatzitheodorou, G. Electrical conductive AsF$_5$-complexes of poly(2,5-thiophenediyl). *Mol. Cryst. Liq. Cryst.* **1982**, *83*, 291–296.

81. Koßmehl, G.; Chatzitheodorou, G. Synthesis and electrical conductivities of poly(2,5~thiophenediyl) salts by the action of nitronium or nitrosonium salts. *Makromol. Chem. Rapid Commun.* **1983**, *4*, 639–643.

82. Heeger, A. J. *Never lose your nerve!* World Scientific Publishing: Singapore, 2016, pp. 181–191.

83. Kobayashi, M.; Chen, J.; Chung, T.-C.; Moraes, F.; Heeger, A. J.; Wudl, F. Synthesis and properties of chemically coupled poly(thiophene). *Synth. Met.* **1984**, *9*, 77–86.

84. Kaufman, J. H.; Chung, T.-C.; Heeger, A. J.; Wudl, F. Poly(thiophene): A stable polymer cathode material. *J. Electrochem. Soc.* **1984**, *131*, 2092–2093.

85. Chung, T.-C.; Kaufman, J. H.; Heeger, A. J.; Wudl, F. Charge storage in doped poly(thiophene): Optical and electrochemical studies. *Phys. Rev. B* **1984**, *30*, 702–710.

86. Moraes, F.; Schaffer, H.; Kobayashi, M.; Heeger, A. J.; Wudl, F. Photoexcitations in poly(thiophene): Photoinduced infrared absorption and photoinduced electron-spin resonance. *Phys. Rev. B* **1984**, *30*, 2948–2950.

87. Moraes, F.; Davidov, D.; Kobayashi, M.; Chung, T. C.; Chen, J.; Heeger, A. J.; Wudl, F. Doped poly(thiophene): Electron spin resonance determination of the magnetic susceptibility. *Synth. Met.* **1985**, *10*, 169–179.

88. Davidov, D.; Moraes, F.; Heeger, A. J.; Wudl, F.; Kim, H.; Dalton, L. R. Electron spin echo modulation and relaxation in polythiophene. *Solid State Commun.* **1985**, *53*, 497–500.

89. Mo, Z.; Lee, K.-B.; Moon, Y. B.; Kobayashi, M.; Heeger, A. J.; Wudl, F. X-ray scattering from polythiophene: Crystallinity and crystallographic structure. *Macromolecules* **1985**, *18*, 1972–1977.

90. Chen, J.; Heeger, A. J.; Wudl, F. Confined soliton pairs (bipolarons) in polythiophene: *In-situ* magnetic resonance measurements. *Solid State Commun.* **1986**, *58*, 251–257.

91. Vardeny, Z.; Ehrenfreund, E.; Shinar, J.; Wudl, F. Photoexcitation spectroscopy of polythiophene. *Phys. Rev. B* **1987**, *35*, 2498–2500.

92. Bredas, J. L.; Wudl, F.; Heeger, A. J. Polarons and bipolarons in doped polythiophene: A theoretical investigation. *Solid State Commun.* **1987**, *63*, 577–580.

93. Vardeny, Z.; Ehrenfreund, E.; Brafman, O.; Heeger, A. J.; Wudl, F. Photoinduced absorption and resonant Raman scattering of polythiophene. *Synth. Met.* **1987**, *18*, 183–188.

94. Jen, K. Y.; Oboodi, R.; Elsenbaumer, R. L. Processible and environmentally stable conducting polymers. *Polym. Mater. Sci. Eng.* **1985**, *53*, 79–83.

95. Jen, K.-Y.; Miller, G. G.; Elsenbaumer, R. L. Highly conducting, soluble, and environmentally-stable poly(3-alkylthiophenes). *J. Chem. Soc., Chem. Commun.* **1986**, 1346–1347.

96. Sato, M.; Tanaka, S.; Kaerivama, K. Electrochemical preparation of highly conducting polythiophene films. *J. Chem. Soc., Chem. Commun.* **1985**, 713–714.

97. Sato, M.; Tanaka, S.; Kaeriyama, K. Soluble conducting polythiophenes. *J. Chem. Soc., Chem. Commun.* **1986**, 873–874.

98. Sugimoto, R.; Takeda, S.; Gu, H. B.; Yoshino, K. Preparation of soluble polythiophene derivatives utilizing transition metal halides as catalysts and their property. *Chem. Express* **1986**, *1*, 635–638.

99. Yoshino, K.; Nakajima, S.; Sugimoto, R. Fusibility of poly thiophene derivatives with substituted long alkyl chain and their Properties. *Jpn. J. App. Phys.* **1987**, *26*, L1038–L1039.

100. Sugimoto, R.; Takeda, S.; Gu, H. B.; Yoshino, K. Optical properties of solution of poly thiophene derivatives as functions of alkyl chain length, concentration and temperature. *Jpn. J. App. Phys.* **1987**, *26*, L1371–L1373.

101. Yoshino, K.; Nakajima, S.; Gu, H. B.; Sugimoto, R. Absorption and emission spectral changes in a poly(3-alkylthiophene) solution with solvent and temperature. *Jpn. J. App. Phys.* **1987**, *26*, L2046–L2048.

102. Yoshino, K.; Nakajima, S.; Park, D. H.; Sugimoto, R. Thermochromism, photochromism and anomalous temperature dependence of luminescence in poly(3-alkylthiophene) film. *Jpn. J. Appl. Phys.* **1988**, *27*, L716–L718.

103. Hotta, S; Rughooputh, S. D. D. V.; Heeger, A. J.; Wudl, F. Spectroscopic studies of soluble poly(3-alkylthienylenes). *Macromolecules* **1987**, *20*, 212–215.

104. Nowak, M. J.; Rughooputh, S. D. D. V.; Hotta, S.; Heeger, A. J. Polarons and bipo-larons on a conducting polymer in solution. *Macromolecules* **1987**, *20*, 965–968.

105. Rughooputh, S. D. D. V.; Nowak, M.; Hotta, S.; Heeger, A. J.; Wudl, F. Soluble conducting polymers: The poly(3-alkylthienylenes). *Synth. Met.* **1987**, *21*, 41–50.

106. Hotta, S.; Soga, M.; Sonoda, N. Novel organosynthetic routes to polythiophene and its derivatives. *Synth. Met.* **1988**, *26*, 267–279.

Chapter 7
Polyisothianaphthene and the Birth of Low-Bandgap Polymers

7.1 Introduction

As outlined in the previous chapters, the family of parent conjugated homopolymers had grown by 1990 to include polyaniline, polypyrrole, polyphenylene, polyphenylene vinylene, polyacetylene, and polythiophene.[1,2] In the process, a variety of synthetic methods had been developed, as well as multiple polymerization methods beyond the simple oxidative polymerizations applied in the early days of polyaniline and polypyrrole.[3-6] By 1980, the final member of the parent polymers, polythiophene, had been introduced,[7] and conductivities in excess of 2000 S cm^{-1} had been achieved for stretch-oriented doped polyacetylene films.[8] While efforts to further increase the conductivities continued, particularly with the efforts of Naarmann, other critical advances were also introduced, such as the production of soluble and processible materials via the inclusion of flexible, alkyl side chains, and the initial introduction of copolymeric materials.

At the same time, efforts began to develop new types of conjugated polymers with the goal of producing materials with potentially higher conductivities and enhanced stability. It was reasoned that this could potentially be accomplished by enhancing the quinoidal resonance structures proposed for polyaromatic materials such as polyphenylene, polypyrrole, and polythiophene (Figure 7.1).[9] While attempts to enhance quinoidal resonance did not lead to materials with new record conductivities, this did result in materials with smaller bandgaps. This, in turn, led to a new class of conjugated polymers, commonly referred to as low-bandgap polymers, which are formally defined as those materials with a bandgap below 1.5 eV.[10-13]

Due to their lower bandgaps, these materials absorb wavelengths in the near-infrared (NIR, λ > 800 nm), with the polymer's optical absorption shifted to longer and longer wavelengths with diminishing bandgap.[11,13-15] As a result, sufficiently low bandgaps would result in materials in which the absorbance would be shifted fully into the NIR. As such materials would no longer absorb

The Origins and Early History of Conjugated Organic Polymers. Seth C. Rasmussen, Oxford University Press.
© Oxford University Press (2025). DOI: 10.1093/9780197638194.003.0007

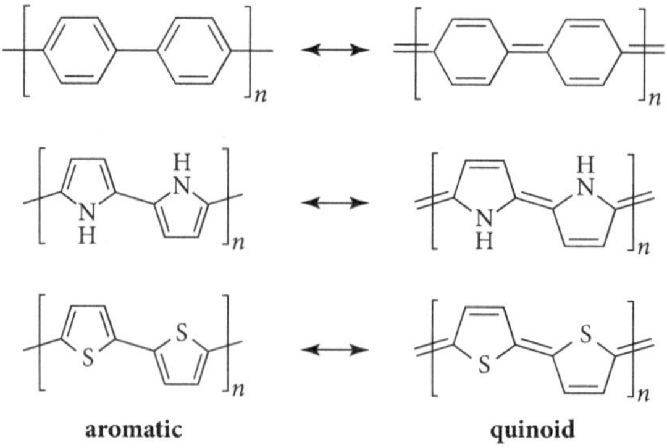

Figure 7.1 Aromatic versus quinoidal resonance forms of polyphenylene, polypyrrole, and polythiophene.

in the visible regime, these polymers would no longer be colored and should, therefore, be transparent.[11,14,15] Smaller bandgaps should also result in greater thermal population of the conduction band, thus increasing the number of intrinsic charge carriers and contributing to enhanced conductivity in the neutral, undoped material.[11-15] Furthermore, as reduction of the bandgap typically involves destabilization of the energy of the highest occupied molecular orbit (HOMO), this would also result in a lower potential of oxidation and thus stabilization of the corresponding oxidized or p-doped state.[11-14] For all of these reasons, low-bandgap polymers have become an important subclass of conjugated polymers, with significant efforts in the design and production of such materials.

The first true low-bandgap polymer to be produced was polyisothianaphthene (PITN, Figure 7.2), a fused-ring derivative of polythiophene. This new material was first reported in 1984 by Fred Wudl in collaboration with Alan Heeger at the recently formed Institute for Polymers and Organic Solids at the University of California, Santa Barbara.[16,17] As such, it is with PITN that we will begin our discussion of the origin and history of these materials.

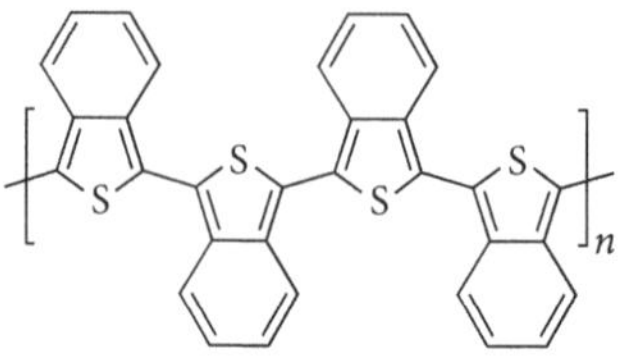

Figure 7.2 Polyisothianaphthene (PITN).

7.2 Wudl, Heeger, and polyisothianaphthene

Fred Wudl (Figure 7.3) was born in Cochabamba, Bolivia, in 1941[18–20] to Jewish parents who were immigrants from Vienna, Austria.[18] The family then emigrated to the United States in 1958. As he was still in the 11th grade at the time, Wudl finished his high school education in the United States but didn't have sufficient grades or the preparation to go directly to university. Thus, he first attended Los Angeles City College for two years, receiving an AB degree in 1961,[18] before transferring to the University of California, Los Angeles (UCLA).[18–22] During the fall of 1962, he was invited by Donald J. Cram (1919–2001) to work in his group as an undergraduate researcher, where he initially worked on a polymer project with Richard F. Smith, a visiting professor from the State University of New York at Geneseo (SUNY Geneseo).[18] Completing his BS in chemistry in 1964,[18–21] Wudl applied for graduate school, beginning with UCLA. To his surprise, he was accepted, even though it was unusual at the time for the department to accept its own undergraduates.[18] After meeting with each of the organic faculty, Wudl elected to remain in Cram's group.[18,19,21,22] After three very enjoyable years, he completed his Ph.D. in 1967[18–21,23] with a dissertation on optically active methyl p-styryl sulfoxide copolymers.[23]

Figure 7.3 Fred Wudl (b. 1941).
Reproduced courtesy of Fred Wudl.

Following his graduate studies, Wudl spent a year as a postdoctoral researcher at Harvard University with Robert B. Woodward (1917–1979), before joining the State University of New York at Buffalo (SUNY Buffalo) as an assistant professor in the fall of 1968.[18–22] It was while at Buffalo that Wudl and his group designed and synthesized tetrathiafulvalene (TTF) and demonstrated the high electrical conductivity of its ion radical salts,[24,25] findings that later contributed to its pairing with tetracyanoquinodimethane (TCNQ) to give the organic metal TTF:TCNQ (see Section 8.2).[18,19] However, SUNY Buffalo was not the environment that he had hoped it would be, which led to his accepting a research staff position in 1972 at Bell Laboratories in New Jersey.[18–21]

It was about a decade later that Wudl received an unexpected call from Alan Heeger.[18] As previously outlined in Section 5.9, Heeger's group at Penn had claimed in the early 1970s that TTF:TCNQ exhibited superconducting fluctuations above 60 K, which was ultimately determined to be flawed. As Wudl was part of the team that proved definitively how Heeger's group made the mistake,[26] this had made him essentially persona non grata with Heeger.[18] Because of this, he was genuinely surprised when Heeger called him one day at his office at Bell Labs with the idea for both of them to move to the University of California, Santa Barbara (UCSB) at the invitation of John Robert Schrieffer (1931–2019).[18] In the end, however, he moved to UCSB in 1982, where he established a lasting and very successful collaboration with Heeger (Figure 7.4).[18–22] In addition to his faculty position, Wudl also served as the associate director of UCSB's Institute for Polymers and Organic Solids (renamed as the Center for Polymers and Organic Solids in 2000).[18,20,21]

In 1997, Wudl moved to his alma mater UCLA to occupy the Courtaulds Chair of Chemistry, as well as establishing the Institute for Exotic Materials as its founding director.[19,21,22] Later, in 2006, he returned to Santa Barbara as co-director of the Center for Polymers and Organic Solids until 2011, as well as participating in the new California NanoSystems Institute.[19,22] Wudl became professor emeritus at UCSB in 2020.

His contributions have been recognized by a number of prestigious awards and honors.[19,20] This included the American Chemical Society (ACS) Arthur C. Cope Scholar Award in 1993, the Wheland Medal in 1994, the ACS Award for Chemistry of Materials in 1996, the Tolman Medal in 2007, the Stephanie L. Kwolek Award from the Royal Society of Chemistry (RSC) in 2010, and the Seaborg Medal in 2014.[19] In addition, he was elected a fellow of the American Association for the Advancement of Science in 1989 and a fellow of the RSC in 2010.

Much of the early focus of the collaboration with Heeger was on polythiophene, which was introduced only two years prior to Wudl's move to UCSB (see Section 6.7). In the process, however, they also began investigations of the

Figure 7.4 Fred Wudl and Alan Heeger in January 2001.
Reproduced courtesy of Fred Wudl.

polymerization of isothianaphthene (or benzo[c]thiophene), with the view that
the quinoidal resonance contributions could potentially lead to higher conductivity than polythiophene. The earliest report of these studies appeared in the
late summer of 1984,[16] with a more substantial paper then following a month
later.[17]

The monomeric isothianaphthene was prepared by the dehydration of a
sulfoxide precursor as illustrated in Figure 7.5. Initial attempts to utilize electrochemical polymerization to produce the corresponding PITN, however,
were unsuccessful. Rather, it was found that standard conditions utilizing
electrolytes consisting of tetrabutylammonium salts of ClO_4^- or BF_4^- (i.e.,
Bu_4NClO_4 or Bu_4NBF_4) resulted in the production of the nonconjugated
poly(dihydroisothianaphthene) (Figure 7.5) when carried out in acetonitrile

(CH$_3$CN).[17] Further study then determined that isothianaphthene could be polymerized though cationic polymerization, which resulted in the observed nonconjugated polymer. As such, it was reasoned that the standard conditions applied favored cationic polymerization over the desired oxidative polymerization.

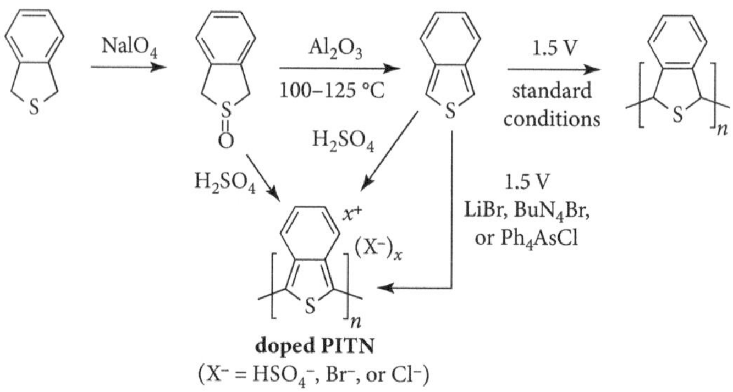

Figure 7.5 Wudl and Heeger's synthetic routes to PITN.

In the process of investigating the cationic polymerization with various Lewis acids, it was found that when methylene chloride solutions of isothianaphthene were treated with H$_2$SO$_4$, a blue-black powder was produced rather than the previous white product.[17] Here, it was reasoned that H$_2$SO$_4$ was acting as both acid catalyst and oxidizing agent, resulting in a doped form of PITN. Furthermore, it was found that treatment of the original sulfoxide with H$_2$SO$_4$ also resulted in the same blue-black material (Figure 7.5).

Returning to the electropolymerization, it was reasoned that if the conditions contained a species that was more nucleophilic than the isothianaphthene monomer, then the propagation step of the cationic polymerization could potentially be interrupted. As such, the electropolymerization was attempted again, finding that the use of electrolytes such as LiBr, Bu$_4$NBr, or Ph$_4$AsCl successfully produced thin films of doped PITN on either Pt or indium tin oxide (ITO) electrodes.[17] From the absorption edge of the neutral film, the polymer bandgap (E_g) was estimated to be ~1 eV. Doped films compensated with either Br$^-$ or Cl$^-$ were found to exhibit conductivities of 0.12–0.40 S cm^{-1}.

This was then followed by another paper in 1985 that focused on the electronic characterization of PITN.[27] The polymer samples were produced via electropolymerization as described earlier using Ph$_4$PCl in CH$_3$CN as the electrolyte. The polymer was found to be partially crystalline with an open honeycomb-like morphology. Neutral PITN films were blue-black in color, which undergo oxidation at ~0.6 V vs. SCE (saturated calomel electrode), becoming an almost transparent greenish yellow. The films seemed to suffer from overoxidation at potentials above ~1.3 V, resulting in a decrease of current response for each additional cycle of the voltammogram. Attempts to n-dope the polymer via reduction were unsuccessful, with no polymer reduction observed within the potential window of the electrolyte solution.

The absorption spectra of thin films reveal a maximum at approximately 1.4 eV (ca. 885 nm). The absorption onset corresponds to a bandgap of 1.0 eV, essentially half the bandgap of polythiophene and the smallest bandgap of any known conjugated organic polymer at the time.[27] The reduction in the energy gap was believed to arise from fusion of the benzene ring onto the thiophene, thereby forcing some quinoid contribution into the polymer ground state.

Chemical or electrochemical doping of PITN diminished the intense interband transition of the polymer, with the absorbance shifted into the infrared. After doping, there is no indication of an energy gap over the measured spectral range, suggesting that the electronic structure of heavily doped PITN is that of a metal.[27] These results are qualitatively similar to those for polythiophene and polyacetylene, where experiments had confirmed a metallic state at high doping levels. The small bandgap of PITN, however, provides an important difference. While the doping of the other polymers gave observable color changes, the doped polymers still exhibit significant absorption in the visible spectrum. In comparison, the absorbance of heavily doped PITN is quite small throughout the visible spectrum, such that thin film samples appear essentially transparent.

The conductivities of a freestanding PITN film and a pressed powder pellet measured via four-point probe were found to be 5 and 7 S cm^{-1}, respectively.[27] *In situ* studies of the effect of doping and compensation were also carried out, which showed that doping with iodine or perchloric acid caused an increase in conductivity by a factor of ~6, to a maximum value of about 50 S cm^{-1}. Compensation with ammonia or hydrazine vapor caused a corresponding increase in resistance as expected.

A final 1986 paper then probed the effect of doping on PITN in greater detail.[28] The blue PITN films were prepared in the same way as in the previous paper, after which they were analyzed using cyclic voltammetry and electrochemical voltage spectroscopy. In general, as the cell voltage was increased, doping of $(ITN)_x$ (i.e., PITN) occurs by the reaction

$$(ITN)_x + xy\text{LiClO}_4 \rightarrow \left[(ITN)_x^{y+}(\text{ClO}_4^-)_y\right] + xy\text{Li}$$

which results in a continuous decrease in the intensity of the interband transition, coupled with an increase of oscillator strength in the infrared. The threshold for charge injection was determined to be circa 2.6 V versus Li, and the doping-induced spectral changes were found to be reversible such that reducing the voltage of the optical cell to ~2.2 V versus Li restores the spectrum of the undoped polymer. Hysteresis was observed in the charge injection–removal cycles, which, based on previous studies of polyacetylene and polythiophene, was viewed to result from structural distortions that occur after charge injection. However, detailed identification of these doping-induced features (e.g., bipolarons or polarons) was not possible in the current study.

7.3 Additional studies on polyisothianaphthene

Due to the large reduction in the bandgap of PITN in comparison with polythiophene, it quickly became the subject of considerable theoretical interest. Such theoretical studies began with a collaboration between Wudl, Heeger, and Jean-Luc Brédas (b. 1954), a computational chemist at the University of Namur, Belgium. This resulted in an initial 1985 paper that reported a theoretical investigation of the geometry and electronic properties of isothianaphthene, with the first ionization potential and the π-$\pi*$ transition calculated to be 1.29 eV and 2.10 eV smaller than in thiophene, respectively.[29] These results were qualitatively consistent with the fact that the bandgap of PITN was found to be half that of polythiophene.

A second collaborative paper in 1986 then focused on band structure calculations.[30] This study concluded that PITN possessed an aromatic-type electronic structure with a calculated bandgap that was 1.17 eV lower than that of polythiophene determined under the same conditions. In addition, it was concluded that the smaller bandgap found in PITN could be understood on the basis of a relationship between the bandgap energy and the effective quinoid contributions to the electronic structure. That is, the fusion of the benzene ring onto the thiophene effectively increases the quinoid contribution to the electronic structure of the polymer ground state, thus reducing the bandgap energy.

Brédas then followed this with a solo contribution in 1987, which attempted to better understand the effect of quinoidal contributions to the polymer bandgap.[31] As with the previous study, it was found that increasing the quinoid character for an aromatic ground state should result in a lowering of the polymer bandgap. However, he notes that this relationship predicts that the corresponding polymers will generally not have a metallic behavior (i.e., zero bandgap) but would correspond to semiconductors with possibly very small bandgaps. Consistent with his previous work, the calculated bandgap for PITN was again calculated to be 1.17 eV lower than that of polythiophene, which was rationalized as the fusion of benzene to thiophene effectively increased the quinoid contributions. In addition to PITN, Brédas also studied the extended analog polyisonaphtothiophene, which resulted in a calculated bandgap of 0.01 eV. While this polymer was not yet known, it was later reported by Wudl in 1988 (see Section 7.4).

Miklos Kertesz and coworkers later determined a similar calculated bandgap of 1.16 eV for PITN in 1990 but considered the structure to be quinoid, rather than the aromatic electronic structure concluded by Brédas.[32] A second paper followed in 1991 that focused on calculating the stability of aromatic versus quinoidal ground states in various polymers, including PITN.[33] Again, the methods applied determined the quinoidal form of PITN to be more stable than the aromatic form.

Another 1990 study by Kasinath Nayak and Dennis S. Marynick also predicted a quinoid ground state for PITN but noted that the relative energies of the aromatic and quinoid forms were so close that both may exist as local minima in the real polymeric system.[34] They also highlight that the details of the polymer chain termination in these systems are extremely important, where termination in a quinoid fashion would result in a quinoid structure for even short chains, but an aromatic termination (as expected in polymers synthesized from aromatic monomers) would likely force the aromatic structure for much longer chain length.

That same year, Kürti and Surján published a short letter that showed the aromatic versus quinoid nature of PITN was highly dependent on chain length and thus one must be careful with calculations based on smaller oligomers. Based upon their methods, it was found that PITN oligomers of eight or fewer repeats were aromatic, while longer chains contained aromatic character localized to the chain ends with quinoidal character through the central structure.[35] It should be pointed out that this model agreed well with the previous conclusions of Nayak and Marynick.

The issue of aromatic versus quinoid structure was also addressed experimentally by two 1989 papers from Hans Kuzmany and coworkers at the University of Vienna.[36,37] Here, the calculated vibrational modes for aromatic and quinoid structures of various polymers were compared with experimental values from Raman scattering. While polymers such as polythiophene and polypyrrole gave results confirming an aromatic form for the ground state of the polymer, the results for PITN did not clearly confirm either an aromatic or quinoid form, and thus both forms must be considered to be contributing to the polymer ground state. In the second paper, however, they conclude that the results for the neutral polymer are in better agreement with the quinoid form, although again the differences between the two forms are small.

In addition to the focus on theoretical studies, modified synthetic methods for the preparation of PITN were also reported by other groups following the initial reports from Wudl and Heeger. The first of these reports came in 1986, in which Kwan-Yue Jen and Ronald Elsenbaumer reported that PITN could be prepared as blue-black powders by the direct oxidation of dihydroisothianaphthene using either oxygen or $FeCl_3$ (Figure 7.6).[38] Soxhlet extraction of the initial $FeCl_3$ polymerized powders with THF resulted in a deep blue powder of the doped polymer compensated with $FeCl_4^-$. Pressed pellets of the powder gave conductivities of 0.5 S cm^{-1}.

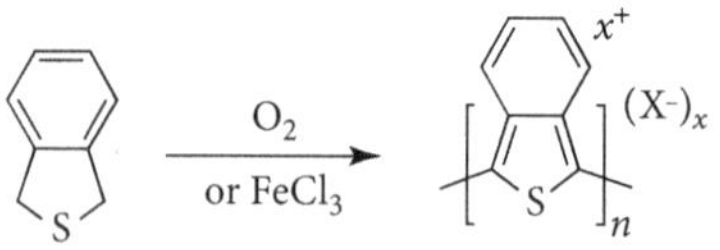

Figure 7.6 Jen and Elsenbaumer's synthetic route to PITN.

doped PITN
$(X^- = FeCl_4^-$ or $OH^-)$

Another report on the electropolymerization of isothianaphthene then came from Simon J. Higgins and coworkers in 1989.[39] Although Wudl and Heeger had previously reported that the use of conventional electrolytes resulted in generation of the nonconjugated poly(dihydroisothianaphthene) (Figure 7.5), the authors here reported that good-quality films of PITN could be produced by these methods, providing care is taken over the control of the potential applied. Thus, good-quality, doped PITN films compensated with BF_4^- could be produced from Et_4NBF_4 via either potential step voltammetry or via potential cycling.

The first functionalized PITN, poly(5-decylisothianaphthene), was then reported in 1992–1993 by Martin Pomerantz and coworkers (we will return to Pomerantz in more detail in Section 7.4).[40,41] Although first mentioned in 1992, all that was given was the polymer bandgap in one of the paper's footnotes.[40] As such, it was not really until 1993 that significant details were reported.[41] The alkyl-functionalized polymer was prepared by $FeCl_3$ oxidation of 1,3-dihydro-5-decylisothianaphthene in a manner analogous to the synthesis of PITN by Jen and Elsenbaumer described earlier. The resulting polymer was then dedoped using NH_3 to give a blue-black material soluble in a variety of organic solvents. Absorption spectra of the neutral polymer in solution gave a maximum at 512 nm, with a shoulder at circa 670 nm.[41] While attempts to cast films of the polymer from solution gave material that apparently scattered considerable light, the absorption onset corresponded to an E_g of ca. 1.0–1.3 eV.[40,41]

7.4 Expanding the family of low-bandgap polymers

After the initial success of PITN, researchers began efforts to develop other polymers based on fused-ring monomeric units. The first new polymer of this type, poly(dithieno[3,4-*b*:3',4'-*d*]thiophene) (Figure 7.7), was developed in 1987 by Carlo Taliani (1945–2015) and coworkers in Italy, with their first paper appearing in early 1988.[42] Taliani was born in Siena, Italy, on March 11, 1945. He earned his doctorate at the University of Bologna in 1968, after which he remained at Bologna as lecturer.[43] He then became a researcher at the Institute of Molecular Spectroscopy of the National Research Council (CNR) in 1972 and assistant professor in the Department of Chemistry at the University of Bologna from 1974 to 1983. He advanced to associate professor for the period of 1983–1989,[43] after which he left the university to become research director at the Italian CNR.[43,44] He also became director of the Institute of Molecular Spectroscopy in 1991.[43,44] He retired in 2010 and passed away on August 12, 2015.[44]

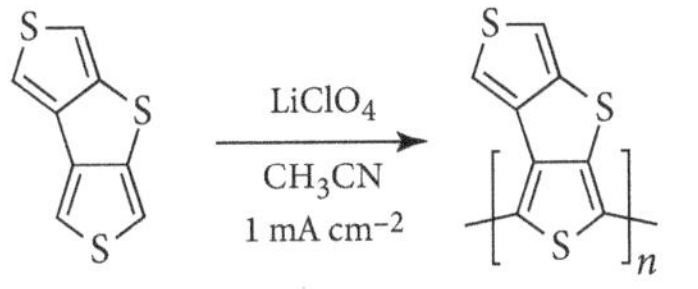

Figure 7.7 Electrochemical synthesis of poly(dithieno[3,4-*b*:3',4'-*d*]thiophene).

It was toward the end of his time at the University of Bologna that his focus turned to this new low-bandgap polymer. The monomer dithieno[3,4-*b*:3',4'-*d*]thiophene was added to a CH_3CN solution of $LiClO_4$ and electropolymerized onto Pt or ITO electrodes at a constant current of 1 mA cm^{-2}.[42] The resulting neutral film exhibited a low-energy transition with a maximum at 590 nm, while the oxidized film was colorless and highly transparent. The conductivity was measured by four-point probe to give a value of 1.0 S cm^{-1}.

This was then followed by a second paper in 1989 that focused on the optical properties of poly(dithieno[3,4-*b*:3',4'-*d*]thiophene) (PDTT), particularly the optical changes involved during doping.[43] The polymer was generated as described in the first paper, after which its optical properties were determined at various applied potentials. The neutral film exhibited a maximum at 2.1 eV (590 nm), with an absorption onset corresponding to a bandgap of 1.1 eV. As doping of the film is induced, a gradual decrease in the intensity of the π-π* absorption is observed, along with the consequent increase of two new bands below 1.8 eV. These new bands were assigned as bipolaron bands, the energy of which shifts to higher energies with increased doping levels (i.e., from ca. 0.53 and 1.1 eV to ca. 0.83 and 1.4 eV). Due to the change in the intensity of the visible transition, opaque neutral film becomes colorless and semitransparent when heavily doped.

This study also attempted to address the fact that the monomeric dithieno[3,4-*b*:3',4'-*d*]thiophene contains four thiophene α-positions potentially capable of coupling during the electropolymerization. Thus, while traditional coupling through the two α-positions of one external thiophene is certainly possible, potential coupling through both external thiophenes could generate a number of possible conjugated pathways (Figure 7.8), as well as potentially introducing multidimensionality to the material. The comparison of X-ray data and theoretical calculations in a second 1989 paper,[46] however, led the authors to propose that the primary contribution to polymer structure was the traditional coupling through only one of the two external thiophenes.

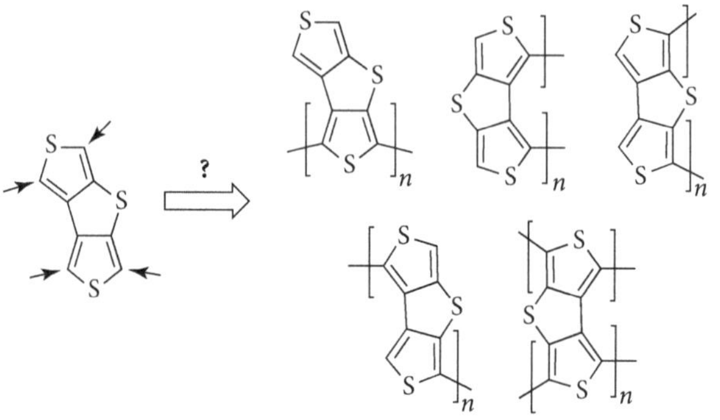

Figure 7.8 Potential coupling motifs for poly(dithieno[3,4-*b*:3',4'-*d*]thiophene).

A later 1993 collaboration with Brédas returned to further address this issue of polymer structure via computational modeling of the various possibles.[47,48] These results not only confirmed that the lowest E_g values resulted from the traditional coupling through only one of the two external thiophenes, but they also addressed the fact that this could result in two possible regiochemistries (Figure 7.9). The HH-TT-type regiochemistry maximizes steric interactions to give a higher calculated E_g of 1.77 eV, while the all-HT-type arrangement allows a more planar conformation and the lower calculated value of 1.27 eV, in good agreement with the experimental E_g of 1.1 eV.

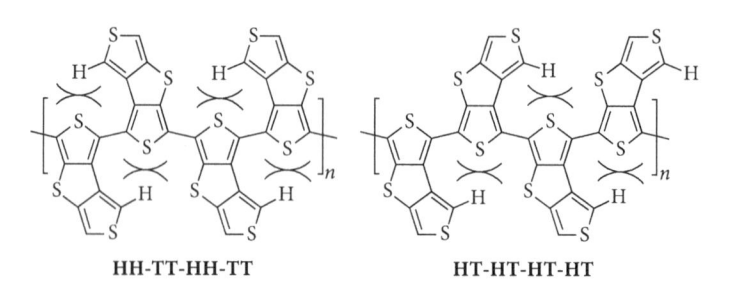

HH-TT-HH-TT   HT-HT-HT-HT

Figure 7.9 Two possible regiochemistries for poly(dithieno[3,4-*b*:3',4'-*d*] thiophene) coupled through one external thiophene.

Another addition to this family was the extended analog of PITN, poly-isonaphtothiophene or poly(naphtho[2,3-c]thiophene), initially introduced by Wudl and coworkers in 1988.[49] Here, attempts to polymerize 1,3-dihydronaphtho[2,3-c]thiophene via both electrochemical and $FeCl_3$ oxidation methods were reported (Figure 7.10), all of which gave polymer films with an absorption onset of 1.4 eV. This value is quite far from the value of 0.01 eV previously calculated for this polymer by Brédas in 1987.[31] However, the bandgap of this polymer was also calculated using Hückel methods by Kertesz and coworkers two years later in 1990, giving a value of 0.28 eV for the aromatic form and 1.5 eV for the quinoid form.[32] That same year Nayak and Marynick used extended Hückel methods to give values of 0.37 and 1.10 eV for the aromatic and quinoid forms, respectively.[34] For both of these later studies, the calculated quinoid form agrees well with the experimental value here.

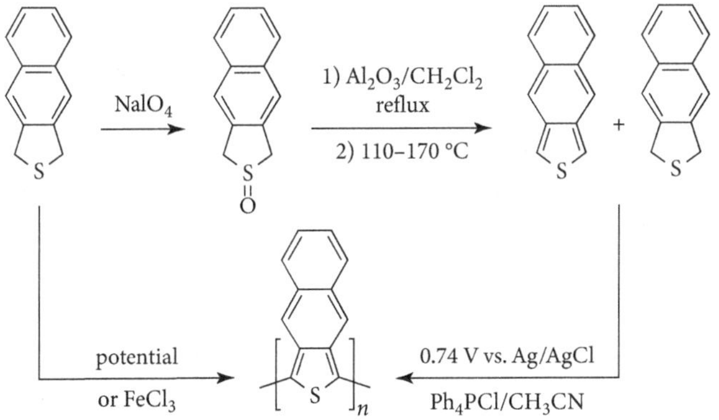

Figure 7.10 Synthetic routes to poly(naphtho[2,3-c]thiophene).

A second paper was then published by Ikenoue in 1990, which focused on the synthesis and polymerization of naphtho[2,3-c]thiophene.[50] The production of naphtho[2,3-c]thiophene was attempted in a manner analogous to isothianaphthene, but the isolated product was determined to be a 1:1 mix of naphtho[2,3-c]thiophene and its 1,3-dihydro compound (Figure 7.10). Electropolymerization of the 1:1 mixture at a potential that would only result in oxidation of naphtho[2,3-c]thiophene (0.74 V vs. Ag/AgCl) resulted in the deposition of a blue-gray film that became transparent gray in the oxidized state. The neutral film exhibited a maximum at ~575 nm with an absorption onset at ~830 nm, corresponding to a bandgap of circa 1.5 eV.[50]

A fourth member of this growing family, poly(2,3-dihexylthieno[3,4-b]pyrazine), was then added by Martin Pomerantz and coworkers in 1992.[40] After receiving a BS in chemistry from the City College of New York in 1959, Pomerantz continued his studies at Yale University, receiving his MS in

1961.[51,52] He then completed his Ph.D. under William von Eggers Doering (1917–2011) in 1964 with a dissertation on cyclopropenes.[52] He then spent a year at the University of Wisconsin as a National Science Foundation Post-doctoral Fellow under Jerome A. Berson (1924–2017),[53] before joining the faculty at the Case Institute of Technology[54] (Case merged with Western Reserve University in 1967 to become the modern Case Western Reserve University). Pomerantz then moved to Yeshiva University in New York in 1971,[55] where he was an Alfred P. Sloan Foundation Research Fellow from 1971 to 1976.[56] He then moved again to join the faculty at the University of Texas at Arlington in 1977,[56] where he remained for the rest of his career.

Pomerantz's interest in the alkylated polythieno[3,4-*b*]pyrazine (Figure 7.11) was partially due to theoretical calculations published by his Arlington colleague Dennis Marynick in 1990 that predicted its bandgap to be 0.70 eV, 0.10 eV less than that predicted for the quinoid form of PITN by the same methods.[34] Marynick also found that the placement of the nitrogens in the pyrazine ring removed unfavorable steric interactions found in the structural analysis of PITN; thus, calculations predicted a more planar structure for polythieno[3,4-*b*]pyrazine. Pomerantz's other interest in this new material was the ability to easily include alkyl groups on the pyrazine ring, thus producing a low-bandgap material that should be soluble in common organic solvents.[40]

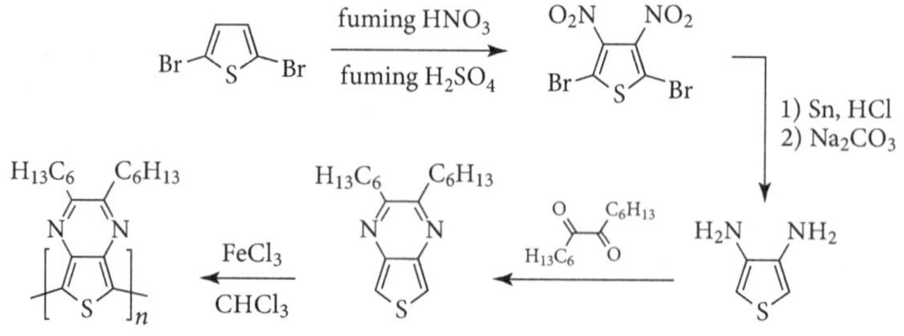

Figure 7.11 Pomerantz's synthetic route to poly(2,3-dihexylthieno[3,4-*b*]pyrazine).

In order to produce the desired material, the monomeric 2,3-dihexylthieno[3,4-*b*]pyrazine was synthesized as shown in Figure 7.11, after which it was oxidatively polymerized via FeCl$_3$.[40] The resulting product was then dedoped with hydrazine hydrate and dialyzed to remove low molecular weight species. This resulted in the isolation of a dark blue-black powder that was soluble in a number of organic solvents. Solution spectra in CHCl$_3$ exhibited an absorption maximum at 875 nm. In comparison, a solvent cast film exhibited a slightly lower energy maximum at 915 nm. The absorption onset of the film corresponded to a bandgap of 0.95 ± 0.10 eV, lower than that of PITN as predicted by the earlier theoretical calculations.

A dark blue-black solution of the polymer in $CHCl_3$ was doped with $NOBF_4$ (22% per monomer unit), resulting in a light yellow solution of the doped polymer.[40] Solution spectra revealed that the original transition at 875 nm had diminished, with the addition of a new transition at circa 1600 nm. This doped material could then be cast into films, after which its conductivity was measured by four-point probe to give a value of 3.6×10^{-2} S cm^{-1}.

A second paper was then reported in 1993.[41] While this report did not provide much in terms of new characterization of the polymer's optical and electronic properties, it did provide full NMR and infrared details for both the monomer and corresponding polymer. Furthermore, it provided full experimental details for the polymerization conditions.

7.5 Conclusions

Although the initial 1984 publications reported the bandgap of PITN to be 1.0 eV,[16,17] it was not until the first of their papers in 1985 that the authors referred to this polymer as a *"small bandgap"* material.[27] Since that point, the terms "small bandgap,"[27,29,32,47,48] "low bandgap,"[40,41,45,46] and "narrow bandgap" came more frequent in the literature. However, these terms were often used somewhat interchangeably and rarely with any formal definition of the terms. While cutoff values of 1.0–1.5 eV were sometimes used by authors to differentiate low-bandgap polymers from their more common counterparts,[12,49] it was not until 1998 that the first working definition was proposed by Pomerantz.[10] Using sound logic, Pomerantz explained that among the family of more commonly studied parent conjugated polymers, polyacetylene has the lowest bandgap ($E_g = 1.5$ eV), yet polyacetylene is still not considered a low-bandgap material. As such, it thus made sense to define low-bandgap polymers as those materials with bandgaps below that of polyacetylene, that is, $E_g < 1.5$ eV.[10,11]

While the initial hope that reduction in the bandgap could lead to new record conductivities or even metallic polymers did not ever come to pass, this did result in a growing family of low-bandgap polymers. Furthermore, these materials did offer the realistic possibility of transparent conducting polymers, as illustrated by some of the examples discussed in this chapter. Lastly, the detailed study of the structural effects that contributed to the lower bandgaps of these materials led to a much greater understanding of the structure–function relationships within conjugated polymers in general, particularly in terms of control of the bandgap energy. Although this chapter only covers the early development of these materials up through the early 1990s, these efforts continued up to the present, and the ability to produce low-bandgap materials remains a significant focus within the field of conjugated and conducting polymers.

References and Notes

1. Rasmussen, S. C. Early history of conjugated polymers: From their origins to the handbook of conducting polymers. In *Handbook of Conducting Polymers*, 4th ed. Reynolds, J. R.; Skotheim, T. A.; Thompson, B.; Eds.; CRC Press: Boca Raton, FL, 2019, pp. 1–35.

2. Rasmussen, S. C. Conjugated and conducting organic polymers: The first 150 years. *ChemPlusChem* **2020**, *85*, 1412–1429.

3. Rasmussen, S. C. The early history of polyaniline: Discovery and origins. *Substantia* **2017**, *1*(2), 99–109.

4. Rasmussen, S. C. The early history of polyaniline - revisited: Russian contributions of Fritzsche and Zinin. *Bull. Hist. Chem.* **2019**, *44*(2), 123–133.

5. Rasmussen, S. C. The early history of polyaniline II: Elucidation of structure and redox states. *Substantia* **2022**, *6*(1), 107–119.

6. Rasmussen, S. C. Early history of polypyrrole: The first conducting organic polymer. *Bull. Hist. Chem.* **2015**, *40*, 45–55.

7. Yamamoto, T.; Sanechika, K.; Yamamoto, A. Preparation of thermostable and electric-conducting poly(2,5-thienylene). *J. Polym. Sci. Polym. Lett. Ed.* **1980**, *18*, 9–12.

8. Park, Y. W.; Druy, M. A.; Chiang, C. K.; MacDiarmid, A. G.; Heeger, A. J.; Shirakawa, H.; Ikeda, S. Anisotropic electrical conductivity of partially oriented polyacetylene. *J. Polym. Sci. Polym. Lett. Ed.* **1979**, *17*, 195–201.

9. Bredas, J. L.; Themans, B.; Fripiat, J. G.; Andre, J. M.; Chance, R. R. Highly conducting polyparaphenylene, polypyrrole, and polythiophene chains: An ab initio study of the geometry and electronic-structure modifications upon doping. *Phys. Rev. B* **1984**, *29*, 6761–6773.

10. Pomerantz, M. Low band gap conducting polymers. In *Handbook of Conducting Polymers*, 2nd ed., Skotheim, T. A.; Elsenbaumer, R. L.; Reynolds, J. R.; Eds; Marcel Dekker: New York, 1998, pp. 277–309.

11. Rasmussen, S. C.; Pomerantz, M. Low bandgap conducting polymers. In *Conjugated Polymers: Theory, Synthesis, Properties, and Characterization*, Skotheim, T. A.; Reynolds, J. R.; Eds.; *Handbook of Conducting Polymers*, 3rd ed.; CRC Press: Boca Raton, FL, 2007, Chapter 12.

12. Rasmussen, S. C. Low-bandgap polymers. In *Encyclopedia of Polymeric Nanomaterials*, Muellen, K.; Kobayashi, S.; Eds.; Springer: Heidelberg, 2015, pp. 1155–1166.

13. Rasmussen, S. C.; Gilman, S. J.; Wilcox, W. D. The eternal quest for practical low bandgap polymers. *Gen. Chem.* **2023**, *9*, 220010.

14. Roncali, J. Synthetic principles for bandgap control in linear π-conjugated systems. *Chem. Rev.* **1997**, *97*, 173–205.

15. Mikie, T.; Osaka, I. Small-bandgap quinoid-based π-conjugated polymers. *J. Mater. Chem. C* **2020**, *8*, 14262–14288.

16. Wudl, F.; Kobayashi, M.; Heeger, A. J. Conducting polymers of thiophene and its benzolog. *Polym. Prep.* **1984**, *25*, 257.

17. Wudl, F.; Kobayashi, M.; Heeger, A. J. Poly(isothianaphthene). *J. Org. Chem.* **1984**, *49*, 3382–3384.

18. Wudl, F. Personal communication, 2021.

19. Bendikov, M.; Martin, M.; Perepichka, D. F.; Prato, M. Fred Wudl. Discovering new science through making new molecules. *J. Mater. Chem.* **2011**, *21*, 1292–1294.

20. ACS Award in the Chemistry of Materials. *Chem. Eng. News* **1996**, *74*(4), 59.

21. Profile. *J. Mater. Chem.* **2005**, *15*, 18–19.

22. Briseno, A. J.; Mallett, J. J.; Schanze, K. S. Celebrating 50 years of organic chemistry applied materials research on the occasion of Fred Wudl's 75th birthday. *ACS Appl. Mater. Interfaces* **2015**, *7*, 27987–27988.

23. Wudl, F. Preparation and survey of resolving properties of optically active methyl *p*-styryl sulfoxide copolymers. Ph.D. dissertation, University of California, Los Angeles, 1967.

24. Wudl, F.; Smith, G. M.; Hufnagel, E. J. Bis-1,3-dithiolium chloride: An unusually stable organic radical cation. *Chem. Commun.* **1970**, 1453–1454.

25. Wudl, F.; Wobschall, D.; Hufnagel, E. J. Electrical conductivity by the bis-1,3-dithiole-bis-l,3-dithiolium system. *J. Am. Chem. Soc.* **1972**, *94*, 670–671.

26. Thomas, G. A.; Schafer, D. E.; Wudl, F.; Horn, P. M.; Rimai, D.; Cook, J. W.; Glocker, D. A.; Skove, M. J.; Chu, C. W.; Groff, R. P.; Gillson, J. L.; Wheland, R. C.; Melby, L. R.; Salamon, M. B.; Craven, R. A.; De Pasquali, G.; Bloch, A. N.; Cowan, D. O.; Walatka, V. V.; Pyle, R. E.; Gemmer, R.; Poehler, T. O.; Johnson, G. R.; Miles, M. G.; Wilson, J. D.; Ferraris, J. P.; Finnegan, T. F.; Warmack, R. J.; Raaen, V. F.; Jerome, D. Electrical conductivity of tetrathiafulvalenium-tetracyanoquinodimethanide (TTF-TCNQ). *Phys. Rev. B* **1976**, *13*, 5105–5110.

27. Kobayashi, M.; Colaneri, N.; Boysel, M.; Wudl, F.; Heeger, A. J. The electronic and electrochemical properties of poly(isothianaphthene). *J. Chem. Phys.* **1985**, *82*, 5717–5723.

28. Colaneri, N.; Kobayashi, M.; Heeger, A. J.; Wudl, F. Electrochemical and opto-electrochemical properties of poly(isothianaphthene). *Synth. Met.* **1986**, *14*, 45–52.

29. Bredas, J. L.; Themans, B.; Andr, J. M.; Heeger, A. J.; Wudl, F. Geometric and electronic structures of isothianaphthene and thieno[3,4-*c*]thiophene: A theoretical investigation. *Synth. Met.* **1985**, *11*, 343–352.

30. Bredas, J. L.; Heeger, A. J.; Wudl, F. Towards organic polymers with very small intrinsic band gaps. I. Electronic structure of polyisothianaphthene and derivatives. *J. Chem. Phys.* **1986**, *85*, 4673–4678.

31. Bredas, J. L. Theoretical design of polymeric conductors. *Synth. Met.* **1987**, *17*, 115–121.

32. Lee, Y.-S.; Kertesz, M.; Elsenbaumer, R. L. Importance of energetics in the design of small bandgap conducting polymers. *Chem. Mater.* **1990**, *2*, 526–530.

33. Karpfen, A.; Kertesz, M. Energetics and geometry of conducting polymers from oligomers. *J. Phys. Chem.* **1991**, *95*, 7680–7681.

34. Nayak, K.; Marynick, D. S. The interplay between geometric and electronic structures in polyisothianaphthene, polyisonaphthothiophene, polythieno(3,4-*b*)pyrazine, and polythieno(3,4-*b*)quinoxaline. *Macromolecules* **1990**, *23*, 2237.

35. Kürti, J.; Surján, P. R. Quinoid vs aromatic structure of polyisothianaphthene. *J. Chem. Phys.* **1990**, *92*, 3247–3248.

36. Faulques, E.; Wallnöfer, W.; Kuzmany, H. Vibrational analysis of heterocyclic polymers: A comparative study of polythiophene, polypyrrole, and polyisothianaphtene. *J. Chem. Phys.* **1989**, *90*, 7585–7593.

37. Wallnöfer, W.; Faulques, E.; Kuzmany, H.; Eichinger, K. Resonance Raman spectroscopy and vibrational analysis of poly(isothianaphthene) and related compounds. *Synth. Met.* **1989**, *28*, C533.

38. Jen, K.-Y.; Elsenbaumer, R. Facile preparation of electrically conductive poly(isothianaphthene). *Synth. Met.* **1986**, *16*, 379–380.

39. Christensen, P. A.; Kerr, J. C. H.; Higgins, S. J.; Hamnett, A. A combined ellipsometric and *in situ* infrared (SNIFTIRS) study of poly(benzo-[c]-thiophene) films. *Faraday Discuss. Chem. Soc.* **1989**, *88*, 261–275.

40. Pomerantz, M.; Chaloner-Gill, B.; Harding, L. O.; Tseng, J. J.; Pomerantz, W. J. Poly(2,3-dihexylthieno[3,4-*b*]pyrazine). A new processable low band-gap polyheterocycle. *J. Chem. Soc., Chem. Commun.* **1992**, 1672–1673.

41. Pomerantz, M.; Chaloner-Gill, B.; Harding, L. O.; Tseng, J. J.; Pomerantz, W. J. New processable low band-gap, conjugated polyheterocycles. *Synth. Met.* **1993**, *55*, 960–965.

42. Bolognesi, A.; Catellani, M.; Destri, S.; Zamboni, R.; Taliani, C. Poly(dithieno[3,4-*b*:3',4'-*d*]thiophene): A new transparent conducting polymer. *J. Chem. Soc., Chem. Commun.* **1988**, 246–247.

43. Prabook. Carlo Taliani. https://prabook.com/web/carlo.taliani/269678 (accessed May 9, 2024).

44. ECME 2015. In memory of Professor Carlo Taliani. www.unistra.fr/index.php/carlo-taliani (accessed Nov. 19, 2022).

45. Taliani, C.; Ruani, G.; Zamboni, R.; Bolognesi, A.; Catellani, M.; Destri, S.; Porzio, W.; Ostaja, P. Optical properties of a low energy gap conducting polymer: Polydithieno[3,4-*b*:3',4'-*d*]thiophene. *Synth. Met.* **1989**, *28*, C507–C514.

46. Bolognesi, A.; Catellani, M.; Destri, S.; Ferro, D. R.; Porzio, W.; Taliani, C.; Zamboni, R.; Ostaja, P. Preparation and properties of a new conducting polyheterocycle: Polydithieno[3,4-*b*:3',4'-*d*]thiophene (PDTT). *Synth. Met.* **1989**, *28*, C527–C532.

47. Quattrocchi, C.; Lazzaroni, R.; Brédas, J. L.; Zamboni, R.; Taliani, C. Theoretical investigation of the structure and electronic properties of poly(dithieno[3,4-*d*:3',4'-*d*]thiophene), a small-band-gap conjugated polymer. *Macromolecules* **1993**, *26*, 1260–1264.

48. Quattrocchi, C.; Lazzaroni, R.; Brédas, J. L.; Zamboni, R.; Taliani, C. Electronic structure and of poly(dithieno[3,4-*d*:3',4'-*d*]thiophene), a small bandgap conjugated polymer. *Synth. Met.* **1993**, *57*, 4399–4404.

49. Wudl, F.; Ikenoue, Y.; Patil, A. O. Synthesis of certain specific electroactive polymers. In *Nonlinear Optical and Electroactive Polymers*. Ulrich, D.; Prasad, P. N.; Eds.; Plenum Press: New York, 1988; pp. 393–400.

50. Ikenoue, Y. Synthesis and characteristic properties of poly(naphtho[2,3-*c*]thiophene) as a series of fused-ring conducting polymers analogous to poly(isothianaphthene). *Synth. Met.* **1990**, *35*, 263–270.

51. The University of Texas at Arlington, Chemistry and Biochemistry. Martin Pomerantz. https://www.uta.edu/academics/faculty/profile?username=pomerant (accessed May 10, 2024).

52. Pomerantz, M. Reactions of some cyclopropenes - 2,4-dimethyltricyclo[1.1.1.0^{2,4}]pentan-5-one. Ph.D. dissertation, Yale University, New Haven, 1964.

53. Jerome A.; Berson, J. A.; Pomerantz, M. Formation and capture of a reactive intermediate related to dimethylpseudoindene. *J. Am. Chem. Soc.* **1964**, *86*, 3896–3897.

54. Pomerantz, M.; Abrahamson, E. W. The electronic structure and reactivity of small ring compounds. I. Bicyclobutane. *J. Am. Chem. Soc.* **1966**, *88*, 3970–3972.

55. Pomerantz, M.; Gruber, G. W. Photochemical reorganization reactions of *o*-divinylbenzene, 3,4-benzotropilidene, 1,2-benzotropilidene, and 1-phenyl-1,3-butadiene. *J. Am. Chem. Soc.* **1971**, *93*, 6615–6622.

56. Pomerantz, M.; Fink, R. Formation and trapping of 1,2,4,5-dibenzotropilidene (10*H*-dibenzo[*a,d*]cycloheptene). *J. Org. Chem.* **1977**, *42*, 2788–2790.

Chapter 8
Retrospective on the Search for Organic Conductors and the Nature of Discovery

8.1 Introduction

The history of conjugated and conducting polymers is often dominated by the awarding of the Nobel Prize in Chemistry to Hideki Shirakawa, Alan MacDiarmid, and Alan Heeger in 2000.[1] While this was an important milestone, the history of these materials far exceeds either this singular award or the work that it is recognizing. As outlined in the previous chapters, this collective history dates back to 1834 and includes a number of landmark events in both the history of synthetic organic dyes and the overall history of organic polymers in general.[2,3]

Furthermore, this history also includes a number of Nobel Prize winners beyond that of Shirakawa, MacDiarmid, and Heeger. As early as 1899, Paul Sabatier studied the polymerization of acetylene over copper, work that directly related to his being awarded the 1912 Nobel Prize in Chemistry "for his method of hydrogenating organic compounds in the presence of finely disintegrated metals."[4] Less than a decade later, in 1906, Richard Willstätter worked to deduce the structure of aniline black (polyaniline) and its various oxidation states, efforts that can viewed as an extension of his work recognized by the 1915 Nobel Prize in Chemistry "for his researches on plant pigments, especially chlorophyll."[5] Perhaps the most important example of other Nobel laureates in the history of conjugated polymers was Giulio Natta, who was the first to successfully generate fully conjugated polyacetylene in 1955. Of course, this was accomplished using metal catalysts previously applied to the polymerization of olefins, for which he shared the 1963 Nobel Prize in Chemistry with Karl Ziegler "for their discoveries in the field of the chemistry and technology of high polymers."[6] Lastly, while technically not a Nobel laureate, John K. Stille made notable contributions to polyacetylene and polyphenylenes in the early and late 1960s, respectively. Of course, Stille is most well-known for his introduction of Still cross-coupling, now used extensively in the synthesis of conjugated polymers, and it is generally agreed that he would have been included in the 2010 Nobel Prize in Chemistry "for palladium-catalyzed cross couplings in organic synthesis" had he not died prematurely in the 1989 crash of United Airlines Flight 232.[7,8]

The Origins and Early History of Conjugated Organic Polymers. Seth C. Rasmussen, Oxford University Press.
© Oxford University Press (2025). DOI: 10.1093/9780197638194.003.0008

In order to place the history of conjugated and conducting polymers outlined in the previous chapters into the greater context of the history of organic electronic materials, this concluding chapter will attempt to outline the evolution of conduction in carbon-based materials, including how the development of conducting polymers falls within this larger picture. This will then be followed by an overview of the early history of conjugated polymer-based devices, thus illustrating the real-world applications of these materials and paving the way for the current field of organic electronics. Lastly, the recognition by the 2000 Nobel Prize in Chemistry "for the discovery and development of conductive polymers" will be discussed, along with an examination of the nature of discovery.

8.2   From carbon black to conducting polymers

When considering carbon-based materials, the most fundamental is simply elemental carbon. As the fullerenes were not discovered until the 1980s, followed later by carbon nanotubes, the early history of carbon's conductivity is limited to diamond and graphite. By the early 1930s, it was known that diamond was essentially nonconductive, but that graphite could exhibit conductivities >50 S cm^{-1}.[9]

Closely related to graphite is the class of materials known as carbon blacks. Such carbon blacks are carbonaceous materials generated via the pyrolysis of organic matter (oil, wood, vegetables, etc.) and are related to other such pyrolysis products such as charcoal, coke, and activated carbon.[10–12] All of these materials differ primarily in terms of the temperatures utilized during the pyrolysis, with higher temperatures giving materials closer to natural graphite. Carbon blacks are the furthest along this spectrum of carbonaceous materials and may be regarded as amorphous precursors of graphite.[10,12] The use of carbon blacks as black pigments extends back to prehistory, with the oldest recorded process for its production dating to 3000 BCE.[11] Again, by the early 1930s, the conductivity of these species had been determined, with carbon blacks giving values of 3–21 S cm^{-1} depending on the method of preparation.[9] As a result, a large number of studies beginning in the 1950s involved the pyrolysis of organic polymers in order to develop conducting materials with more defined and controllable compositions.[10,13–15] As might be expected, the results depended on the specific polymer used, but conductivity generally increased with the pyrolysis temperatures applied.[15]

The first report of conduction in a nonpyrolyzed organic compound was reported in 1948, when Daniel D. Eley (1914–2015) measured the electronic characteristics of metal-free phthalocyanine (Figure 8.1).[16] The resistance of the phthalocyanine single crystals exhibited the temperature dependence typical of

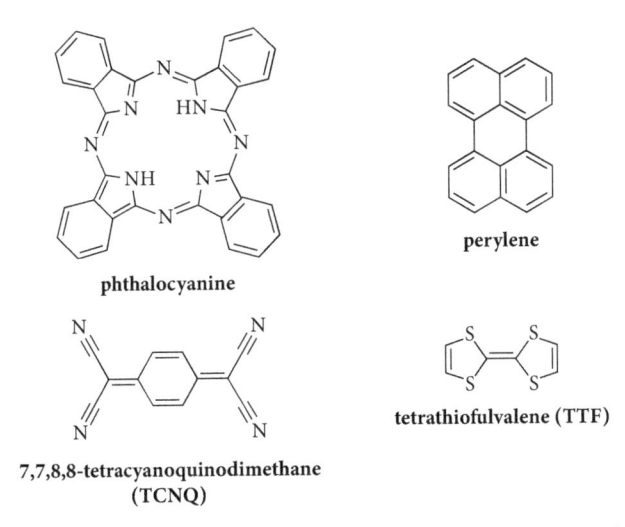

phthalocyanine

perylene

**7,7,8,8-tetracyanoquinodimethane
(TCNQ)**

tetrathiofulvalene (TTF)

Figure 8.1 Various small molecules used in organic conducting materials.

semiconductors, with a room temperature resistance of circa 5.4×10^{10} Ω cm (corresponding to a conductivity of 1.8×10^{-11} S cm^{-1}). It was thus concluded that metal-free phthalocyanine appeared to be an intrinsic semiconductor.

Another class of conductive carbon-based materials based on graphite intercalation compounds was then introduced by Alfred Ubbelohde (1907–1988) in 1951.[17] The earliest examples consisted of graphite intercalated with either bromide or potassium to give conductivities up to 1.5×10^3 S cm^{-1}. By 1969, more conductive examples had been produced to give conductivity values of 10^4–10^5 S cm$^{-1,18}$ at which point Ubbelohde started referring to these species as *synthetic metals*, a term introduced by Herbert N. McCoy (1870–1945) in 1911 to describe metallic substances comprised of nonmetallic elements.[19] Although intercalated graphite compounds represent the first true case of a synthetic metal, other carbon-based species followed.[20]

A few years later, it was found that solid-state complexes between polycyclic aromatic compounds and halogens exhibited greater electrical conductivity, although often with limited stability.[21] Of these, the bromine complex of perylene (Figure 8.1) was found to be relatively stable with conductivities of 0.01–1 S cm^{-1}. As with phthalocyanine, the temperature dependence of this perylene–bromine complex was typical of an intrinsic semiconductor.

Such complexes, commonly known as *charge-transfer complexes*, were generally viewed as the formation of an ionized structure resulting from the interaction of donor and acceptor molecules.[22] In addition to the donor-halogen species such as the perylene–bromine complex discussed previously, this class of organic conductors soon also included organic donor–organic acceptor combinations as well. As such, this became a very active area of research, with hundreds of examples known by the mid-1960s.[22]

An organic acceptor of particular interest was 7,7,8,8-tetracyanoquinodimethane (TCNQ, Figure 8.1), which was introduced by D. S. Acker and coworkers in 1960.[23] It was quickly shown that TCNQ could be reduced via treatment of metals or organic amines to give conductive species, with the organic salts 5,8-dihydroxyquinolinium (TCNQ) and quinolinium (TCNQ)$_2$ exhibiting conductivities of 0.7 S cm^{-1} and 100 S cm^{-1}, respectively. While early charge–transfer complexes of TCNQ generally exhibited conductivities in the range of 10^{-11} S cm^{-1}, a new TCNQ complex exhibiting conductivity in the metallic regime was reported by Dwaine O. Cowan (1935–2006) and coworkers in 1973.[24] This new charge–transfer complex utilized the organic donor tetrathiofulvalene (TTF, Figure 8.1), which had been introduced by Fred Wudl in 1970,[25] to give room temperature conductivities as high as 652 S cm^{-1}.[24] Furthermore, a semiconductor-to-metal transition was determined at 66 K, with a maximum conductivity of 1.47 × 10^4 S cm^{-1} at this temperature. As a consequence, TTF-TCNQ was added to the growing family of synthetic metals[20] and went on to become a major focus of research in both physics and chemistry, with Heeger studying the metal physics of this charge–transfer complex prior to his work with MacDiarmid (see Section 5.9).

The development of conjugated and conducting polymers then overlapped with the development of both intercalated graphites and charge–transfer complexes, beginning with initial conductivity measurements in the period of 1958–1961 for the neutral forms of both polyacetylene[26] and polyphenylene.[27] In both cases, the conductivities of these materials were found to be in the range of 10^{-10} and 10^{-12} S cm^{-1} and exhibited a temperature dependence consistent with an intrinsic semiconductor. By 1961, the ability to produce highly crystalline polyacetylene increased the conductivity of the neutral polymer to circa 7 × 10^{-5} S cm^{-1}.[28]

The earliest known example of the electronic characterization of a doped conjugated polymer appears in 1961, with reported conductivities of 10^{-5} to 10^{-2} S cm^{-1} for a sample of aniline black (Table 8.1).[29] Unfortunately, little is given about the exact form of the sample, such as the predominate oxidation state of the polymer backbone (i.e., emeraldine?) or the associated counterions. More significant, however, are two 1963 reports, one for I$_2$-doped polyphenylene[30] and the other for I$_2$-doped polypyrrole,[31] both of which show improvements in conductivity compared with the initial 1961 report of aniline black (Table 8.1).

Although both polymers are doped with the same species, an important difference between these two materials lies in the method of doping. In the case of the polyphenylene, a neutral polymer was first prepared and then treated with iodine, which is the typical view of doping a conjugated polymer. As such, the

Table 8.1 Conductivities of various doped conjugated polymers from 1961 to 1977

Year	Polymer	Conductivity (S cm^{-1})	Reference
1961	aniline black	10^{-5} to 10^{-2}	29
1963	I$_2$-doped polyphenylene	4×10^{-5}	30
1963	I$_2$-doped polypyrrole	0.005–0.09	31
1966	emeraldine sulfate	10^{-5}–10	32
1968	BF$_3$-doped polyacetylene	0.0013	33
1968	polypyrrole sulfate	7.54	34
1969	emeraldine sulfate	100	35
1977	Br$_2$-doped *trans*-polyacetylene	0.5	36
1977	I$_2$-doped *trans*-polyacetylene	38	36
1977	I$_2$-doped *trans*-polyacetylene	160	37
1977	AsF$_5$-doped *trans*-polyacetylene	220	37
1977	AsF$_5$-doped *cis*-polyacetylene	560	37

conductivity was seen to increase with increasing amounts of I$_2$, giving a maximum conductivity of 4×10^{-5} S cm^{-1}.[30] In comparison, the polypyrrole material was produced via the thermal polymerization of tetraiodopyrrole, which produced a cross-linked polypyrrole via the elimination of I$_2$. The I$_2$ released thus reacted with the polymer during its polymerization to give an I$_2$-doped material as the initial product. Here, higher conductivity values of 0.005–0.09 S cm^{-1} were determined for the initial product, after which attempts to remove the I$_2$ resulted in decreases in the conductivity.[31] Both of these cases were viewed in terms of polymeric charge–transfer complexes, analogous to the perylene–bromine complex discussed earlier.

Following these initial reports are several more over the period of 1966–1969 (Table 8.1), most of which are produced via oxidative polymerization and thus generate doped materials as the polymer products. The single exception to this is the doping of neutral polyacetylene with BF$_3$ vapor.[33] In all cases, one can see a fairly steady increase in conductivity over the time period, with a maximum of 100 S cm^{-1} achieved for polyaniline (emeraldine sulfate) by 1969.[35]

Interestingly, there seem to be no additional reports between 1970 and 1976. As such, the next studies are the polyacetylene reports recognized by the 2000 Nobel Prize, which includes doping with Br$_2$, I$_2$, and AsF$_5$.[36,37] As can be seen in Table 8.1, however, while the values reported in the first 1977 paper are always presented as especially noteworthy, these are actually a step back in magnitude compared with the two previous reports. The results of the second paper are a significant advancement, however, and represent the first doped conjugated polymers that exhibit conductivities firmly within the metallic regime (i.e., >100 S cm^{-1}). As such, doped polyacetylene was soon added to the family of known synthetic metals, with other doped conjugated polymers added by 1991.[20]

8.3 Early applications and the first devices

Although conjugated and conducting polymers provide plenty of content for academic studies aiming to advance our understanding of electronic materials, their importance for society in general depends on how they can be developed into real-world applications. As covered in Chapter 2, the earliest applications of these materials took advantage of the intense coloring of conjugated polymers, resulting in the development of various commercial polyaniline-based fabric dyes for printing on cotton in the mid-1800s (i.e., emeraldine, azurine, aniline black). Of course, once the conductive nature of these polymers was realized, this opened up a whole new set of possible applications as electronic materials.

The first example to take advantage of the conductive nature of these materials was the application of polyacetylene to battery technologies. As early as 1980, MacDiarmid, Heeger, and coworkers were developing doped polyacetylene films as cathode-active materials in lightweight rechargeable storage batteries.[38] The initial battery configuration utilized a 6% predoped polyacetylene film (i.e., $[CH(ClO_4)_{0.06}]_x$) as the cathode and a lithium metal electrode as the anode, both of which were immersed in a propylene carbonate solution of $LiClO_4$. For this configuration, a 0.5 cm^2 (~3 mg) polymer film provided an open-circuit voltage (V_{OC}) of 3.7 V and the short-circuit current (I_{SC}), was 23 mA. An energy density of 176 W-hr/kg was estimated based on the weight of the polymer film employed and the weight of lithium consumed during a partial discharge.[38]

A second 1981 paper then expanded battery configurations in which polyacetylene films could be used for both the anode and cathode.[39] For example, a p-doped film such as $[CH^{+0.05}(ClO_4)_{0.05}]_x$ could be used as the cathode, while an n-doped film such as $[Li^+_{0.05}(CH^{0.05-})]_x$ acted as the anode. Upon connecting the anode and cathode a V_{OC} of 2.9 V and an I_{SC} of 1.9 mA are achieved. Systems using other anions and cations were also investigated, after which the next significant advancement was the move to all-polymeric solid-state batteries by the end of 1981.[40] Further studies of polyacetylene batteries continued through the early 1980s but never really advanced to any commercial products.

Applications following these early investigations into batteries moved away from efforts to utilize the high conductivities of conducting polymers but, rather, emphasized the semiconducting properties of neutral conjugated polymers. This began with the application of conjugated polymers to photovoltaic devices (i.e., solar cell) beginning in the early 1980s. The earliest of these reports was in

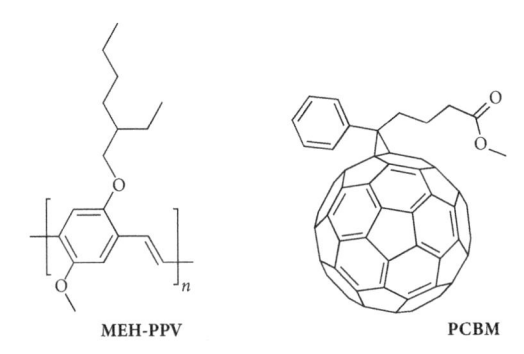

Figure 8.2   Poly[2-methoxy-5-(2-ethylhexyloxy)-1,4-phenylene vinylene] (MEH-PPV) and phenyl-C_{61}-butyric acid methyl ester (PCBM).

1982 from B. R. Weinberger and coworkers at Chronar Corporation, which fabricated neutral polyacetylene devices in which the polymer was deposited onto polycrystalline CdS substrates and then coated with a graphite-based material for the second electrode.[41] Under a simulated solar irradiation (70 mW/cm²), devices exhibited an I_{SC} of 63 µA/cm², a V_{OC} of 0.31 V, and a fill factor of 0.21, all giving an internal conversion efficiency of 0.1%.

A second report then came in 1984 from Gérard Tourillon and Francis Garnier, as previously discussed in Section 6.5.[42] The cells here were comprised of neutral, electrochemically grown polythiophene films sandwiched between a gold electrode and a 10 nm semitransparent Al overlayer. While promising power efficiencies were achieved at low illumination (Pi < 1 µW cm⁻²), these decreased rapidly to circa 0.01% at higher light intensities. They followed this with a second 1986 report in which poly(3-methylthiophene) was electrochemically grown onto an electrode comprised of a thin platinum or gold layer (15 nm) deposited onto a glass substrate.[43] The polymer film utilized was either neutral or 5% doped. A semitransparent (10 nm) Al electrode was then evaporated onto the top of the polymer layer. At a light power of 10 µW cm⁻², an I_{SC} of 12 µA/cm², a V_{OC} of 0.35 V, and a fill factor of 0.25 were achieved.

Heeger and coauthors then utilized the poly(phenylene vinylene) (PPV) derivative poly[2-methoxy-5-(2-ethylhexyloxy)-1,4-phenylene vinylene] (MEH-PPV, Figure 8.2) to fabricate photodiode devices in 1994.[44] These devices utilized the neutral polymer encased between transparent indium tin oxide (ITO) and a calcium electrode. Under 20 mW/cm², the devices produced a V_{OC} of 1.05 V and an I_{SC} of 1.1 µA/cm². The sensitivity and the quantum yield at −10 V were 5 × 10 mA/W and 1.4% electrons per photon.

One of the most significant advances in conjugated polymer OPV devices then came the following year when Heeger, Fred Wudl, and coworkers blended

MEH-PPV with a new, soluble methanofullerene derivative (phenyl-C_{61}-butyric acid methyl ester or PCBM; Figure 8.2) together to produce an active layer with a 1:4 ratio.[45] The resultant bicontinuous network of the electron-rich MEH-PPV and electron-poor PCBM allowed charge transfer between the two species and thus more effective separation of charge, resulting in devices with a power conversion efficiency (PCE) of 2.9%. These new forms of OPV devices consisting of donor–acceptor blends became known as bulk hetero-junction (BHJ) devices and continue to remain the device standard for organic photovoltaics.[46]

Another potential application was then introduced in 1990 with the discovery of electroluminescence in PPV by Richard H. Friend (b. 1953), Andrew B. Holmes (b. 1943), and coworkers, thus paving the way for organic light-emitting diodes (OLEDs) with semiconducting conjugated polymers as the active light-emitting layer.[47] Such polymer-based OLEDs are also sometimes referred to as polymer light-emitting diodes or PLEDs. The initial PLED devices reported in 1990 utilized the polymer sandwiched between a transparent indium oxide bottom contact and an Al top contact. Such devices exhibited a turn-on voltage of just under 14 V, with the spectrum of emitted light very similar to that measured in photoluminescence, with a peak near 2.2 eV (ca. 560 nm). The quantum efficiencies for these devices were up to 0.05%.

The following year, Braun and Heeger reported OLEDs fabricated from the soluble PPV derivative MEH-PPV (Figure 8.2).[48] Using a glass substrate coated with ITO, the MEH-PPV was deposited by spin casting from tetrahydrofuran (THF) or xylenes solutions to give films with thicknesses near 120 nm. Indium or calcium was then used as a top metal contact. A 13 V forward bias resulted in room temperature electroluminescence near 2.1 eV with a hint of a second peak above 1.9 eV.

Friend, Holmes, and coworkers then returned with additional papers in 1992, beginning with a report on the application of random copolymeric PPV derivatives.[49] These materials utilized the copolymerization of phenylene vinylene and 2,5-dimethoxyphenylene vinylene segments (Figure 8.3), the content of which could be modified by the monomer feed ratio used. By controlling the copolymerization in this way, some control of the color of the electroluminescent output could thus be achieved. It was also found that the device efficiency was affected by the monomer feed ratio, with a 9:1 ratio of unfunctionalized to dimethoxy units resulting in an increase to 0.3%. A second paper then detailed the effects of introducing an electron-transport layer between the emitting polymer and the top metal electrode.[50] The electron transport layer utilized in this study was a dispersion of 2-(4-biphenylyl)-5-(4-tert-butylphenyl)-1,3,4-oxadiazole in poly(methyl methacrylate), resulting in enhanced efficiencies up to 0.8% for basic PPV devices.

The following year then brought a number of new studies, beginning with a report from Karg and coworkers at the Universität Bayreuth in Germany that reported PPV devices on either ITO or Au, with top metal electrodes of Al, Mg, or Ca.[51] The primary focus here was the study of the devices at both room temperature and at 77 K, from which it was concluded that the electroluminescence mechanism was temperature dependent. At room temperature, it was concluded that the Schottky barrier exists, and that emitted light is due to radiative recombination of injected electrons with holes of the p-type PPV. These efforts were continued in a second paper while also introducing data indicating that the forward threshold potential for visible electroluminescence was dependent on a combination of polymer synthesis conditions, film thickness, the top electrode material (Al, Mg, Ca), and the specific device fabrication conditions.[52] Optimum conditions could provide threshold voltages as low as 2 V at room temperature.

This was followed by another contribution from Friend, Holmes, and coworkers that summarized many of the advances achieved to date, as well as providing some new innovations.[53] In particular, they were able to show that the efficiency of the devices could be enhanced with both the choice of the top metal electrode and the introduction of an electron-conducting hole-blocking layer. In this way, they were able to achieve efficiencies of 1% for basic PPV devices.

A second paper then detailed a new, soluble, cyano-substituted PPV (Figure 8.4) that provided a uniform red emission under forward bias.[54] As the lowest unoccupied orbital of this new polymer lies at lower energy than those of PPV, electrodes made from stable metals such as Al can be used for electron injection, rather than the Ca electrodes normally needed to achieve higher efficiencies. Devices were thus fabricated from ITO coated with a PPV layer that helps to localize charge at the interface between the PPV and the new polymer, increasing the efficiency of recombination. This, combined with Al electrodes resulted in high internal efficiencies up to 4%. A second PPV derivate (Figure 8.4) was also investigated that provided an orange-yellow emission and efficiencies of circa 2%.

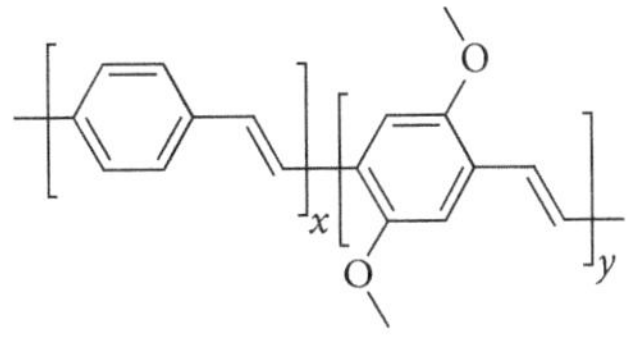

Figure 8.3 Random copolymeric PPV analog comprised of phenylene vinylene and 2,5-dimethoxy-phenylene vinylene segments.

Research on these polymer-based OLED devices continued to ramp up such that Friend, Holmes, and coworkers published a review of the progress to date in the beginning of 1994,[55] amazing considering that these devices had only been studied for four years at that point. The rapid advancement in these

red emission   orange-yellow emission

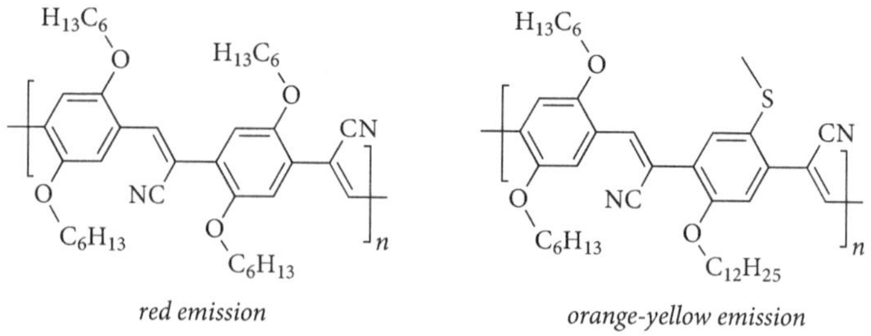

Figure 8.4 New cyano-substituted PPV derivatives.

OLED devices also led to the founding of Cambridge Display Technology by Friend and Holmes in 1992 (along with colleagues Donal Bradley, Jeremy Burroughes, and Chloe Jennings). Prior to this, in 1990, Heeger and Paul Smith had established the UNIAX Corporation in Santa Barbara in order to develop potential applications of conjugated polymers.[56] Within its first few years, however, UNIAX refocused its efforts largely on OLED technologies. UNIAX and Cambridge Display Technology thus represented the first companies established in order to commercialize conjugated polymer technologies, although others followed in the next decade. These developing applications, along with efforts to commercialize them, then served as the genesis of the new field of *organic electronics*,[57] or the development of devices based on organic electronic materials.

8.4 Nobel Prize 2000

As the most well-known event in the history of conjugated and conducting polymers, it is worth providing what details are available about the 2000 Nobel Prize in Chemistry. After the announcement of the Nobel Prize in Medicine on October 9, the selections for the prizes in both physics and chemistry were announced the following day. Thus, it was on October 10, 2000, that The Royal Swedish Academy of Sciences announced via a press conference and an associated press release that the Nobel Prize in Chemistry for 2000 was to be awarded jointly to Alan J. Heeger, Alan G. MacDiarmid, and Hideki Shirakawa *"for the discovery and development of conductive polymers."*[58] Further justification for the award was given in the accompanying Advanced Information released by the Academy, which stated:[59] "The choice is motivated by the important scientific position that the field has achieved and the consequences in terms of practical applications and of interdisciplinary development between chemistry and physics."

Table 8.2 Members of the 2000 Swedish Nobel Committee for Chemistry

Name	Field	University	City
Bengt Nordén[a]	physical chemistry	Chalmers University	Gothenburg
Björn Roos	theoretical chemistry	Lund University	Lund
Carl-Ivar Brändén	molecular biology	Karolinska Institute	Stockholm
Ingmar Grenthe	inorganic chemistry	Royal Institute of Tech.	Stockholm
Per Ahlberg	organic chemistry	Univ. of Gothenburg	Gothenburg
Astrid Gräslund[b]	biophysics	Stockholm University	Stockholm
Hakan Wennerström[c]	medical and physiological chemistry	Lund University	Lund
Torvard Laurent[c]	theoretical physical chemistry	Uppsala University	Uppsala
Gunnar von Heijne[c]	theoretical chemistry	Stockholm University	Stockholm

[a] Chairman of the Committee.
[b] Secretary of the Committee.
[c] Adjoint member.
Source: Data collected from T. Frängsmyr, Ed. Les PrixNobel. The Nobel Prizes 2000. Nobel Foundation: Stockholm, 2001.

The contributions of the Nobel laurates to the development of conductive polymers as a research field was also stressed in the press release, as was the importance of the various practical applications that resulted from these materials.[58] The prize included a monetary amount of nine million Swedish Krona (roughly $900K in 2000), which would also be shared equally among the three awardees. Unfortunately, there is no information currently available on the nominators for the 2000 Nobel Prize, as the statutes of the Nobel Foundation stipulate that such details cannot be released until at least 50 years have passed.

The 2000 Swedish Nobel Committee for Chemistry was made up of nine professors from seven different Swedish universities (Table 8.2), although all were from only four cities in Sweden.[60] However, as these cities included three of the four largest population centers in Sweden (Stockholm, Gothenburg, and Uppsala), this is perhaps not too surprising. Bengt Johan Fredrik Nordén (b. 1945) of Chalmers University of Technology served as the Chairman of the Committee, with Ruth Astrid Olivia Gräslund (b. 1945) of Stockholm University serving as the Committee Secretary.[59,60] The committee included three adjoint members, which only had one-year terms in comparison with the three-year terms of normal members and thus were specific to each Nobel committee. As can be seen in Table 8.2, over half the committee were from fields with strong ties to physics (i.e., physical chemistry, theoretical chemistry, biophysics), which most likely played a role in both the stated motivation concerning the interdisciplinary development between chemistry and physics and the awarding of the prize for a topic shared equally between chemistry and physics. However, it has also been noted that the 2000 Nobel Prize in Medicine also had a strong connection to physics.[61] As such, while this could be coincidence, the perceived importance of physics in the selections in 2000 may also have gone beyond any particular committee.

According to Heeger, the awardees were notified by the president of the Swedish Academy of Sciences via phone on October 10, 2000, shortly before the official announcement was made in Stockholm during a press conference held at 3:15 pm.[62] Due to the time difference, Heeger received his call at 5:45 am in Santa Barbara, only 30 minutes before the Stockholm press conference. The formal awarding of the Nobel Prize medals and diplomas then occurred two months later during ceremonies in Sweden at the Stockholm Concert Hall on December 10, 2000.[63] The ceremonies began with a brief introductory speech on the 2000 Nobel Prize in Chemistry given by the chair of the Committee for Chemistry, Bengt Nordén,[64] after which the Nobel laurates received their prize from Carl XVI Gustaf, the king of Sweden.[62] These ceremonies were followed by the Nobel Banquet, held on the lower floor of the City Hall. During the banquet each of the three Nobel laurates gave a brief speech.

Prior to the award ceremonies, the three laurates gave Nobel Lectures at Stockholm University on December 8, 2000.[65-67] Alan Heeger began with a lecture entitled "Semiconducting and Metallic Polymers: The Fourth Generation of Polymeric Materials," which attempted to give a broad overview of the field, with particular focus on the underlying physics and potential applications.[65] Alan MacDiarmid then continued with his lecture entitled "'Synthetic Metals': A Novel Role for Organic Polymers," which focused primarily on his more recent efforts with polyaniline.[66] Finally, Hideki Shirakawa presented a lecture entitled "The Discovery of Polyacetylene Film: The Dawning of an Era of Conducting Polymers," which focused primarily on his polyacetylene work prior to the collaboration with Heeger and MacDiarmid.[67] Various versions of all three of these Nobel Lectures were then published in the journals *Angewandte Chemie International Edition*,[68-70] *Reviews of Modern Physics*,[71-73] and *Synthetic Metals*.[74-76]

8.5 The nature of discovery

Considering the history laid out in the previous chapters, one must question exactly what was meant by the wording of the 2000 Nobel Prize in Chemistry that credits Heeger, MacDiarmid, and Shirakawa as the discoverers of conductive polymers. In evaluating the meaning of this claim, we first must consider what exactly is meant by the term *conductive polymers*. As previously discussed in Chapter 1, the terms *conjugated polymers* and *conducting polymers* are often used interchangeably, even if not technically correct. As such, this could refer to conjugated polymers in general. However, as all of the primary parent conjugated polymers besides polythiophene were known prior to the 1977 polyacetylene research, this would make little sense, and it would not be logical to assign discovery of conjugated polymers to the Nobel laureates.

In his introduction of his Nobel Lecture, MacDiarmid tried to clarify that the conductive polymers recognized by the award are different from those described by the more general term *conducting polymers*:[66] "This class of polymer is completely different from 'conducting polymers' which are merely a physical mixture of a nonconductive polymer with a conducting material such as a metal or carbon powder distributed throughout the material." Still, MacDiarmid does not refer to these as conductive polymers but, rather, "intrinsically conducting polymers," which he describes as polymers whose conductivity is enhanced through the process of doping to give the properties of a metal while retaining the mechanical properties commonly associated with a conventional polymer.[66] This language is confusing, however, as an intrinsic semiconductor refers to an undoped semiconductor, which is not consistent with the description used here. Nevertheless, the term *conducting polymers* is still far more common than the alternate term *conductive polymers*, and it must be concluded that these terms are being used interchangeably.

Of course, Heeger, MacDiarmid, and Shirakawa admitted as early as late 1978 that others had previously shown many orders of magnitude increases in conductivity by treating various materials, including polyacetylene, with an electron-withdrawing species.[77] At the same time, however, they dismissed this previous work by stating that the final room temperature conductivity obtained was still very small. Of course, this overview of previous work conveniently excluded conjugated polymers generated via oxidative polymerization, which directly produced the final doped, conducting forms, and only included cases in which neutral materials were treated with oxidants to give the final doped materials. In their defense, one could argue that the relationships between these two types of materials was not completely understood in 1978, but this was certainly not the case by 2000. Either way, doped conjugated polymers certainly predated the work on conducting polyacetylene films, as outlined in Section 8.2 (see Table 8.1).

Another point of consideration is the extent of conductivity. The dismissal of some previous examples was due to their low conductivity values, and Mac-Diarmid's description of his intrinsically conducting polymers emphasizes the properties of metals. Nordén's introductory speech at the Nobel ceremonies also emphasized the properties of metals.[64] As such, the conductive polymers with whose discovery the Nobel laureates are credited could refer to doped conjugated polymers that exhibit conductivities in the metallic regime. By 1969, Rene Buvet and Marcel Jozefowicz had achieved conductivities as high as 100 S cm^{-1} for doped polyaniline,[35] but it was the doped polyacetylene work recognized by the Nobel Prize that was the first to surpass this achievement and produce polymers with conductivities firmly in the metallic regime (i.e., >100 S cm^{-1}). As such, crediting Heeger, MacDiarmid, and Shirakawa with the discovery of polymers capable of this level of conductivity would be valid. However, at no point is this distinction stated anywhere in the award documentation, and the common use of the term *conducting* (or *conductive*) *polymers* does not have any such associated magnitude of conductivity.

A final point to consider is the physical nature of the polymers in question. Both the Nobel press release[58] and the advanced material[59] emphasize conductive plastics, as does Nordén in his introductory speech of the Nobel ceremonies.[64] In fact, the press release specifically states:[58] "Yet this year's Nobel Laureates in Chemistry are being rewarded for their revolutionary discovery that plastic can, after certain modifications, be made electrically conductive." Of course, the official wording of the award states *conductive polymers*, not *conductive plastics*. Still, the relationships between plastics and polymers can be poorly understood, even by practicing chemists, such that the two terms can be incorrectly viewed as referring to the same thing.[3] Evidence of this can be seen in the Nobel material where the statement "We are used to polymers— that is, plastics" can be found in the advanced material,[59] while the statement "Plastics are polymers" is found in the press release.[58] In reality, the term *polymer* refers to a molecular species, while the term *plastic* refers to the nature of a material, its plasticity, which is a product of polymer morphology (i.e., packing and ordering), rather than its molecular structure directly. As such, most common plastics are comprised of assemblies of polymers, but any particular polymer can exist in both plastic and nonplastic forms. However, even if the committee meant for conductive polymers to refer to conductive plastics, doped polyacetylene was not the first report of a conducting plastic, with freestanding, conductive plastic films of doped polypyrrole previously reported in 1968.[34]

As the determination of exactly what the Nobel committee meant by the term *conductive polymers* is inconclusive, one thus must continue with the view that this simply referred to the class of organic materials commonly known as *conducting polymers*. As such, one must then ask what is actually meant by "discovery" in the official wording of the award. Of course, the topic of scientific discovery has been a recurring one in the history and philosophy of science,[78–83] with it becoming a part of the wider endeavor of exploring creative thinking and creativity.[79] For many, however, the act of discovery is simply viewed to be a leap of insight, what is often described as the "eureka moment."[79,80] However, major discoveries typically require more than just a single moment of insight and require scientists to continue pursuing the various elements involved until the discovery is complete.[80]

As highlighted by the *Stanford Encyclopedia of Philosophy*,[79] it was during the course of the 19th century that the act of having an insight, the alleged "eureka moment," was distinguished from the later aspects of articulating, developing, and testing that insight. An important contribution to this process was the work of William Whewell (1794–1866), as he distinctly separated the eureka moment, or "happy thought," from other aspects of scientific inquiry. Thus, he viewed the complete process of discovery to consist of three elements: the happy thought, the articulation of that thought, and its testing or verification.[79] Commonly, however, the general view of discovery is often limited to either the eureka moment alone, or to just the eureka moment and its articulation.

Still, according to the historian and philosopher of science Thomas S. Kuhn (1922–1996), in order to discover something, one must be aware of the discovery while also knowing what it is that one has discovered.[81] The sticking point thus becomes: How much must one know about what was discovered? MacDiarmid also highlighted this view of discovery when asked about the nature of the contributions of Pyun and Shirakawa on the discovery of polyacetylene films, to which he stated:[84]

> Of course, your remark touches an interesting aspect of this discovery and, generally speaking, of scientific discoveries. Who is the discoverer? Is it the person who does something mechanically, or is it the person who realizes its significance? Quite often the person who does the mechanical operation in the lab is also the person that realizes its significance. Sometimes it is not the same person.

Because of this dual requirement, Kuhn did not consider discovery to be a simple act, but an extended, complex process, which ultimately results in a paradigm shift.[79] Thus, as with Whewell, Kuhn felt that the process of discovery involves several aspects, beginning with observations of an anomalous phenomenon. Discovery then continues with attempts to conceptualize the anomaly, and changes in the paradigm such that the anomaly can be accommodated.[79] This condition of a paradigm shift is also echoed by historian of science Robert Fox (b. 1938), who points out that a discovery "normally entails the abandonment, or at least the qualification, of earlier theories or observations."[82] This can be especially difficult if the discovery requires the marginalization of entrenched ideas or previously respected authorities.

Of course, the more complex nature of scientific discovery outlined previously can complicate the narrative of popular interpretations of scientific history, which often highlight great discoveries by select figures of great stature, particularly such events easily commemorated in an anniversary. The reality of science is that most discoveries are much more nuanced, complicated, and communal.[83] As such, it is not always a simple task to assign discovery to particular figures at a particular time.

In terms of the discovery of conducting polymers, as examples of conducting polymers predate the polyacetylene work of the 1970s (see Table 8.1), is there some aspect of the later work that could be said to illustrate a greater understanding of the underlying process of enhancing conductivity via the addition of oxidizing species? As early as 1963, it was understood that the conductivity was dependent on the amount of the dopant (i.e, I_2) reacting with the polymer.[30,31] Furthermore, Weiss in 1963, as well as Smith and Berets in 1968, had at least a partial understanding that this was enhancing the p-type nature of the material by increasing the number of holes in the polymer.[31,33] As such, it is unclear what aspects of the polyacetylene work might have been viewed by the Nobel committee as a significantly advanced understanding of the process or how this overturned earlier theories or observations. There is no doubt that the Nobel laureate's work on polyacetylene achieved much higher conductivities than previous efforts, including the first in the metallic regime, and ultimately resulted in a greater understanding of the doping process, yet it is difficult to point to a clear resulting paradigm shift associated with their contributions.

Of course, in the minds of the Nobel laureates, they clearly view themselves as the discoverers of these materials and the originators of the field of conjugated and conducting polymers. As Heeger stated in 2005:[85] "We were the pioneers. This is not to say that many other people didn't make important contributions." Later, he expanded on the issue of discovery in his 2016 autobiography, stating:[86]

Had we really created this new field? Were we really first? The answer to such questions was clearly and eloquently first stated by Sir Issac Newton: "If I have seen further, it is by standing on the shoulders of giants." Many chemists and physicists had touched upon the subject of conjugated polymers. There were reports in the literature of modest electrical conductivities. We were aware of some of these and ignorant of others. I will not attempt to list those earlier "giants" because I would surely leave out important names that should have been included. Our interdisciplinary approach, breath and solidity of our results, the new theoretical concepts that we introduced, and the firm scientific foundation that we build all contributed to what turned out to be the creation of the field of semiconducting and metallic polymers.

8.6 Conclusions

As previously stated in the introduction, the goal of the current volume is to highlight the extent of work carried out from the 19th century to the 1980s, while also properly give credit to all of the various figures that contributed to the origin and development of this important field of electronic materials. While not comprehensive, the previous chapters do cover the early periods of most of the major parent conjugated polymers, as well as attempts to develop low-bandgap polymers (< 1.5 eV) starting in 1984. In addition to highlighting all of the overlooked heroes throughout this history, the hope was also to strip away the myth and legend concerning these materials, the most prevalent of which is the mistaken belief that polyacetylene is the first known example of a conducting polymer.

Another common view is that all other conjugated polymers developed from polyacetylene as the prototype material. Of course, as illustrated in the previous chapters, the history of conjugated and conducting polymers is much older than generally thought, with over 120 years of history that predate the first report of linear polyacetylene. Furthermore, this is not a sporadic history of disconnected events, but a continuous run of publications that clearly tracks the development of these materials and our growing understanding of their properties. This development and the connections between these materials can be illustrated by the synthetic methods utilized for their production across the historical record, providing a clear line of development for most conjugated polymers that has little to do with polyacetylene (Figure 8.5).

The clear influence of polyaniline on the development of polypyrrole can be seen, both in the similarity of their initial names (i.e., aniline black vs. pyrrole black) and the fact that both polymers are produced by oxidative polymerization. Another line of development can then be created starting with polyphenylene, the first conjugated polymer not made by oxidative polymerization. From there, lines of development can be extended to both the polyphenylene derivative poly(phenylene vinylene) (PPV) and polythiophene, which was initially produced via methods developed for polyphenylene. At the same time polythiophene was also later produced using the oxidative polymerization methods developed for polypyrrole, so polythiophene can be viewed as the intersection of the two primary lines of conjugated polymer development. As clearly illustrated, however, polyacetylene is not part of either line of development, as it is produced by chain-grown methods with Ziegler–Natta catalysts, a method unrelated to any of the other members of the larger class of

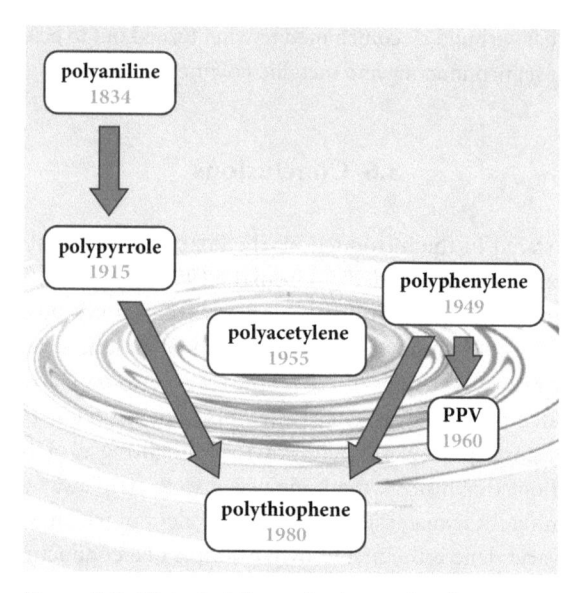

Figure 8.5 Historical lines of polymer development.

conjugated polymers. Nevertheless, polyacetylene still relates to the other polymers via its material properties and its ability to generate conducting polymers via redox doping, thus having a ripple effect on the rest of the family, in which a better understanding of polyacetylene and its conductivity can be applied to advancing these processes in other conjugated polymers.

As highlighted by Heeger in the quote earlier, it is also generally viewed that while these materials might have been studied at earlier points, the field of conjugated polymers did not begin until after the polyacetylene work of the 1970s. However, as the history outlined in the previous chapters shows, an earlier field did exist, although not as significant or well organized. This is clear in that several figures worked on multiple types of conjugated polymers, rather than isolated studies on a single system. This included people like Stille (polyphenylene and polyacetylene), Buvet and Jozefowicz (polyphenylene and polypyrrole), Kossmehl (PPV and polythiophene), and Yamamoto (polyphenylene and polythiophene). Furthermore, the work of prominent figures like Kossmehl and Yamamoto span the time frame both before and after the supposed start of the field in the 1970s. With that said, it is quite true that the field changed significantly after the 1970s. Not only did it grow significantly, but the earlier field was largely the domain of chemistry. In contrast, the high conductivities of doped polyacetylene then attracted the physics community that had been previously focused on other highly conductive organic species like TTF-TCNQ, thus resulting in a much more multidisciplinary community.

As such, it is clear that the work of Heeger, MacDiarmid, and Shirakawa initiated a new phase in the study of these materials and sparked the rapid growth of a niche area of scientific study into the wider community of modern conjugated materials. In addition, the contributions of MacDiarmid and Heeger did not end with their initial work on polyacetylene, with both men remaining central figures in the further growth of the field for decades afterward. Furthermore, this included not just the study of these materials, but also efforts to develop real-world applications that ultimately led to the field of organic electronics.

In the end, the field of conjugated materials owes much to them, and in no way is this current deep dive into the history of these materials an attempt to marginalize their contributions. Rather, the hope is that this will better allow us to recognize the very deep history of these materials that spans centuries, while also properly placing the more familiar contributions within the much larger history of these materials. Not only does this properly credit all of the important early contributions that helped shape the fields of both conjugated polymers and organic electronics, but it also reveals that these activities make up a more significant part of the overall histories of chemistry and physics than is often recognized.

References and Notes

1. Nobelprize.org. The Nobel Prize in Chemistry 2000. Nobel Prize Outreach AB 2024. https://www.nobelprize.org/prizes/chemistry/2000/summary (accessed April 23, 2024).
2. Rasmussen, S. C. Conjugated and conducting organic polymers: The first 150 years. *ChemPlusChem* **2020**, *85*, 1412–1429.
3. Rasmussen, S. C. Revisiting the early history of synthetic polymers: Critiques and new insights. *Ambix* **2018**, *65*, 356–372.
4. Nobelprize.org. The Nobel Prize in Chemistry 1912. Nobel Prize Outreach AB 2024. https://www.nobelprize.org/prizes/chemistry/1912/summary (accessed April 23, 2024).
5. Nobelprize.org. The Nobel Prize in Chemistry 1915. Nobel Prize Outreach AB 2024. https://www.nobelprize.org/prizes/chemistry/1915/summary (accessed April 23, 2024).
6. Nobelprize.org. The Nobel Prize in Chemistry 1963. Nobel Prize Outreach AB 2024. https://www.nobelprize.org/prizes/chemistry/1963/summary (accessed April 23, 2024).
7. Nobelprize.org. The Nobel Prize in Chemistry 2010. Nobel Prize Outreach AB 2024. https://www.nobelprize.org/prizes/chemistry/2010/summary (accessed April 23, 2024).
8. Lenz, R. W. In memory of John Kenneth Stille. *Macromolecules* **1990**, *23*, 2417–2418.
9. Cordebas, R. Conductivity of carbon. *Rev. Gen. Electr.* **1932**, *31*, 547–556.

10. Weiss, D. E.; Bolto, B. A. Organic polymers that conduct electricity. In *Physics and Chemistry of the Organic Solid State*, Vol. II. Fox, D.; Labes, M. M.; Weissberger, A.; Eds.; Interscience Publishers: New York, 1965, pp. 67–120.

11. Dannenberg, E. M.; Paquin, L.; Gwinnell, H. Carbon black. In *Kirk-Othmer Encyclopedia of Chemical Technology*, 4th ed.; Howe-Grant, M., Ed.; John Wiley & Sons: New York, 1992, Vol. 4, pp. 1037–1074.

12. Okamoto, Y.; Brenner, W. *Organic Semiconductors*. Reinhold Publishing Corporation: New York, 1964, pp. 82–124.

13. Gutmann, F.; Lyons, L. E. *Organic Semiconductors*. John Wiley & Sons, Inc.: New York, 1967, pp. 471–484.

14. Okamoto, Y.; Brenner, W. *Organic Semiconductors*. Reinhold Publishing Corporation: New York, 1964, pp. 125–158.

15. Brennan, W. D.; Brophy, J. J.; Schonhorn, H. Electrical conductivity in pyrolyzed polyacrylonitrile. In *Organic Semiconductors. Proceedings of an Inter-industry Conference*. Brophy, J. J.; Buttrey, J. W.; Eds.; The Macmillan Co.: New York, 1962, pp. 159–168.

16. Eley, D. D. Phthalocyanines as semiconductors. *Nature* **1948**, *162*, 189.

17. Johnston, R. J.; Ubbelohde, A. R. Some physical properties associated with "aromatic" electrons. Part III. The pseudo-metallic properties of potassium-graphite and graphite-bromine. *J. Chem. Soc.* **1951**, 1731–1736.

18. Murray, J. J.; Ubbelohde, A. R. Electronic properties of some synthetic metals derived from graphite. *Proc. Roy. Soc. A* **1969**, *312*, 371–380.

19. McCoy, H. N. Synthetic metals from non-metallic elements. *Science*, **1911**, *34*, 138–142.

20. Rasmussen, S. C. On the origin of "synthetic metals": Herbert McCoy, Alfred Ubbelohde, and the development of metals from nonmetallic elements. *Bull. Hist. Chem.* **2016**, *41*, 64–73.

21. Akamatu, H.; Inokuchi, H.; Matsunaga, Y. Electrical conductivity of the perylene-bromine complex. *Nature* **1954**, *173*, 168–169.

22. Gutmann, F.; Lyons, L. E. *Organic Semiconductors*. John Wiley & Sons, Inc.: New York, 1967, pp. 548–553, 720–731.

23. Acker, D. S.; Harder, R. J.; Hertler, W. R.; Mahler, W.; Melby, L. R.; Benson, R. E.; Mochel, W. E. 7,7,8,8-Tetracyanoquinodimethane and its electrically conducting anion-radical derivatives. *J. Am. Chem. Soc.* **1960**, *82*, 6408–6409.

24. Ferraris, J.; Cowan, D. O.; Walatka, Jr, V.; Perlstein, J. H. Electron transfer in a new highly conducting donor-acceptor complex. *J. Am. Chem. Soc.* **1973**, *95*, 948–949.

25. Wudl, F.; Smith, G. M.; Hufnagel, E. J. Bis-1,3-dithiolium chloride: An unusually stable organic radical cation. *Chem. Commun.* **1970**, 1453–1454.

26. Natta, G.; Mazzanti, G.; Corradini, P. Polimerizzazione stereospecifica dell'acetilene. *Atti Accad. Naz. Lincei Rend. Cl. Sci. Fis. Mat. Nat.* **1958**, *25*, 3–12.

27. Jozefowicz, M.; Buvet, R. Préparation de poly-p-phénylènes oligomères. *C. R. Acad. Sc.* **1961**, *253*, 1801–1803.

28. Hatano, M.; Kanbara, S.; Okamoto, S. Paramagnetic and electric properties of polyacetylene. *J. Polym. Sci.* **1961**, *51*, S26–S29.

29. Parini, V. P.; Kazakova, Z. S.; Berlin, A. A. Polymers with conjugated bonds and with heteroatoms in the conjugated bond hain XIX. Some properties of aniline black. *Vysokomol. soedin.* **1961**, *3*, 1870–1873.

30. Originally published as Mainthia, S. B.; Kronick, P. L.; Ur, H.; Chapman, E. F.; Labes, M. M. Electronic conductivity of complexes of poly-p-phenylene. *Polym. Prepr.* **1963**, *4*(1), 208–212. Data was later republished in Labes, M. M. Conductivity in polymeric solids. *Pure Appl. Chem.* **1966**, *12*, 275–285.

31. McNeill, R.; Siudak, R.; Wardlaw, J. H.; Weiss, D. E. Electronic conduction in polymers. *Aust. J. Chem.* **1963**, *16*, 1056–1075.

32. Yu, L. T.; Jozefowicz, M. Conductivité et constitution chimique de semi-conducteurs macromoléculaires. *Rev. Gen. Electr.* **1966**, *75*, 1014–1018.

33. Berets, D. J.; Smith, D. S. Electrical properties of linear polyacetylene. *Trans. Faraday. Soc.* **1968**, *64*, 823–828.

34. Dall'Olio, A.; Dascola, G.; Varacca, V.; Bocchi, V. Resonance paramagnètique èlectronique et conductiviè d'un noir d'oxypyrrol èlectrolytique. *C. R. Acad. Sci., Ser. C* **1968**, *267*, 433–435.

35. Jozefowicz, M.; Yu, L. T.; Perichon, J.; Buvet, R. Proprietes nouvelles des polymeres semiconducteurs. *J. Polym. Sci. Part C: Polym. Symp.* **1969**, *22*, 1187–1195.

36. Shirakawa, H.; Louis, E. J.; MacDiarmid, A. G.; Chiang, C. K.; Heeger, A. J. Synthesis of electrically conducting organic polymers: Halogen derivatives of polyacetylene, $(CH)_x$. *J. Chem. Soc., Chem. Commun.* **1977**, 578–580.

37. Chiang, C. K.; Fincher, C. R., Jr.; Park, Y. W.; Heeger, A. J.; Shirakawa, H.; Louis, E. J.; Gau, S. C.; MacDiarmid, A. G. Electrical conductivity in doped polyacetylene. *Phys. Rev. Lett.* **1977**, *39*, 1098–1101.

38. Nigrey, P. J.; MacInnes, D. Jr., Nairns, D. P.; MacDiarmid, A. G.; Heeger, A. J. Lightweight rechargeable storage batteries using polyacetylene, $(CH)_x$ as the cathode-active material. *J. Electrochem. Soc.* **1981**, *128*, 1651–1654.

39. MacInnes, D., Jr.; Druy, M. A.; Nigrey, P. J.; Nairns, D. P.; MacDiarmid, A. G.; Heeger, A. J. Organic batteries: Reversible n- and p- type electrochemical doping of polyacetylene, $(CH)x$. *J. Chem. Soc. Chem. Commun.* **1981**, 317–319.

40. Chiang, C. K. An all-polymeric solid state battery. *Polymer* **1981**, *22*, 1454–1256.

41. Weinberger, B. R.; Akhtar, M.; Gau, S. C. Polyacetylene photovoltaic devices. *Synth. Met.* **1982**, *4*, 187–197.

42. Glenis, S.; Horowitz, G.; Tourillon, G.; Garnier, F. Electrochemically grown poly-thiophene and poly(3-methylthiophene) organic photovoltaic cells. *Thin Solid Films* **1984**, *111*, 93–103.

43. Glenis, S.; Tourillon, G.; Garnier, F. Influence of the doping on the photovoltaic properties of fhin films of poly-3-methylthiophene. *Thin Solid Films* **1986**, *139*, 221–231.

44. Yu, G.; Zhang, C.; Heeger, A. J. Dual-function semiconducting polymer devices: Light-emitting and photodetecting diodes. *Appl. Phys. Lett.* **1994**, *64*, 1540–1542.

45. Yu, G.; Gao, J.; Hummelen, J. C.; Wudl, F.; Heeger, A. J. Polymer photovoltaic cells: Enhanced efficiencies via a network of internal donor-acceptor heterojunctions, *Science* **1995**, *270*, 1789–1791.

46. Dastoor, P. C.; Belcher, W. J. How the west was won? A history of organic photo-voltaics. *Substantia* **2019**, *3*(2) Suppl. 1, 99–110.

47. Burroughes, J. H.; Bradley, D. D. C.; Brown, A. R.; Marks, R. N.; Mackay, K.; Friend, R. H.; Burnst, P. L.; Holmes, A. B. Light-emitting diodes based on conjugated polymers. *Nature* **1990**, *347*, 539–541.

48. Braun, D.; Heeger, A. J. Visible light emission from semiconducting polymer diodes. *Appl. Phys. Lett.* **1991**, *58*, 1982–1984.

49. Burn, P. L.; Holmes, A. B.; Kraft, A.; Bradley, D. D. C.; Brown, A. R.; Friend, R. H.; Gymer, R. W. Chemical tuning of electroluminescent copolymers to improve emission efficiencies and allow patterning. *Nature* **1992**, *356*, 47–49.

50. Brown, A. R.; Bradley, D. D. C.; Burroughes, J. H.; Friend, R. H.; Greenham, N. C.; Burn, P. L.; Holmes, A. B.; Kraft, A. Poly(p-phenylenevinylene) light-emitting diodes: Enhanced electroluminescent efficiency through charge carrier confinement. *Appl. Phys. Lett.* **1992**, *61*, 2793–2795.

51. Karg, S.; Riess, W.; Dyakonov, V.; Schwoerer, M. Electrical and optical characterization of poly(phenylene-vinylene) light emitting diodes. *Synth. Met.* **1993**, *54*, 427–433.

52. Karg, S.; Riess, W.; Meier, M.; Schwoerer, M. Characterization of light emitting diodes and solar cells based on poly-phenylene-vinylene. *Synth. Met.* **1993**, *54*, 4186–4191.

53. Holmes, A. B.; Bradley, D. D. C.; Brown, A. R.; Burn, P. L.; Burroughes, J. H.; Friend, R. H.; Greenham, N. C.; Gymer, R. W.; Halliday, D. A.; Jackson, R. W.; Kraft, A.; Martens, J. H. F.; Pichler, K.; Samuel, I. D. W. Photoluminescence and electroluminescence in conjugated polymeric systems. *Synth. Met.* **1993**, *57*, 4031–4040.

54. Greenham, N. C.; Morattit, S. C.; Bradley, D. D. C.; Friend, R. H.; Holmes, A. B. Efficient light-emitting diodes based on polymers with high electron affinities. *Nature* **1993**, *365*, 628–630.

55. Baigent, D. R.; Greenham, N. C.; Grtiner, J.; Marks, R. N.; Friend, R. H.; Moratti, S. C.; Holmes, A. B. Light-emitting diodes fabricated with conjugated polymers - recent progress. *Synth. Met.* **1994**, *67*, 3–10.

56. Heeger, A. J. *Never Lose Your Nerve!* World Scientific Publishing: Singapore, 2016; pp. 195–205.

57. Moliton, A.; Hiorns, R. C. The origin and development of (plastic) organic electronics. *Polym. Int.* **2012**, *61*, 337–341.

58. Nobelprize.org. The Nobel Prize in Chemistry 2000: Press release. Nobel Prize Outreach AB 2024. https://www.nobelprize.org/prizes/chemistry/2000/press-release (accessed April 28, 2024).

59. Nobelprize.org. The Nobel Prize in Chemistry 2000: Advanced information. Nobel Prize Outreach AB 2024. https://www.nobelprize.org/prizes/chemistry/2000/advanced-information (Accessed April 28, 2024).

60. Frängsmyr, T., Ed. *Les PrixNobel. The Nobel Prizes 2000.* Nobel Foundation: Stockholm, 2001, p. 8.

61. Schewe, P. P. 2000 Nobel Prizes announced in physics, chemistry, medicine. *APS News* 2000, *9*(11), https://www.aps.org/publications/apsnews/200012/nobel.cfm (accessed April 28, 2024).

62. Heeger, A. J. *Never Lose Your Nerve!* World Scientific Publishing: Singapore, 2016, pp. 151–163.

63. Nobelprize.org. The Nobel Prize Award Ceremony 2000. Nobel Prize Outreach AB 2024. https://www.nobelprize.org/prizes/chemistry/2000/award-video (accessed April 28, 2024).

64. Nordén, B. The Nobel Prize in chemistry. In *Les Prix Nobel. The Nobel Prizes 2000.* Frängsmyr, T., Ed. Nobel Foundation: Stockholm, 2001, pp. 21–23.

65. Heeger, A. J. Semiconducting and metallic polymers: The fourth generation of polymeric materials. In *Les Prix Nobel. The Nobel Prizes 2000.* Frängsmyr, T., Ed. Nobel Foundation: Stockholm, 2001, pp. 144–181.

66. MacDiarmid, A. G. "Synthetic metals": A novel role for organic polymers. In *Les Prix Nobel. The Nobel Prizes 2000*. Frängsmyr, T., Ed. Nobel Foundation: Stockholm, 2001, pp. 191–211.

67. Shirakawa, H. The discovery of polyacetylene film: The dawning of an era of conducting polymers. In *Les Prix Nobel. The Nobel Prizes 2000*. Frängsmyr, T., Ed. Nobel Foundation: Stockholm, 2001, pp. 217–266.

68. Shirakawa, H. The discovery of polyacetylene film: The dawning of an era of conducting polymers (Nobel Lecture). *Angew. Chem. Int. Ed.* **2001**, *40*, 2574–2580.

69. MacDiarmid, A. G. "Synthetic metals": A novel role for organic polymers (Nobel Lecture). *Angew. Chem. Int. Ed.* **2001**, *40*, 2581–2590.

70. Heeger, A. J. Semiconducting and metallic polymers: The fourth generation of polymeric materials (Nobel Lecture). *Angew. Chem. Int. Ed.* **2001**, *40*, 2591–2611.

71. Heeger, A. J. Nobel Lecture: Semiconducting and metallic polymers: The fourth generation of polymeric materials. *Rev. Modern Phys.* **2001**, *73*, 681–700.

72. MacDiarmid, A. G. Nobel Lecture: "Synthetic metals": A novel role for organic polymers. *Rev. Modern Phys.* **2001**, *73*, 701–712.

73. Shirakawa, H. Nobel lecture: The discovery of polyacetylene film—the dawning of an era of conducting polymers. *Rev. Mod. Phys.* **2001**, *73*, 713–718.

74. Shirakawa, H. The discovery of polyacetylene film. The dawning of an era of conducting polymers. *Synth. Met.* **2002**, *125*, 3–10.

75. MacDiarmid, A. G. Synthetic metals: A novel role for organic polymers. *Synth. Met.* **2002**, *125*, 11–22.

76. Heeger, A. J. Semiconducting and metallic polymers: The fourth generation of polymeric materials. *Synth. Met.* **2002**, *125*, 23–42.

77. Chiang, C. K.; Druy, M. A.; Gau, S. C.; Heeger, A. J.; Louis, E. J.; MacDiarmid, A. G.; Park, Y. W.; Shirakawa, H. Synthesis of highly conducting films of derivatives of polyacetylene, $(CH)_x$. *J. Am. Chem. Soc.* **1978**, *100*, 1013–1015.

78. Rasmussen, S. C. *Acetylene and Its Polymers. 150+ Years of History*. Springer Briefs in Molecular Science: History of Chemistry. Springer: Heidelberg, 2018, pp. 127–130.

79. Schickore, J. Scientific discovery. In *The Stanford Encyclopedia of Philosophy* (Winter 2022 ed.). Zalta, E. N.; Nodelman, U., Eds.; https://plato.stanford.edu/archives/spr2014/entries/scientific-discovery (accessed April 29, 2024).

80. Koshland, D. E., Jr. The Cha-Cha-Cha theory of scientific discovery. *Science* **2007**, *317*, 761–762.

81. Kuhn, T. S. Historical structure of scientific discovery. *Science* **1962**, *136*, 760–764.

82. Fox, R. The nature of discovery. *Notes Rec.* **2014**, *68*, 319–321.

83. Awkward first dates. *Nature* **2017**, *550*, 7.

84. Hargittai, B.; Hargittai, I. *Candid Science V: Conversations with Famous Scientists*. Imperial College Press: London, 2005, pp. 401–409.

85. Hargittai, B.; Hargittai, I. *Candid Science V: Conversations with Famous Scientists*. Imperial College Press, London, 2005, pp. 411–427.

86. Heeger, A. J. *Never Lose Your Nerve!* World Scientific Publishing: Singapore, 2016, pp. 143–144.

Index